THE INCREDIBLE HUMAN JOURNEY

ALICE ROBERTS

BLOOMSBURY

LONDON · BERLIN · NEW YORK

To Jonathan Musgrave and Kate Robson-Brown,
wise friends and mentors

First published in Great Britain 2009

Text © 2009 by Alice Roberts
Illustrations © 2009 by Alice Roberts
Maps © 2009 by Dave Stevens

The quotation on page vii by D. J. Cohen taken from *The Origins of Pottery and Agriculture* by Yoshinori Yashuda is reprinted by permission of Roli Books. The quotations on pages 144, 149 and 334 from *The Songlines* by Bruce Chatwin, published by Jonathan Cape, are reprinted by permission of The Random House Group and Aitken Alexander Associates.

The moral right of the author has been asserted.

Bloomsbury Publishing Plc, 36 Soho Square, London W1D 3QY

By arrangement with the BBC

The BBC logo is a trademark of the British Broadcasting Corporation and is used under licence. BBC logo © BBC 1996.

ISBN 978 0 7475 9839 8

10 9 8 7 6 5 4 3 2 1

Typeset by seagulls.net
Printed in Great Britain by Clays Limited, St Ives plc

The paper this book is printed on is certified by the © 1996 Forest Stewardship Council A.C. (FSC). It is ancient-forest friendly. The printer holds FSC chain of custody SGS-COC-2061

www.bloomsbury.com

Contents

Try to imagine the world in which humans have lived for the overwhelming majority of our existence, a world without cities, settled villages, or even permanent residences, a world without farmed fields and crops, without possessions larger than those which we could easily carry with us, and with everything we needed in daily life – all of our tools, weapons and clothing – produced by ourselves or by those within our small social bands. We did not grow food or have others grow it for us, but instead exclusively relied upon our own knowledge of the surrounding natural environment to survive, foraging for plant foods and scavenging, hunting or fishing for meat.[1]

D. J. Cohen

INTRODUCTION

We are very familiar with the idea that humans are everywhere; that wherever you go in the world you will probably find people there already. We are an unusual species in that we have a near-global distribution. And although people around the world may look quite different from each other, and speak different languages, they can nevertheless recognise each other as distant cousins.

But where and when did our species first appear? What are the essential characteristics of our species? And how did people end up being everywhere? These are rephrasings of fundamental questions. Who are we? What does it mean to be human? Where do we come from? For thousands of years, such questions have been explored through philosophy and religion, but the answers now seem to lie firmly within the grasp of an empirical approach to the world and our place within it. By peering deep into our past and dragging clues out into the light, science can now provide us with some of the answers to the questions that people have always asked.

They are questions that have always captivated me. As a medical doctor and anatomist (I lecture in anatomy on the medical course at Bristol University), I am fascinated by the structure and function of the human body, and the similarities and differences between us and other animals. We are certainly apes; our anatomy is incredibly similar to that of our nearest relations, chimpanzees. I could put a chimpanzee arm bone, or humerus, in an exam for medical students and they wouldn't even notice that it wasn't human.

But there are obviously things that mark us out – not as special creations, but as a species of African ape that has, quite serendipitously, evolved in ways that enabled our ancestors to survive, thrive and expand across the whole world. There are aspects of anatomy that are entirely unique to us; unlike our arms, our spines, pelvis and legs are *very* different from those of our chimp cousins, and no one would mistake a human skull for that of another African ape. It's a very distinctive shape, not least because we have such enormous brains for the size of our bodies. And we use our big brains in ways that no other species appears to.

Unlike our closest ape cousins, we make tools and manipulate our environments to an extent that no other animal does. Although our species evolved in tropical Africa, this ability to control the interface between us and our surroundings means that we are not limited to a particular environment. We can reach and survive in places that should seem quite alien to an African ape. We have very little in the way of fur, but we can create coverings for our bodies that help to keep us cool in very hot climates and warm in freezing temperatures. We make shelters and use fire for warmth and protection. Through planning and ingenuity, we create things that can carry us across rivers and even oceans. We communicate, not just through complicated spoken languages but through objects and symbols that allow us to create complex societies and pass on information from generation to generation, down the ages. When did these particular attributes appear? This is a key question for anyone seeking to define our species – and to track the presence of our ancestors through the traces of their behaviour.

The amazing thing is – it *is possible* to find those traces, those faint echoes of our ancestors from thousands and thousands of years ago. Sometimes it could be an ancient hearth, perhaps a stone tool, that shows us where and how our forebears lived. Occasionally we find human remains – preserved bones or fossils that have somehow avoided the processes of rot and decay and fragmentation to be found by distant descendants grubbing around in caves and holes in the ground, in search of the ancestors.

I've always been intrigued by this search, by the history that can be reconstructed from the few clues that have been left behind. And at this point in time, we are very lucky to have evidence emerging from several different fields of science, coming together to provide us with a compelling story, with a better understanding of our real past than any humans have ever had before. From the study of bones, stones and the genes within our living bodies comes the evidence of our ancestors, of who we are, of where we came from – and of how we ended up all over the world.

When the BBC offered me the opportunity to follow in the footsteps of the ancients, to delve into the past, to meet people, see artefacts and fossils for myself and visit the places that seem most sacred to those searching for real meanings, I couldn't wait to get started. I took a year off from teaching anatomy and looking at medieval bones in the lab, and set off on a worldwide journey in search of our ancestors.

The Human Family Tree

My journey would take me all around the world, starting in Africa and then following in the footsteps of our ancestors into Asia and around the Indian coastline, all the way to Australia, north into Europe and Siberia, and, eventually, to the last continents to be peopled: the Americas.

Modern humans are just the latest in a long line of two-legged apes, technically known as hominins. We've grown used to thinking of ourselves as rather special, and a quick glance at the human family tree shows us that we're in a rather unusual position at the moment, being (as far as we know) the only hominin species alive on the planet. Going back into prehistory, the family tree is quite bushy, and there were often several species knocking around at the same time. By 30,000 years ago, it seems there were only two twigs left on the hominin family tree: modern humans and our close cousins, the Neanderthals. Today, only we remain.

The ancestral home of hominins is Africa, although some species, including our own, have made it out into other continents at various times. Whether we actually met up with our 'cousins' on our ancient wanderings is something I'll be investigating in this book. Certainly, it seems that there was some overlap in Europe, and that for a good few thousand years modern humans and Neanderthals were sharing the continent.

It may sound strange but it's actually quite difficult to know exactly how *many* different species of ancient hominins there were. It's something that prompts a huge amount of debate. The world of palaeontology – the branch of science that peers into the past to examine extinct and fossil species – is inhabited by 'lumpers' and 'splitters'. Lumpers use very wide definitions of species to group lots of fossils together under one species name. Splitters, as the term suggests, divide them up into lots of different species. But which group is right? It's hard to know, and this is one of the debates that enlivens this science. Both lumpers and splitters are looking at the same evidence – but making different interpretations.

Trying to decide if two populations really are different enough from each other to be labelled as separate species is more difficult than it seems. Some species can even interbreed and produce fertile, hybrid offspring. But, basically, species are populations that are diagnosably different from each other, in terms of their genes or their morphology (the way their bodies are constructed) – or both.

With long-dead fossil animals, all palaeontologists have to go on is the skeleton, and sometimes only fragments of bones. So the problem of species definition becomes even more difficult. Looking at the skeletons of living animals, they can get an idea of the range of morphological variation in a species (because, even within a species, animals come in a variety of shapes and sizes). They can also measure the level of morphological difference between species. This gives them a benchmark for how similar skeletons are within a species, and how different they need to be to be classed as separate species. Then the palaeontologist can apply that standard to sort fossil animals into species. It's a knotty problem, and it's not really that surprising that different palaeontologists, each of whom may have spent a lifetime studying the fossils, can reach different conclusions.

Indeed, some palaeontologists shy away from talking about ancient 'species' at all. The eminent physical anthropologist William White Howells suggested we call these groupings 'palaeodemes' ('ancient populations') instead. But we can recognise evolving lineages in the fossil record, and giving distinct populations a genus (e.g. *Homo*) and species (e.g. *sapiens*) name provides a useful handle and helps when we're trying to reconstruct family trees.[1]

Within palaeoanthropology, the discipline that looks at fossil hominins in particular, taxonomies range from extreme lumping, with some researchers calling all hominins from the last million years *Homo sapiens*, to extreme splitting, with other scientists finding room for eight or more species. Chris Stringer, palaeoanthropologist at the Natural History Museum in London, has recognised four species during and since the Pleistocene (in the last 1.8 million years): *Homo erectus, Homo heidelbergensis* (the putative common ancestor of modern humans and Neanderthals), *Homo sapiens* and *Homo neanderthalensis*,[2] although the recent discovery of the tiny 'Hobbit' skeletons in Indonesia requires us to make room for *Homo floresiensis* as well.

Throughout this book I use the word 'human' in an inclusive but nevertheless precise sense to mean any species in the genus *Homo*, whereas 'modern human' refers to our own species, *Homo sapiens*. In the same way, 'Neanderthals' are *Homo neanderthalensis*.

Each one of these human species made it out of Africa, to Eurasia. *Homo erectus* got all the way to Java and China by about one million years ago. About 800,000 years ago, another lineage formed and expanded: *Homo heidelbergensis* fossils have been found in Africa and Europe. The European branch of this population went on to give rise to Neanderthals, about 300,000 years ago. Modern humans sprang from the African population, some around 200,000 years ago – and *their* descendants spread across the globe.

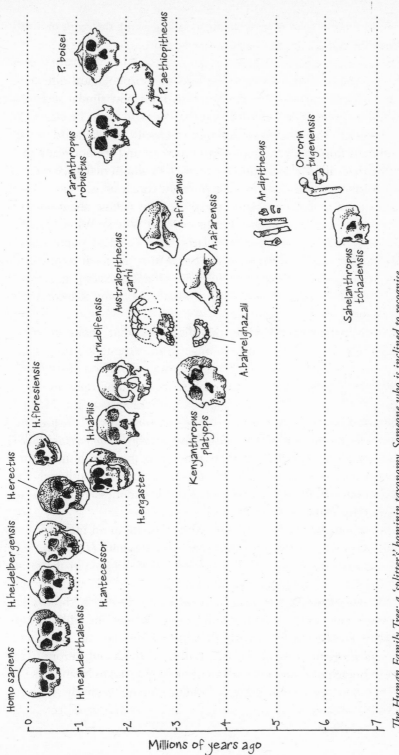

Homo sapiens

H.heidelbergensis

H.erectus

H.neanderthalensis

H.floresiensis

H.antecessor

H.habilis

H.ergaster

H.rudolfensis

Australopithecus
garhi

A.africanus

A.afarensis

Kenyanthropus
platyops

Paranthropus
robustus

P. boisei

P. aethiopicus

A.bahrelghazali

Ardipithecus

Orrorin
tugenensis

Sahelanthropus
tchadensis

Millions of years ago

0   1   2   3   4   5   6   7

The Human Family Tree: a 'splitter's' hominin taxonomy. Someone who is inclined to recognise lots of different species might cram this many hominins into the last seven million years.

This is a version of events that is now accepted by most palaeoanthropologists, and is supported by the weight of fossil evidence and genetic studies. It is known in the jargon as the 'recent African origin' or 'Out of Africa' model. But although this is now the majority view, it is not the only theory about how modern humans evolved and ended up everywhere. Some palaeoanthropologists still argue that archaic species like *Homo erectus* and *heidelbergensis*, having spread from Africa into Asia and Europe, then 'grew up' into modern humans in all of these continents. At the end of the twentieth century, a great debate raged over which version, the recent African origin or the 'Regional Continuity' (also called 'Multiregional Evolution'), was a more accurate representation of events. Since then, the evidence – genetic, fossil and climatological – has stacked up quite impressively in favour of a recent African origin,[3, 4] but there is still a minority of scholars who argue for multiregionalism. There are also some palaeoanthropologists, who, while accepting a recent African origin, suggest that modern humans may have interbred with other archaic species as they spread into other continents, for instance mixing with the Neanderthals in Europe.[2]

Reconstructing the Past

Generally speaking, someone who studies ancient human ancestry is a called a **palaeoanthropologist**. Palaeoanthropology is a sphere of enquiry that essentially started off with fossil-hunting, but which today draws on many other disciplines (so that, for example, people who now end up contributing to palaeoanthropology may come from fields as disparate as genetics and climate science).

When Charles Darwin wrote *The Descent of Man* in 1871, not a single early human fossil had been discovered, but he nonetheless tentatively suggested that Africa might just be the homeland of the human species:

> In each great region of the world the living animals are closely related to the extinct species of the same region. It is, therefore, probable that Africa was formerly inhabited by extinct apes closely allied to the gorilla and chimpanzee; and as these two species are now man's nearest allies, it is somewhat more probable that our early progenitors lived on the African continent than elsewhere.

Then fossils started to appear. For a long time, the study of fossils formed the basis of palaeoanthropology, supplemented with comparisons with the anatomy of living humans and our closest relatives, the African apes: chimpanzees and gorillas. Scientists in this specific field might call themselves **physical** or **biological anthropologists**. Much of their work is focused on bones; after all, that's normally all that is preserved in fossils.

As well as looking at the physical remains of our ancestors, palaeo-anthropology also draws on the clues left behind – the traces of the material culture of past people, in other words **archaeology**. Palaeolithic archaeologists are, by necessity, experts in recognising and interpreting stone tool types. Some engage in experimental archaeology, testing out methods of making and using ancient tools and other cultural items. The insights from such practical work can be profound.

The earth itself contains 'memories' of past climate and geography held in sediments and layers of ice. Unlocking these secrets has armed palaeoanthropologists with powerful tools for reconstructing the human family tree, and for understanding the environments in which our ancestors lived. **Geologists** now join the fray, as people who understand how landscapes are formed, how sediments are laid down, how caves are made, and dating experts often come from this field. The study of both fossils and archaeological remains has benefited hugely from advances in dating techniques, meaning that we can now pin fairly precise ages on clues from the deep past. The study of climate change in the past is called **palaeoclimatology.**

As well as digging for physical remains in the ground, there are clues to our ancestry held in the DNA (deoxyribonucleic acid, the stuff of life) of everyone alive today. **Geneticists** involved in palaeoanthropology often come from a background of medical genetics – where genes responsible for particular diseases or conditions are tracked down. But differences in our genes can also be used to reconstruct past histories. An exciting new development is the possibility of obtaining ancient DNA from fossil bones – providing another way of approaching the species question.

Linguists have also tried to reconstruct human histories, by looking at language families. However, most linguists feel that languages cannot be reliably traced back more than 10,000 years, although, as we shall see, there are some interesting insights emerging from studies combining linguistics with genetics.

In my journey around the world I have visited many communities of indigenous people in different continents. Many of those I have met have been given different names at different times by outsiders, some of which carry racist or, at the very least, derogatory overtones. I have always tried to use terms to describe people that they themselves are happy with, which is why, for example, in the first part of 'African Origins', I refer to the people of the Kalahari as 'Bushmen', a name they use in English to refer to themselves. Similarly, people of mixed European and sub-Saharan African ancestry in South Africa refer to themselves as 'coloureds'; the Evenki of Siberia call themselves by that name, and the same goes for the Semang and Lanoh tribes of Malaysia, the Native Americans of Canada and North America, and the Aboriginal Australians.

The Ice Age

This story of ancient human migrations across the world is set almost entirely within the later stages of what geologists call the Pleistocene period, or the Ice Age. In its entirety, the Pleistocene lasted from 1.8 million years ago to 12,000 years before the present. Although our species appears only in the late Pleistocene, by the end of that period modern humans had made their way into every continent (except Antarctica). In some chapters we will also dip our toes into the Holocene, the period that followed the Pleistocene or Ice Age, and in which we're still living today.

As we look deep into the past, over vast stretches of time, the apparent stability of geography and climate that we perceive as individuals melts away and we see instead a picture of changing climate, with sea levels and whole ecosystems in flux. The population expansions and migrations of our ancient ancestors were governed by climate change and its effect on the ancient environment. Reconstructing past climates, or palaeoclimates, is an exciting field of science that draws on ancient clues that have been 'frozen in time' as well as our understanding of the relationship between the earth and the sun.

The earth's orbit is not a perfect circle, and so there are times (spanning thousands of years) when the earth is nearer the sun, and warmer, and other times when it is further away, and consequently colder. These cycles last around 100,000 years. As well as this, the tilt of the earth's axis varies, on a 41,000-year cycle, affecting the degree of difference between the seasons. The earth also wobbles a little around its axis, on a 23,000-year cycle. There are times when the factors affecting tilt and orbit work together to create exceptional chilliness – a glacial period. At other times, the factors come together to produce a very warm period, called an interglacial. This theory was developed by the Serbian mathematician Milutin Milankovitch in the early twentieth century.[5, 6]

During the 1960s and 1970s, researchers were able to pin down ice ages with increasing levels of accuracy using deep-sea cores, samples drilled from the seabed. Those cores contain the shells of tiny marine animals, called foraminifera, and the carbonate in their shells contains different isotopes of oxygen. The two isotopes of relevance here are ^{16}O, the lighter, 'normal' kind, and ^{18}O, a heavier version. Both are present in the ocean, but water that evaporates from the oceans contains more of the lighter kind. This means that water precipitating from the atmosphere – as rain, hail, snow or sleet – also contains more of the lighter ^{16}O than the seas. And it's *that* water, falling on to land or ice caps, which becomes frozen into large ice sheets during an ice age. That means there's more of the heavier ^{18}O left behind in the seas, and more of it gets incorporated into those tiny shells, during an ice age.[7] So marine cores, which can be dated using uranium series dating and by looking at the way the earth's magnetic pole has switched in the past, hold an amazing record of past climate and ice ages.

Formations in limestone caves – in stalagmites, stalactites or flowstone, or in the useful, all-embracing jargon, 'speleothem' (from the Greek for cave deposit) – also contain a record of past climate, depending on the proportions of oxygen isotopes that are present in the water that forms them. At any one time, the ratio of heavy and light oxygen isotopes in that water depends on global temperatures – and how much water is locked up as ice, as well as on local air temperatures and the amount of rainfall. While deep-sea cores are useful for looking at global climate, speleothem is very useful for investigating how climate has varied in a specific locality. Another indicator of past climates is pollen: soil samples containing pollen can be analysed to show the range of plants that were living in a particular area.

The Pleistocene was a period marked by repeated glaciations and ending with the last ice age. As ice sheets grew and then shrank back, sea levels would fall and rise. With differences of up to 60 million km^3 in the amount of water locked up as ice, sea levels fluctuated by up to 140m.[7] The oxygen isotopes trapped in deep-sea cores and speleothem can be used to draw up a series of alternating cold and warm stages called 'oxygen isotope stages', often abbreviated to OIS. Just looking at the last 200,000 years, there have been three major cold periods (corresponding with oxygen isotope stages or OIS 2, 4 and 6) interspersed with four warmer periods (OIS 1, 3, 5 and 7). But the Pleistocene really was one long, cold ice age. Interglacials account for less than 10 per cent of the time.[7]

At the moment we're enjoying a nice, warm interglacial, oxygen isotope stage 1. The last full-glacial period, OIS 2, lasted from 13,000 to 24,000 years ago. The peak of this most recent cold phase, around 18,000 to 19,000 years ago, is known as the 'Last Glacial Maximum' (or 'LGM'). OIS 3, from 24,000 to 59,000 years ago, was a bit warmer and more temperate, though still much colder than the present, and is called an 'interstadial'. OIS 4, from 59,000 to 74,000 years ago, was another full-glacial period, though nowhere near as cold as OIS 2.[4, 8] OIS 5, the last (sometimes called the 'Eemian' or 'Ipswichian') interglacial, was a warm, balmy period, lasting from about 130,000 to 74,000 years ago. Before that, there was another glacial period, OIS 6, starting at about 190,000 years after the preceding interglacial, OIS 7.

This level of detail might seem a bit excessive, but our ancient ancestors were very much at the mercy of the climate (as we still are today). For instance, there was a population expansion during the wet warmth of OIS 5, and a crash, or 'bottleneck', during the cold dryness of OIS 4. And sea levels fluctuated according to how much water was locked up in ice: during cold, dry periods, sea levels were significantly lower – by as much as 100m – than during warm, wet periods. Between 13,000 and 74,000 years ago (i.e. during OIS 2–4), the world was a drier, colder place than it is today. Although the map of the world was generally very similar, there was more land exposed; many of today's islands would have been joined to the mainland, and in places where the coast slopes gently the shoreline would have been much further out than it is today. This is of particular significance to archaeologists looking for traces of our ancestors along those ancient coasts – which are now submerged.

Stone Age Cultures

Archaeologists classify periods differently from geologists, depending on what humans were doing at the time. During the Stone Age, humans (including *Homo sapiens* and their ancestors) were making stone tools. This is before metals – copper, tin, iron – were discovered and used. In fact, in the scheme of things, metal-working is a very recent invention.

The Stone Age is traditionally divided up into the **Palaeolithic** (old stone age – roughly corresponding with the Pleistocene period), **Mesolithic** (middle stone age) and **Neolithic** (new stone age). These stages happened at different times in different places, so it can become quite confusing. The categories are also based on European prehistory, where much early

Ages and Stages

Age (Kyr) cycle	Geological period	Stage of glacial	Archaeological stage	
			Western Eurasia	Africa
13–present	Holocene	Interglacial (OIS 1) Warm	Metal Ages Neolithic Mesolithic	
13–24	Pleistocene	Glacial (OIS 2) Cold	Upper Palaeolithic	Later Stone Age
24–59		Interstadial (OIS 3) Warmish		
59–74		Glacial (OIS 4) Cold	Middle Palaeolithic	Middle Stone Age
74–130		Interglacial (OIS 5) Warm		
130–190		Glacial (OIS 6) Cold		

Table showing the relationship between geological periods, oxygen isotope stages and what humans were getting up to at the time.

archaeological work was carried out. But in terms of global archaeology, western Europe is a bit of a backwater, even a cul-de-sac[9] and so the terminology that has grown up there is sometimes rather unhelpful when we're trying to understand what was happening in the rest of the world. However, the categories at least provide us with a vocabulary and some kind of framework to help us think about the deep past.

Each stage is characterised by different styles and ways of making stone tools, but also by differences in the broader lifestyles of people. Put very simply (really too simply, as we shall see later), the Palaeolithic lifestyle was that of a nomadic hunter-gatherer, the Mesolithic saw a trend towards settling down, and the Neolithic saw the beginning of settled villages, cities, agriculture, pottery and organised religion.

Throughout the Palaeolithic, and in the Mesolithic, too, our ancestors were nomadic. They left barely a trace of their passing – no buildings, and very little in the way of possessions – and those material possessions that they did have were often made of what we now think of as biodegradable materials, so they have long since disappeared. When we find stone tools, we are often looking at something that was just a part of a more complex piece of equipment. Sometimes there are hints from polished areas of the stone tool suggesting how it might have been tied to something. Very rarely are the conditions right for organic materials – like pieces of wood or animal hide – to be preserved. When you consider the scarcity of the remains, it's quite amazing that we *can* find the occasional trace and, from this, reconstruct part of our collective (pre)history.

During the Palaeolithic period, there are changes in the types of stone tools people were making, and the period is divided up into Lower, Middle and Upper Palaeolithic (or, in Africa, the Early, Middle and Later Stone Age). Stone tools start to appear in the ground, in what is grandly termed the 'archaeological record', around 2.5 million years ago, made by early members of our own genus, *Homo*. These are crude, pebble tools, and the toolkit or stone tool 'technology' is called *Oldowan* after the sites excavated by Mary Leakey in the Olduvai Gorge. These basic tools continued to be made for hundreds of thousands of years. Our early ancestors were not great innovators! But we have to grant them some skill. In the wild, chimpanzees make tools out of easily modified materials like sticks or grass stems, and use stones to crack nuts; chimpanzees in captivity can be taught to *make* stone tools, but their products are still not as good as those Oldowan tools.[10]

The next stone toolkit to come along is called the *Acheulean*. And this toolkit is not found only in Africa. In fact, it is named after the site of St Acheul in France, where a characteristic 'hand axe' was discovered in the nineteenth century. Acheulean tools are found in Africa from about 1.7 million years ago, but it's not until 600,000 years ago that they are found in Europe. The tool from St Acheul is actually quite late: it dates to between 300,000 and 400,000 years ago. By 250,000 years ago, this technology had disappeared. Slightly strangely, this hand axe technology never reached East Asia. The fossil record suggests that people – probably *Homo erectus* – first made their way out of Africa around a million years ago, so it's unlikely that the East Asian pebble-tool-makers were direct descendants of the Oldowan people in Africa; they are more likely to have been, culturally, 'Acheuleans' who gave up making hand axes as they moved east.[10]

Hand axes are pointed, teardrop-shaped tools, flaked on both sides. It seems that nobody knows much about how these tools were used: were they designed for use in the hand, or hafted on to a shaft? Many archaeologists prefer simply to call them 'bifaces' (a general term for tools flaked on two sides). Acheulean bifaces are much more refined (though still big, chunky things) than Oldowan tools. Some of the bifaces are quite beautifully symmetrical, and many archaeologists have suggested that their form is governed by aesthetics as well as function. It's a tempting but ultimately conjectural idea, as there is no other evidence for any art at this time. And, once again, there seems to be extreme conservatism in tool-making throughout this period: there was very little invention. Over the huge time span of the Acheulean – from 1.7 million to 250,000 years ago – that culture hardly changed.[10]

But then a new sort of culture appeared. South of the equator in Africa, it's called the Middle Stone Age (MSA). Similar tools in North Africa, Europe and western Asia are called Middle Palaeolithic, or Mousterian, the latter a name that comes from the Neanderthal site of Le Moustier in south-west France. These labels carry with them a great deal of historical baggage, and the distinction between Africa and Eurasia is not particularly helpful. What we can say is that these tools seem to have been made by the archaic species *Homo heidelbergensis*, as well as by its (probable) daughter species: *Homo sapiens* and Neanderthals.

The MSA/Middle Palaeolithic differs from the Acheulean in that bifaces disappear from the toolkits. And tools are often made from stones that have first been shaped into a tool blank, or 'prepared core' – although, actually, the distinction isn't that easy as this technique was also used in the Acheulean. From studies of the wear on MSA/Middle Palaeolithic tools, it seems that the people were regularly mounting, or 'hafting', stone points on to shafts (although, as I mentioned, it is possible though not proven that so-called Acheulean bifaces were hafted). The new generation of tools, and ways of making them, were much more varied than the preceding stone tool technologies. There were other developments during this stage: people started to collect reddish, iron-rich rocks, perhaps for use as pigment; the first hearths appear; they had control of fire; and they started to bury their dead. From the composition of their bones, it also seems that people started to eat more meat in this period. Although people had hunted before, judging from artefacts such as the 400,000-year-old Schöningen spears from Germany, archaeologists believe that it is in the MSA/Middle Palaeolithic that hunting – not just scavenging – became routine.[10]

About 40,000 years ago, there was another change: to what is called the Later Stone Age (LSA) in Africa, and the Upper Palaeolithic in Eurasia. A huge and varied range of stone tools emerged, and people were also regularly making things out of bone. They were also using 'true' projectile weapons – spear-throwers with darts and bow and arrow (as opposed to just hand-cast spears)[11] – and they were building shelters, fishing and burying their dead with a degree of ritual that had not really been seen before. They also created magnificent art – particularly in Europe. Although this was probably not the first art (as there's much earlier evidence of pigment use in Africa), the painted caves of Spain and France are quite exceptional. From the fossils that have been discovered alongside archaeological finds, it is generally believed that the Later Stone Age and Upper Palaeolithic were made by just one species: *Homo sapiens*, modern humans. Us. Some palaeoanthropologists believe that the appearance of this new phase marks the relatively sudden beginning of truly 'modern' human behaviour[10] but others think that it is possible to see traces of fully modern behaviour much earlier, even before 100,000 years ago. They also suggest that this behaviour developed gradually, mirroring the physical, biological transition to a modern form.[9, 12]

The continuing debate goes to show that it's actually very difficult trying to tease out how and when this behavioural transition, to something we can truly recognise as modern human, happens. As far as stone tools are concerned, it's hard to find a clear signature of the earliest modern human tools. To begin with, the earliest modern humans were manufacturing exactly the same type of toolkits as their parent and sister species, *heidelbergensis* and Neanderthals; they all made bog-standard, Middle Stone Age tools. But there *is* a distinct MSA toolkit, from the northern Sahara, that has been attributed to modern humans. Similar

Basic Guide to Toolkits

LSA & Upper Palaeolithic tools
(from about 50,000 years ago)

Gravette
point

Solutrean
point

antler
spear-thrower

MSA & Middle Palaeolithic tools
(appears about 250,000 years ago)

Levallois
point

point

flake

Acheulean tools
(from 1.7 million years ago)

bifaces

Oldowan tools
(from about 2.5 million years ago)

hammerstone

bifacial
chopper

flake
scraper

discoid

core scraper

to other MSA toolkits in many ways, the *Aterian* includes stemmed or 'tanged points' (perhaps spear- or arrowheads). At a site in Morocco, further evidence of 'modern behaviour' has been discovered, alongside Aterian tools, in the form of shell beads.[13] Even so, before the LSA and the Upper Palaeolithic, it is difficult to discern the presence of modern humans based on stone tools alone. So fossilised skeletons become like holy grails for those seeking evidence of the earliest modern humans.

Dating Fossils and Archaeology

It is important to understand something about the techniques archaeologists can now draw on to date their discoveries. Dating is at the very centre of some of the biggest controversies and knottiest problems in palaeoanthropology.

Relative dating often involves judging the age of something by its position in the ground. So, for instance, you might judge something to be Iron Age if it lay underneath a Roman mosaic but on top of a Bronze Age burial. A more scientific approach, sometimes referred to as 'absolute dating', involves some means of measuring the age of the object itself, or at least the layer that it is buried in. Absolute dating techniques relevant to the period we're considering include radiometric and luminescence dating.

Radiometric techniques work by measuring the levels of different radioactive isotopes in materials. Radioactive isotopes decay over time, from one form to another. If the proportions of those forms can be measured, and the rate of decay is known, then the date of the object can be calculated.

The best-known radiometric dating technique is radiocarbon dating. The radioactively unstable $C14$ isotope decays to stable $C12$ over time. As $C14$ is present in the atmosphere, plants trap it when they photosynthesise, and animals eating plants also obtain it. This means that a living plant or animal contains a proportion of $C14$ to $C12$ that matches the ratio in the atmosphere. But when the plant or animal dies, it stops taking in any more $C14$; the $C14$ it already contains gradually decays to $C12$. So, by knowing the rate of decay, and then by measuring the proportions of the carbon isotopes in an organic object, whether that's a piece of wood, some charcoal or a bone, you can work out when that organic thing was last alive.

The precision of radiocarbon dating has recently improved with the use of accelerator mass spectrometry (AMS), which has also pushed the useful limit of radiocarbon dating back to 45,000 years ago. Accuracy has also improved with pre-treatment of samples to remove contamination from modern carbon, and with calibration, to take account of the fact that the amount of C14 in the atmosphere has changed over time (dates given in this book are calibrated, calendar years, rather than 'radiocarbon years'). Radiocarbon dates published before these advances – before 2004 – need to be treated with caution. Generally speaking, when archaeological materials have been redated using the new, improved techniques, the dates turn out to between 2000 and 7000 years older than the previous estimates. An added advantage of AMS radiocarbon dating is that it requires only a minute sample from a precious archaeological object. AMS radiocarbon dating is the best way of dating organic things – as long as they are less than 45,000 years old.[14] Beyond that, and if we're interested in early modern humans and their forays out of Africa, going back more than 50,000 years ago, we have to look to other methods.

Two other radiometric dating techniques that can be used to date rocks are uranium series and potassium-argon dating. Uranium series dating uses radioactive isotopes of uranium and thorium, which decay to stable lead isotopes. It depends on soluble isotopes being precipitated and then changing to insoluble forms, so it can be applied to speleothem and coral. Potassium-argon (and argon-argon) dating is used to date volcanic rocks. Argon can escape from molten rock, but is trapped in solidified lava. So if archaeological finds or fossils are found between layers of speleothem (in limestone caves) or between layers of volcanic tuff from ancient volcanic eruptions, these techniques can provide a date, or at least a date range, for the discoveries.

A relatively new technique that is proving incredibly helpful in Palaeolithic archaeology is luminescence dating. It is used to pin a date on the last time that grains of quartz or feldspar were exposed to either heat or light. It can be used to date the layers of sediment that an object is buried in, or sometimes even to date an object itself if it was heated – for instance, a piece of pottery or a hearth stone. Luminescence dating is a very powerful tool, useful for pinning an age on objects that are just a few years old, all the way to things that have existed for a few million years.[15]

The way that luminescence dating works is, I think, quite mind-blowing. When grains made of natural quartz crystals (i.e. grains of sand) are exposed to ionising radiation – from naturally occurring radioactive elements like uranium as well as cosmic rays – electrons get trapped in tiny flaws inside their crystal structure. Light or heat makes the crystal release its electrons. But once a quartz grain is buried, it starts to accumulate electrons again ... until somebody comes along and digs it up. Samples for luminescence dating have to be kept completely in the dark when they are collected.

Back in the lab, the quartz grains are sorted out from the sample, under very dim, red light. Then they are exposed either to heat (in thermoluminescence, or TL, dating) or light (in optically stimulated luminescence, or OSL, dating). Then the crystals within them release their trapped electrons – making them glow. By measuring this luminescence, while knowing levels of natural radiation at the place where the quartz grains were buried (from other sediment samples and measures of cosmic radiation), the length of time that the crystal was buried can be estimated.[15]

Another method that measures levels of trapped electrons, again resulting from bombardment with radiation within sediments, is electron spin resonance (ESR). This technique works well for ageing tooth enamel, which is also a crystalline material – so it is very useful for dating hominin fossils.[16]

Genetic Studies

Quite recently, another branch of science has begun to provide important clues about our ancestry, about how we are all related to each other and even about the way the world was colonised. This time, however, the evidence is not buried in the ground but inside *us*, because the DNA contained in each cell of each of our bodies holds a record of our ancestry. Getting samples of DNA is surprisingly simple – and painless. DNA can be collected from volunteers using just a cheek brush or saliva swab. These samples contain cells, and inside those cells is the precious DNA.

While everyone's DNA is mostly identical, there are some differences. There have to be, otherwise we'd all look exactly the same, clones of each other. Some genes govern our appearance, while others control the machinery of life. There are differences in those genes, too. You can't tell just by looking at someone, but they might have a different blood group from you, a slightly different enzyme for breaking something down inside their cells. The differences in these active genes and the protein products they make are constrained by natural selection. If a mutation happens in an important gene, it could make the protein product of that gene work better, worse, or perhaps have no effect at all.

If the effect of a mutated gene is deleterious, it may be that the individual carrying it won't be able to survive at all, or, perhaps, won't live long enough to pass their genes on. So the mutant gene will disappear from the mix of genes in the population, or 'gene pool'. If the product of a mutant gene proves advantageous, it could be that the individual with it gains a better chance of survival, and is therefore very likely to pass the new version of that gene on to their offspring. So, gradually, over many generations, a really advantageous gene can spread through a population. If the mutation is neutral, then it is pure chance whether it sticks around or disappears from the gene pool.

But there are also long stretches of our DNA that mean nothing to the cell; they are the bits between genes which are never 'read' to produce proteins. Sometimes they contain parts of old, disused genes, or historical bits of genetic material inserted into the chromosomes by viruses. These unused sections are not subject to natural selection like the working genes. Alterations appearing by random mutations in these regions will not be weeded out in the same way. This means that they are quite useful for tracing genetic lineages.

Most of our DNA is coiled up in chromosomes inside the nuclei of our cells; there is also a little bit of DNA in tiny capsules inside the cell. These are mitochondria, the tiny 'power stations' of the cell, taking fuel – sugar – and burning it to produce energy. The genes in the mitochondrial DNA have a very specific but incredibly important job, controlling energy transformation inside the cell. To a large extent, because they're so hidden away, they too are protected from pruning by the grim reaper of natural selection. Mutations accumulate more rapidly in mtDNA than in the nuclear DNA.[17] So this means that mtDNA is particularly useful for reconstructing genetic family trees. Geneticists can assume that there is a standard rate of mutation within the mitochondrial DNA, and that, unless they really hamper the work of the mitochondria, those mutations will persist.

The other important thing about mitochondrial DNA is that it does not get mixed up at each generation like the nuclear genes. Gametes (eggs and sperm) contain only half the number of chromosomes contained in every other cell in your body. But when the sperm and the egg are first made, it's not simply a question of dividing the pairs of chromosomes up – before

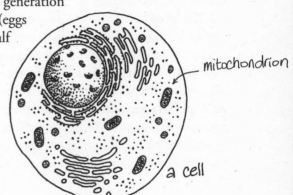

mitochondrion

a cell

that happens, each pair of chromosomes swaps DNA with its partner, in a process called recombination. That means that the twenty-three chromosomes left in the gamete contain new mixes of DNA that weren't there in the father or mother.

Sexual reproduction – with this shuffling of genes at each generation – means that genetically 'new' and different individuals keep being created. This in turn creates variability within the gene pool. This variation is incredibly important: it means that, if circumstances change, if the environment changes around us, there will be some individuals who may be able to survive better than others. Biology cannot predict what changes may be needed at some far-off point in the future, but species that have developed this way of 'future-proofing' themselves, through sexual reproduction, have been successful in the past – and so we still do it today. But for geneticists trying to trace genes back through generations, it is a nightmare, because the genes keep jumping about.

Mitochondrial DNA, on the other hand, does not get involved in recombination, and stays, chaste and untouched, inside the mitochondria – which we all inherit from our mothers. The sperm from our father contributes just its nucleus with twenty-three chromosomes at fertilisation. The egg also contains twenty-three chromosomes, as well as all the other cellular machinery – including mitochondria. This means that all your mitochondria, and the DNA they contain, are inherited from your mother. And she got hers from her mother, and so on. Geneticists can therefore trace back maternal lineages using mitochondrial DNA (mtDNA). Of the nuclear DNA in our chromosomes, there is actually one bit that doesn't recombine, and that is part of the Y chromosome (which only men have). So this can be used to trace back paternal lineages.

In fact, other genes, in the nuclear DNA, *can* be traced back, although the history of these genes is more convoluted and much harder to track back through time than the non-recombining bit of the Y chromosome or mtDNA. Techniques for analysing DNA, for reading the sequence of nucleotide building blocks, are getting better and faster, almost by the day. Many labs are not looking at just individual genes from mtDNA or nuclear DNA but are now attempting to read *all* the DNA – mapping entire mitochondrial and even nuclear genomes. It's an exciting time.

In terms of investigating our ancestry, it's those tiny differences in our DNA, nuclear or mitochondrial, that are important. The traditional approach to studying genetic variation was 'population genetics', where frequencies of different gene types are compared between different populations. The problem with this approach is that it is particularly subject to distortions, as people migrate and populations mix. The approach of building 'family trees' of genes, from mtDNA, the Y chromosome and other nuclear DNA, makes for a much clearer picture of our inter-relatedness and our ancestry. The branch points of the trees correspond with the appearance of specific mutations.[18]

There are obviously ethical issues involved in the collecting of DNA: it should be done only with the consent of the individual concerned, it should be used only for the purposes initially laid out, and should not be passed on to any third parties. Some people have worried that genetic analyses of human variation might be used for a racist agenda, but these fears can be allayed as this science actually contains a strong anti-racist message. As eminent geneticist Luigi Luca Cavalli-Sforza has put it: 'Studies of human population genetics and evolution have generated the strongest proof that there is no scientific basis for racism, with the demonstration that human genetic diversity between populations is small, and perhaps entirely the result of climatic adaptation and random [genetic] drift'.[18]

The Illusion of the Journey and the Folly of Hero-Worship

This book is about several different sorts of journey. There are the physical journeys made by our ancestors as they spread around the world. Then there is a more abstract, philosophical journey, with the gradual transformation of body and mind to something we can recognise as fully modern human. And then, there is my own physical and mental journey. I spent six months travelling around the world, meeting all sorts of experts and indigenous people, and experiencing the huge range of environments that humans manage to survive in today, from the frozen taiga of Siberia to the blazing aridity of the Kalahari.

When we cast our minds back and imagine our ancestors, sometimes surviving against the odds, and managing to make their way into and survive in the most extreme environments, it may inspire us with humility, awe and great admiration. And it certainly *is* an awe-inspiring story: from the origin of our species in Africa, to the colonisation of the globe.

But it's all too easy to start thinking of this journey as a heroic struggle against adversity, and to imagine our ancestors setting out with the explicit intention of colonising the world. In fact, this 'human journey' is a metaphor – as it wasn't really a journey at all – and they had no such goal in mind. I think that words like 'journey' and 'migration' are useful metaphors for describing how populations have moved across the face of the earth over vast stretches of time, but it's very important to realise that our ancestors were *not* on some kind of quest to get on and colonise the world. Certainly, they were nomadic and they would have moved around the landscape as the seasons changed, but most of the time they would not have been purposefully moving on from one place to another. It's just that, as populations (of humans or other animals) expand, they spread out. But I think it's acceptable to use words like journey and migration to mean something more abstract, a diaspora happening over thousands of years. So there was no quest, and no heroes. We may feel humbled by the survival of our species, though set about by vicissitudes, and we may marvel at our ancestors' ingenuity and adaptability – but we must remember, through it all, that they were just *people* – like you and me.

1. African Origins

Matay making beadwork.

Skhul & Tabun Caves

Salalah, Wadi Darbat

Nile

Herto

Addis Ababa

Omo Kibish

Omo River

Lake Turkana

Norikiushin

Enkapune ya Muto

Mumba

● places visited

○ places mentioned

Nhoma

Windhoek

Diepkloof

Cape Town

Blombos

Klasies River & Boomplaas

Pinnacle Point

Meeting Modern-Day Hunter-Gatherers: Nhoma, Namibia

I was sitting at a wooden table, under a thatched roof, somewhere out in the bush, in Namibia. A small but noisy flock of grey louries was flying about in the trees around the camp, calling 'go-away!' loudly. Beyond, the landscape of scrub and grassland stretched off, unbroken, into the distance. I was excited about being in Africa: it was the starting point of my journey and where the story of the human colonisation of the world begins.

I had flown into Windhoek and from there taken a small plane to fly out into the Kalahari Desert. We approached our destination, somewhere on the northern edge of the Nyae Nyae conservancy area, and circled, looking for the landing strip: a bare patch of dusty ground in the middle of the bush.

We landed, kicking up great clouds of dust and sending the crowd of inquisitive children that had gathered at the end of the airstrip running, screaming away. They returned to watch as we taxied, came to a halt, leapt out and started to unload. A few of the boys were holding long, straight sticks, and they were trimming off side twigs and whittling them down with knives.

It was extremely hot and very dry. Small trees and bushes were dotted about, with swathes of pale golden grass between. It was a short drive to Nhoma, a Bushman village situated on a ridge looking out across this landscape. There was nothing else for miles around.

I got out at a lodge near the village, and met Arno Oostuysen, who introduced me to field guides Bertus (himself a Bushman) and Theo (a young South African spending a year at Nhoma). We walked through the bush on sandy paths into the village. There were shelters, about twenty of them, around the edges of a clearing. The shelters were simply built, with branches anchored in the sandy ground and bent in to form a dome shape, covered with sheafs of grass.

Theo said there were 110 people living in the village, and that they were mostly members of just two extended families. This was a 'matrilocal' society: men married out and went to live with other Bushman bands in neighbouring villages, while women stayed in the village of their birth.

He took me to meet one of the older men of the village. The Bushmen do not have leaders as such, but this man had the hunting rights over the land around, and it was courteous to thank him for letting me visit.

These people looked completely different from the black Namibians I had seen in Windhoek. The Bushmen were short and very lightly built, and relatively pale-skinned. They had the tightest curls of black hair on their heads, and open, wide faces, with high cheekbones. I looked at some of their profiles, and, from the side, their faces were quite flat below the nose: they did not have the jutting jaws of other sub-Saharan Africans. They had very narrow shoulders and a very pronounced curve in the lumbar spine which placed the hips far back.

Some people were sitting in groups, gathered in the shade under the trees. One woman was making ostrich-shell beads. Having carefully chipped pieces of shell into tiny discs, she was now drilling holes into them. She had a plank of wood about half a metre long on the ground in front of her. She placed the shell discs into small holes worn into the wood by previous beads, then drilled them by twirling a long, pointed stick between her hands. She drilled into one side, then turned them over and drilled the other. The beads would then be polished before they were ready to be threaded into a necklace or bracelet.

I walked over to where three other women were sitting on colourful cloths spread out on the ground, piles of small glass beads between their legs. They were threading beads to form colourful bands which would make bracelets, necklaces and headdresses. Some of the women had rows of traditional small, black, linear scars on their thighs and faces. Children of various ages were sitting with them, watching the women at their beadwork. I sat down and watched. After a while I indicated that I'd like to try making something, and one of the women started me off, with two rows of yellow beads on a new length of thread, which she then handed over to me to continue. I was given my own small pile of beads and I got to work. It was very calm, almost meditative. Patterns started to emerge in the beadwork. More children came over to see what I was doing and to inspect my slowly growing band of beads. All the time, there were soft conversations going on, and every now and then the children would start up a song that would spread throughout the group. The words sounded very strange to me. There were sounds like vowels and consonants I was used to, but there were also clicks. Some words seemed to be almost entirely made up of clicks.

The language was one of the things that brought me here, to this isolated village in the Kalahari. Click languages are unique to populations in southern Africa and Tanzania – including the Bushmen (San) of Namibia and Botswana, and the Khoi Khoi (Khwe) of South Africa (sometimes these people are collectively referred to as Khoisan). Historically, these populations have had different lifestyles: the Bushmen being traditional foragers (i.e. hunter-gatherers) and the Khoi Khoi traditional herders.[1] Although their languages differ, they are bound together by these common sounds: characteristic clicks made by snapping the tongue away from the teeth or the hard palate. For many years anthropologists and linguists have suspected that the languages shared by these now widely separated tribes might indicate a very ancient, common ancestry.[2]

After a while, one of the little girls spoke to me in English.

'What is your name?' she asked, carefully pronouncing the words. I told her, and asked what hers was. 'Matay,' she said.

I asked Matay the name of the woman who was teaching me beadwork. She was called 'Tci!ko' – pronounced 'Jeeko', with a click on the second consonant.

I took some of the coloured beads and laid them out on the cloth between us. 'This is red … yellow … and green,' I said. 'What do you call them?'

Matay understood and tried to teach me the words for the three colours but I found it really difficult. There were clicks in the middle of words, sort of on top of other consonants as well. And not only that, but the clicks were not all the same!

An elderly woman – she must have been at least seventy – came over to see what I was doing. Her face was deeply lined and she had only a few teeth left. The other women and children treated her with great deference. She held out her hand and I gave her the fragment of beadwork I had made. She turned it over and inspected the threading carefully, then handed it back to, beaming with approval. The other women nodded and smiled as well. Although I felt quite isolated by my lack of understanding of the language, this was a form of communication that did not need words.

Later in the day I had a click-speaking lesson from Bertus. We sat on low wooden stools in the doorway of one of the shelters. Inside the hut, bags and clothes hung from wire hooks, and there were a few reeds with guinea-fowl feathers attached. (This was a traditional game called *djani* – or the 'helicopter game' – a winged toy, with a short thong-and-resin weight at the bottom. Thrown into the air, the reed would spiral down like a sycamore key, and the player would try to catch it with a long stick. One of the hunters later gave me a demonstration.)

Bertus explained that there were four clicks in Ju/'hoansi (pronounced something like 'Voo–nwasee', with a click in the middle). The Ju/'hoansi are a group of Bushmen living around the border between northern Namibia and Botswana.[3] They are the same group that anthropologists used to call '!Kung'. (Although the indigenous people of the Kalahari know themselves by their individual, tribal names (such as Ju/'hoansi), they are happy to be known collectively, and refer to themselves in English, as Bushmen. Fifty years ago, this was a term that many anthropologists avoided, as it was one given to those communities by the first European settlers and it was considered to be derogatory. Anthropologists therefore sought another, more acceptable term, and they settled for a word used by other southern Africans to describe the Bushmen: 'San'. This word has become common but itself has derogatory overtones as it apparently means 'cattle thief'.)

Bertus went through the four clicks, accentuating the shape of his mouth and repeating each of them so that I could see where he was placing his tongue in relation to his teeth and soft palate. So here they are:

1. / The dental click: made by placing the tip of the tongue against the front teeth then snapping it away. It's like a tutting noise.
2. ≠ The alveolar click (I found this difficult because it sounded so similar to the first one, but the tongue is placed in a slightly different place: just behind, rather than against, the front teeth).
3. ! The alveolar-palatal click: the tongue is placed against the hard palate just where it starts to dome up – behind the 'alveolar' ridge. (For those who are interested, *alveolus* means 'cup', and so the word 'alveolar' applied to the upper and lower jaws refers to the part of the jaws that bears the cup-like sockets for the teeth.) When the tongue is pulled down quickly, it makes a loud 'nop!'
4. // The lateral click: the tongue is positioned as for the second click, but only the side is released. 'Like if you call a dog,' said Bertus. To me, it sounded like a noise you might make to gee up a horse.

Genetic studies of people speaking click languages, looking at both mitochondrial DNA and the Y chromosome, have shown that the Ju/'hoansi have remained quite isolated from populations around them, while other Bushman groups and the Khwe people have mixed extensively with Bantu populations. The Ju/'hoansi have unique, deep-rooted genetic lineages. A recent study found that the branches leading to different groups of click-speaking people appeared very early in the family tree of modern humans. It's impossible to prove, but geneticists suggest that click languages may have been around for tens of thousands of years, originating before any movement of people out of Africa.[1]

For years anthropologists have debated whether or not the Ju/'hoansi have roots going back into the Later Stone Age, or perhaps even earlier, in the area that they still inhabit.[3] In the 1950s, anthropologist Lorna Marshall found that the Bushmen believed their ancestors to have lived in the region forever. The genetic lineages of the Bushmen suggest that they are indeed descendants of very ancient inhabitants of the area.

The children thought my attempts to speak Ju/'hoansi hilarious. There were plenty of children in the village, of all ages. And the older kids of nine or ten were looking after the toddlers; some were carrying tiny ones round in shawls tied to their backs. A few small boys were pushing round model cars, made of coat hanger wire, with such great attention to detail that some even had wire 'treads' on the wheels to make authentic tyre tracks, and grills on the front. One was unmistakably – in skeletal form – a Toyota Hilux. Theo said that each time a new vehicle arrived at the village it was only a couple of days before an accurate wire model of it appeared on the scene.

A few of the boys were playing a game on the edge of the village, with the thin sticks I had seen them whittling earlier, at the airstrip. They were throwing the sticks, hard, at a small grass-covered mound: the sticks ricocheted off and sped away like arrows. There didn't seem to be any particular aim to the game, and no one was scoring, winning or losing. Theo said this was a way of doing things that permeated Bushman society: they were not competitive. Anthropologists visiting Bushmen communities remarked on the lack of inter-group violence as well.[2]

Two men were sitting beside one shelter, attending to their hunting equipment. One had a small bowl of water with sinews soaking in it, and he was wrapping these shiny, silver ligaments around the ends of his bow, to help keep the string (also sinew) in place. The other hunter was inspecting his arrows, holding each one up and looking down the shaft to check that it was straight. They were composite arrows, with a shaft made of a thick grass stalk, and a detachable end or foreshaft. Theo explained that this was designed to break off from the rest of the shaft once the arrow had found its quarry; the remaining, shortened arrow would be less likely to get knocked out as the animal ran on. About 10cm of the shaft just behind the gleaming, beaten, steel wire arrowhead was covered in a black substance. Theo cautioned me not to touch it: it was poison. The hunters used the larvae of *Diamphidia* beetles to anoint their arrows with a deadly mixture. Hunting would involve tracking an animal until the hunters were close enough to fire poisoned arrows into it, then they would carry on, tracking it at leisure, until the quarry weakened from the poison and exhaustion and the hunters could approach it and deliver the *coup de grâce*. Theo introduced me to the hunters. They were brothers-in-law, !Kun and //ao, and they were going to take me out – right out – into the bush the following day.

That night I slept in my tent at the lodge and was woken up regularly throughout the night by the incredibly loud, explosive noise of Zambezi teak pods popping. And if it wasn't the teak, it was the sand camwood. And if it wasn't the camwood, it was all sorts of weird noises that were quite frightening when you didn't know what they were. But I was in a safari tent with a door, and I was quite secure. So each time I woke up, I snuggled down a little further under my duvet (it was quite chilly) and was soon asleep again.

The following day we set off early, driving 13km away from the village to the nearest waterhole. In the Kalahari there are small rain showers in November and December, and then plenty of rain with summer thunderstorms from January to April. The rest of the year is almost entirely rainless. I was visiting Namibia during the hottest, driest time of the year. Bush fires are common, and in fact there had

//ao
mending
his bow

been one quite close to the village; Arno had driven out to help fight it the night before. The smoke was still rising into the sky behind the ridge. Theo said that Bushmen used to start fires on purpose so that vegetation would send out new shoots and attract the game, and also because it cleared the bush and made it easier to track game. This practice was now banned in Namibia, because of the danger to the people and animals of the Kalahari.

———

The Kalahari Basin is a huge area, extending across the borders of four countries: South Africa, Angola, Botswana and Namibia. It is 'semi-desert': very sandy and dry but still supporting a wealth of plants and animals. What it really lacks is surface water, and, weirdly, this is what seems to have facilitated the survival of the Bushmen and their way of life. In less arid parts, Bantu have moved in to farm the land – and the Bushmen have moved out. But in more arid areas, the Bushmen are well adapted to life in this environment, which seemed so incredibly harsh and unforgiving to me.

Their diet includes the storage organs of plants that have adapted to the dry landscape: bulbs, tubers and roots. And they are expert trackers of the animals that leave their footprints in the sandy soil of the bush. There are no streams in the southern Kalahari: water beneath the ground surface upwells to feed waterholes: some temporary, just there during the wet season, and some permanent, lasting right through the dry season as well. The Bushmen are tied to the waterholes: between these muddy, life-giving pools are vast stretches of uninhabitable desert. And the waterholes are also where the animals go to drink, under cover of darkness. Starting at the waterhole, !Kun and //ao would be able to track antelope that had been drinking there the night before.

Picking up the tracks of an oryx, the hunters set off at a determined pace, an extremely fast walk that occasionally broke into a gentle jog. As they passed raisin bushes, the hunters would grab the small orange fruits to eat. //ao offered me one: it was tough but sweet and tangy. Every now and then I heard a loud clucking noise coming from low down in the bush: the cry of the southern yellow-billed hornbill. We continued, and I tried to follow the spoor as well. Sometimes, the cloven-hoofed prints were obvious on the sandy game trail, but then the oryx would leave the track and head into the bush. We would strike out, through the sparse grass and evil, low-lying thorns, and I couldn't believe the hunters were still on the trail. But then they'd point out a snapped twig, a crushed leaf, a pile of droppings or another hoof print, and I could see that they were still right on track. This happened again and again. I was amazed at their intuitive ability to track an animal that had passed this way many hours before. At one point, the hunters stopped dead. They had found a large raisin bush – *Grewia flava* – with long, straight branches. !Kun took his axe and chopped down four long branches. This was a precious resource: it was the wood used to make bows and spears. Then we were back on the trail of the oryx.

But not long after – perhaps half an hour – the hunters stopped again. They knelt down and I thought they were inspecting tracks, but instead there was a pile of nuts, like large almonds, on the ground. They were both scooping up handfuls of them, shaking them free of dirt and sand, and putting them into !Kun's antelope-skin knapsack (a whole skin, sewn together with the legs forming the handles), and I helped them. These were Manketti nuts, much prized by the Ju/'hoansi. (I later cracked one between two stones and it was delicious: a bit like a brazil nut.) The neat pile seemed strange – we were a long way from the Manketti groves. But elephants eat the nuts too, although they can't digest them: the nuts we'd collected had been left behind after a pile of elephant dung had disintegrated. !Kun and //ao were pragmatic in their approach to hunting. While on the tracks of a potential quarry, they would collect wood, berries and nuts as well.

As they were moving and tracking, the hunters would speak softly to each other. I could still hear the clicks very clearly. The geneticists who investigated the ancestry of the click-speaking people conceded that these languages might have survived for tens of millennia by chance. But perhaps clicks had been retained because they provided a great way of communicating while hunting.[1] This hypothesis is pretty much impossible to test. All I can say is, having been out with the Bushmen, when they were whispering to each other the clicks were still crystal clear. I don't think there would be much hilarity generated from a Ju/'hoansi game of Chinese whispers. I can add my own extension of this hypothesis, completely untested (at the time of writing), that clicks, being high-frequency sounds, wouldn't travel as far as other vocalisations. So a language with clicks might provide a clear means of communication between hunters, moving through the bush close to each other, while being unheard by a more distant quarry.

We had started early in the morning, when the sun was low and the bush was just starting to lose the chill of the night, but the sun soon turned the day from cool to warm to blazing hot. I knew I was sweating, even though I felt quite dry. As soon as sweat appeared on my skin, it evaporated, but I still looked sweatier than the hunters. They were wearing very little, which probably helped in this respect. I had opted for long linen trousers to save my legs from thorns and insect bites, while being cool. As I'd anticipated some running would be in order, I was wearing a sports top and a vest to protect my very white midriff from the harsh sun. I had decided my already tanned shoulders and arms would be safe under a decent layer of high-factor sunscreen. The hunters were wearing very little in comparison: just bead-embroidered loincloths and headbands. But they were also shorter, leaner and much more slightly built than me. Small stature and build makes for a larger surface area to volume ratio: relatively more skin surface for sweat to lose heat from. The idea that Bushmen may be physically adapted to endurance running (or walking) in hot conditions, by virtue of their smallness and consequent ability to lose heat effectively, without sweating much, seemed to be borne out by how much we were all drinking. I could feel dehydration stalking me that morning as we tracked the oryx. I kept up with !Kun and //ao, but I got very hot and drank much more water than they did. While each of the hunters had brought with them just half a litre of water, I had three times as much in my Camelback.

In international sports African athletes dominate distance running. Investigations have shown that there may be a number of reasons for this. Elite African runners have a greater fatigue resistance than non-Africans – able to run for some 20 per cent longer before fatigue sets in. This seems to be due in part to a difference in the composition of the muscles. Body mass is also an important factor: larger and heavier runners, in a hot environment, will not lose heat as quickly as smaller runners, and will reach the point of overheating to exhaustion much quicker. A study run by a group of sports scientists, including Tim Noakes from Cape Town University, showed that, even in cool conditions, larger Caucasian (European) athletes sweated more and had higher heart rates than smaller, African runners. In hot conditions, the Caucasian athletes ran slower than their smaller African colleagues. This is fascinating because it suggests that experienced athletes 'know' their limits when it comes to heat exhaustion and adjust their pace accordingly. The African runners ran, on average, 1.5km/h faster than the Caucasians, without overheating.[4]

I'm certainly not a trained athlete, so this wasn't a fair comparison, but after three hours of walking and running in the bush I had sweated a lot, and I had drunk all of my water. !Kun and //ao hadn't even touched theirs. And the tracks had become confused. The oryx tracks were crossed by a kudu being chased by a hyena. The hunters decided it was time to circle back. I was glad they knew the way. With the sun high in the sky, I had lost all sense and means of judging direction. As we made our way back, I saw a truly enormous bird launch itself from a small tree and heavily flap off through the bush: a Kori bustard.

It seems that endurance running was generally important to our early ancestors. While the Bushmen seem optimally adapted to running in the heat, we all have anatomical features in our bodies that suggest we evolved to perform well in endurance running. There are things about the way our bodies are designed that allow us to store energy in tendons and ligaments as we run, providing us with an efficiency gain. You don't need to be a habitual runner to have these adaptations: they are firmly there, in the blueprint of all our bodies.

Take a look at your feet. They are very strange for the feet of an ape (which, strictly speaking, we are). They're designed specifically for standing, walking and running on. We have lost the ability to grasp things with our big toe that our close cousins, chimpanzees and gorillas, still possess. Instead, it was more important for our big toe to have been brought into line with the others, to form a stable platform. We also have arches in our feet: a long arch down each side, and an arch across the foot as well. These arches are held in place by ligaments and tendons – which stretch. When you're running, each time your foot hits the ground these ligaments and tendons act like springs, stretching, and then giving energy back as your foot lifts off. Our Achilles tendons, attaching to the lever of the calcaneum, or heel bone, are massive: another stretchy spring. Humans also have very long legs, which makes for a good, long stride length. And we've had long legs for quite a while. The first hominins were australopithecines; they walked around on two legs, but their limb proportions were like those of chimps: long arms and short legs. Early species of *Homo* also had chimpanzee-like limb proportions, but by the time *Homo erectus* came along, at around 1.9 million years ago, long legs had appeared as part of the human package. We also have strong back muscles to stop us pitching forward while we're running, and very large bottom muscles. Gluteus maximus swings the leg back at the hip joint; it's hardly used at all during walking, but comes into its own when we run.[5, 6]

Although the small size of the Bushmen makes them able to lose heat easily, and therefore suits them to running in the heat, all humans have adaptations to heat loss which may also be related to keeping cool during endurance running. We have very little body hair, so we lose heat both by convection and evaporation of sweat. And we have lots and lots of sweat glands.

All these features make us good runners. Compared with other animals – four-legged ones – we are not very fast sprinters at short distances, but we are actually excellent endurance runners. And we are unique among primates in developing this capability of endurance running. Over long distances, trained humans can even outrun horses and dogs.

Some explanations of human anatomy and locomotion have suggested that the apparent adaptations to running are actually just by-products of a body designed for *walking* on two legs. Certainly, long legs improve efficiency in walking as well as running. But the springiness of the leg and foot, and those big bum muscles, aren't really used in walking – but they are great for running. American anthropologists Dennis Bramble and Daniel Lieberman gathered the physical evidence and published an article in the journal *Nature* in 2004, arguing that, while walking was undoubtedly a fundamentally important way of getting about, for us and our ancestors the role of running had been overlooked. They suggested that the human body has evolved to cover long distances: walking and running.

But why run at all when walking uses up less energy? Well, in the days before weapons like bows and arrows, endurance running might have enabled our ancestors to get close enough to an animal to throw weapons at it from close range, or even to run it to exhaustion. It's difficult to imagine how hunting large mammals could have been achieved in those early days without *some* running. I mean, imagine trying to walk an antelope to exhaustion. Equally, endurance running could have helped in scavenging: getting to a carcass before carnivores.[5] Although running uses more energy than walking, it's still *worth* it if it increases the chances of getting an animal to eat. For instance, a three-hour endurance run costs around 900kcal (about 30 per cent higher than walking the same distance). If the hunter manages to kill a 200kg duiker, he'd get around 15,000kcal in return. If he brought down something bigger, say a 200kg wildebeest, the return could be a stonking 240,000kcal, for the same effort.[6]

So you could imagine a scenario where, over tens of thousands of years, early humans who had bodies that were better at endurance running – through random but lucky mutations in their genes – had an advantage, passed their genes on … and so our bodies today bear witness to that early way of life. However we make a living today, we have legs and bottoms befitting a persistence hunter.

Although theories of form and function in human evolution should not stand or fall depending on ethnographic evidence (why should any humans today be behaving in exactly the same way as our earliest ancestors?), it's interesting to look at how modern hunter-gatherers hunt. The Bushmen who still hunt today, like !Kun and //ao, use a combination of endurance walking and running over great distances. While they use bows and poisoned arrows, so don't need to get as close to the animal as would have been necessary before this development, they still have to cover a *lot* of ground, quite swiftly, in pursuit of their quarry. And even though they have bows and – poisoned – arrows, the technique itself is still persistence hunting.[6]

It's hard to know how much meat hunter-gatherers like the Bushmen actually eat. Theo thought that it was quite rare for the hunters to bring down an oryx or kudu, but, when they did, the whole village would feast. Nothing belonged to any one person: sharing was the unwritten law among the Bushmen. Smaller animals were caught on a more regular basis, with snares – and hooks. Hanging from the trees in the village were long sticks ending with vicious-looking hooks: springhare probes. Hunters would push them down the burrows to catch these large rodents.

Being able to hunt in the middle of the day gave the Bushmen a unique advantage, a particular niche, away from other predators and scavengers. (Presumably this would also have been a key advantage for our early ancestors as well. It's impossible to know if early *Homo* had the ability to follow the tracks of animals, but it seems reasonable to assume that they did. Surely they must have also worked out what it meant when vultures gathered on the horizon.)[6] While the lions, leopards and hyenas hunted at night, the Bushmen of the Kalahari could track and hunt in the noonday sun. But when dusk fell, it was time to get back to the safety of the village. Bushman villages are always situated a safe distance away from waterholes.

But I wasn't going home to the village that night. Whereas the Bushmen thought it wise to leave the waterholes to the predators, I was going to try sleeping out in the bush. I had planned to stay out there – just 20m from the waterhole – with a bedroll (two sleeping bags in a canvas bag), basic provisions and a video camera, to make a diary. I wasn't entirely alone: Theo (with gun) and cameraman Rob were both going to sleep another 20m away from me. But that was far enough away for me to feel alone. As dusk gathered around us, and barking geckos set up their staccato chorus in the fading light, we collected thorn branches and arranged them around our respective bedrolls to keep the hyenas at bay. I'm not kidding. As darkness fell, I was startled to hear a hyena howling at the waterhole, just metres away. It was an eerie sound. We shone our torches out into the darkness but saw nothing. Theo and Rob barricaded me into my small camp then retired to their own thorny kraal.

For about an hour I sat on my bedroll, listening intently to the sounds of the night around me. The long grass was dry and rustly, and I could quite clearly hear the sound of LARGE animals walking through it. I caught myself holding my breath. I had no idea what (though I knew hyenas were out there) or how many creatures were walking around my sleeping place to the waterhole. I heard hyenas again: blood-chilling howls. Theo had described these animals as the most fearless creatures of the bush. Lions, leopards and elephants would often run away if startled or shouted at by a human being. Hyenas stood their ground.

Then, in the stillness, I heard a strange sound: quite quiet, but rhythmic, coming from the direction of the waterhole. It took me a while before I worked out what it sounded like. It was the sound of lapping. Like a huge cat with a massive saucer of milk. Was that the sound of a leopard drinking? I was too scared to turn on my torch.

Eventually, in a quiet interlude, I plucked up the courage to move around, turn on a small headlamp and get into bed. My bedroll was tucked under the leaning branch of a tree. Theo said I was less likely to get stepped on by an elephant that way. I could hear rustlings close by, and in the light of my headlamp I saw a tiny mouse and straw-coloured stick insects in the grass around me. I shone a larger torch around but couldn't see anything moving outside my circle of thorns.

Lying down and turning the headlamp off, I saw a bat flying low over me. He came back again and again, flying just inches from my face, so close I could hear his wings flapping. He was a benign presence. I knew what he was, and I knew what he was up to: hoovering up flying insects from the air. It was the other, rustling, mysterious sounds that gave me goosebumps and brought tears pricking into my eyes. Tucked up in my sleeping bag, I felt incredibly vulnerable. My legs were enclosed, swaddled so that I couldn't leap up and run away. And I was low, on the ground. I felt very exposed, and I just had to hope that Theo knew his stuff and the thorn branches really were enough to protect me from the wildness all around.

I looked up through the branches of the tree above at the southern stars. Eventually I started to fall asleep, and I welcomed it. I was fatigued by international travel and tired in my bones after going tracking with the hunters. It was good, deep sleep, breathing clean air in the wilderness.

I woke suddenly. It reminded me of being woken, as a child, by cats fighting on the roof. A terrifying sound ripped through the chill night air. A howling, screeching, growling sound. More than one animal.

I lay on my back, absolutely still. I was frightened to the core and my instinct was to lie as still as possible – and listen. My breathing had become halting and heavy, and I made a conscious effort to make my breaths as quiet as possible, while I tried to work out what was going on out there in the darkness. Was it lions? Leopards? Hyenas? The noise went on for what was probably only half a minute, but it echoed in my ears in the silence afterwards. I lay very still, looking at the stars and questioning the wisdom of this venture. It was a long time before I got back to sleep.

When I next awoke, the sky was a light grey above me. Dawn was approaching fast. I felt so much more courageous now that my eyes worked to reveal my environment. I extricated myself – still moving slowly and as quietly as possible – from the bedroll and stood up to look around. There were rustling noises, but nothing big lurking nearby. The birds had started singing. The darkness and terror and chill of the night were gone. The sky was turning from grey to pink, and I could feel the air around me getting warmer by the minute.

I walked over to Theo and Rob's camp, and we drank some warmish coffee before heading to the waterhole to check out the evidence of the previous night's activity. On one trail, there were prints of at least one hyena – a very large female. And there were the wider, shorter paw prints of leopards: a female leopard and her cub. At the waterhole itself there were many muddy, deep prints of hyenas. Theo told me that's what the awful noise had been: a few hyenas, he thought at least four, squabbling at the waterhole. Close to the water were prints of a – very brave – roan antelope. As we walked back up a trail towards my camp, Theo pointed out more tracks: African wildcat, and the

hyena footprint near the waterhole

massive paw prints of a big, male leopard – just metres away from where I'd been sleeping.

We're used to thinking of ourselves as dominant animals. Top of the pecking order. Right at the top of the food chain. Sleeping out in the Namibian bush was a truly frightening, awe-inspiring and humbling experience.

Before I left Nhoma, I talked to Arno about the lodge, the village and what he thought about the future of the Bushmen. Arno and his wife Estelle had set up the camp eight years before, with nine huts, toilets and washing facilities. And the three hundred or so tourists who stayed there each year were also catered for, eating dinners prepared by some of the Bushmen in the large open-sided thatched building. Arno had installed a pump at what had been a semi-permanent waterhole, so that there was water available there for humans and animals throughout the year. This meant that the Bushmen did not have to move around the landscape, following water and animals, and that Nhoma became a permanent village. This wasn't out of the ordinary; there are very few truly nomadic Bushmen left now. Some had criticised the set-up at Nhoma village for failing to preserve a traditional way of life, and for encouraging the Bushmen to settle down; not only that, tourism provided the Bushmen with an income, enabling them to buy Western clothes and maize-based foods to supplement their bush food. But for Arno it was important that the people had some autonomy, and a choice in their own future. Tourism was allowing, even encouraging, the traditional way of life to continue for a while, but Arno did not think it was sustainable.

'How much longer do you think the Bushmen will hunt?' I asked.

'Oh – maybe fifteen years,' he replied. 'There are only eleven hunters left in the whole village. The children go to school and they want other things.'

The Bushmen – and their way of life – had survived for hundreds of years after the incursion of agriculturalists into the area. In most places where foragers come into close contact with food producers, the hunters lose their independence and end up right at the bottom of the social pile.

In the late nineties, archaeologists from Cape Town University excavated at Cho/ana in northern Namibia, with Bushmen as part of the team. They found four levels of archaeology: the first (uppermost) level related to the recent historical period, with occupation by Bushmen, black Africans and Europeans. The archaeologists found items like bottle glass, plastic beads and bullets, as well as local materials like nuts and ostrich eggshell. But they also found stone tools in this top layer which suggested the Bushmen had been making lithics until very recently (even though Lorna Marshall and others had found that the Bushmen had no folk memory of using stone tools). The second level of archaeology contained local materials as well as pottery from Mbukushu people who had moved into the area some three hundred years ago. In the third level, Divuyu pottery indicated that the Bushmen had contact with the farming people of the Tsodilo Hills in modern-day Botswana, going back around 1500 years. Level four, dating to between three and four thousand years ago, contained natural materials and stone tools, but no pottery, so the archaeologists presumed that this layer pre-dated contact with agriculturalists. Based on this evidence, and interviews with Bushmen elders, the archaeologists concluded that the Ju/'hoansi had been independent hunter-gatherers for millennia, and that they had continued with their way of life, while more exotic materials had been incorporated into their culture with increased contact with the outside world.[3] But it seemed that this traditional society may now, finally, be in danger of becoming swamped.

In the 1950s ethnographer Lorna Marshall had spent several years with the Bushmen, and in her 1976 book, *The !Kung of Nyae Nyae*,[2] she had written:

> I personally wish the !Kung could have remained as they were, remote, self-sustaining, independent, and dignified; but that is wishful thinking. Our modern society does not allow people to remain remote. Furthermore, many of the !Kung themselves want change; they want to have land and cattle like the Bantu.

The Bushmen were not ignorant. They may not have had televisions but they knew there was a wider world out there. I tended to agree with Arno. It was sad in many ways that this way of life was disappearing, but these were *people*, not museum exhibits – they had to be able to make their own choices. I felt very privileged to have visited Nhoma and learnt about their traditions, and I left wondering if I'd ever go back and, if I did, whether the Bushmen would still be there.

African Genes: Cape Town, South Africa

My first destination in Africa had provided me with insights into life in a contemporary hunter-gatherer culture, and the fragile nature of this way of life in the modern world. I had learnt about the deep roots of Bushmen ancestry and click languages, adaptations to endurance running in humans, and enjoyed an impromptu crash course in African beadwork. Flying from Nhoma to Windhoek to Cape Town, I was in search of more genetic revelations about African ancestry.

On a sunny spring day in Camps Bay, I met up with Professor Raj Ramesar, of the Cape Town University, to find out about the results of an ambitious study of the diversity of mitochondrial DNA among Capetonians.

In 2007, 326 Capetonians had volunteered to have their mtDNA sampled by the African Genome Education Institute, using salivary swabs to collect precious cells containing their genetic fingerprints. Cape Town is a cosmopolitan city. As well as a wide-ranging mix of Africans, people from other continents have also made Cape Town their home over the centuries. It's a classic melting pot of cultures – and genes.

The results of the Cape Town study bore out that diversity. Everyone taking part in the study was asked what ethnic group they considered themselves to belong to. Of those people who considered themselves either white or coloured, 8 per cent had a maternal ancestry traceable to West Africa. This fits with waves of Bantu-speaking people moving down from the Niger region into southern Africa, bringing agriculture with them, starting around 3000 years ago.[1] Among just the 'white' people, 3 per cent had black African mtDNA markers, and 10 per cent of black Africans had maternal lineages traceable not to Africa, but to Europe. Twenty per cent of black Africans had markers that went back to some of the earliest Africans: very early branches of the human family tree.

Ten of the original volunteers had assembled on the day I met Raj, and their maternal ancestry could be traced back to lineages originating throughout Africa, Europe and Asia. Firstly, Raj explained that the genetic markers used to construct the lineages actually represented tiny differences in DNA between people. 'As different as we may look on the outside, we are practically identical at a genetic level,' said Raj. 'But the minor differences are what we use to plot population migrations from one part of the world to another.'

Then he revealed the results of the mtDNA tests to the volunteers. Some who felt themselves to be African through and through found they had maternal lineages on Asian and European branches of the tree. Of course, mitochondrial DNA reveals only a small bit of our ancestry: just the 'motherline'. But nevertheless, there is still something quite profound about being able to trace part of your heritage back that far.

Although mtDNA traces only *one* line down through the centuries and millennia, among the many ancestors we all gather as we follow our family trees back through time, the study threw up some surprises for the participants involved. The results show quite graphically what a subjective and shifting concept ethnicity is. While similarities and differences between human populations are fascinating, the idea of 'race' doesn't make sense in biology – it's a concept pulled together from a ragbag selection of physical characteristics, culture and religion and attachment to place of birth. Ultimately though, however much we may feel that we come from a particular place, our genes show that our ancestry is far more diverse – and more interesting.

Rather boringly and predictably, I thought, my mtDNA was of European vintage. I was on the 'I' branch. But once I'd got over the lack of exotic genes in my motherline, I have to admit feeling a sense of wonderment at the revelation. Someone had managed to take some minute cells from my mouth, delve into my mitochondrial genome and pull out this fact about my ancestry. I know my family tree only back to my great-grandparents, going back perhaps two hundred years, but here was this piece of information, passed down to me from one of my very ancient female ancestors. It means my maternal ancestry goes back to the second wave of modern humans migrating into Europe around 26,000 years ago.[2]

On the one hand, we can all trace elements of our ancestry back to a great range of places and times. But the mitochondrial DNA family tree also shows very clearly that, if we go back far enough, all the lineages converge – in Africa. In 1987, three geneticists at the University of California, Rebecca Cann, Mark Stoneking and Allan Wilson, published a seminal paper in *Nature*, reporting on an analysis of mtDNA from 147 people, and showing how their maternal lineages could be traced back to one woman, in Africa, some 200,000 years ago.[3] Since then, mtDNA from thousands and thousands of people has been analysed, and the tree has grown bushier, but the origin has stayed firmly anchored. If you traced your ancestry back far enough, along the 'motherline' (your mother's mother, then her mother... and keep going), then eventually you'd get to a point where you reach a common female ancestor of *everyone* alive on the planet today. It's not surprising that the geneticists have named her mitochondrial, or African, Eve.

How do we know she was African? Well, the highest density of branches, in other words, of different types of mtDNA, is in Africa. This is extremely strong evidence that all of us, our species, originated in Africa. And it's not just mitochondrial DNA that reveals this pattern: the Y chromosome, and genes on other chromosomes, also show greater genetic diversity among indigenous Africans compared with Asians or Europeans.⁴ All this genetic diversity points to Africa being the homeland of *Homo sapiens*. People have lived there longer than anywhere else, so there has been more time for mutations to accumulate and different lineages to sprout in Africa than on any other continent. In 2008, the results of the most detailed study of human genetic variation to date were published in *Nature*. Part of the study involved looking at more than half a million points in the nuclear genome, where the individual building blocks of DNA – or nucleotide bases – are known to vary, and comparisons were made between twenty-nine populations from around the globe. Analysis of the differences produced a tree, with branches spreading out from East Africa.⁵

This genetic piece of the puzzle has not come as a surprise to most palaeoanthropologists; rather, it confirmed what they already suspected from the fossil evidence, because the oldest anatomically modern human remains come from Africa. However, there can always be arguments over the veracity of lineages reconstructed from fossils, because of the patchy and inconsistent nature of the evidence. The story we tell based on palaeontological, and indeed archaeological evidence, is woven together from fragments of information. The likelihood of material evidence, whether of people themselves or cultural objects, surviving for thousands of years and then being found, is minute. Many things made by Palaeolithic societies would have been organic, biodegradable, and so would have little chance of surviving. Very often, pieces of worked stone and waste flakes are the only objects left for us to find. Most skeletons will not become fossilised. They will get crushed or trampled or broken up and leave no trace. For a body to be preserved, it needs to get covered by mud fairly quickly, before it is ripped to pieces by scavengers. And the sediments it is sealed in need to be chemically and physically just right for any bones to be preserved. In many cases, water moving through soil will leach out mineral from the bones, and bacteria will eat away at the protein part of the bones, until virtually nothing is left. In a few lucky cases, the shape of the bone is preserved, while the substance of the bone changes as it exchanges minerals with the surrounding sediment – and it turns to stone, in a word: fossilises.

Even if something is preserved for all that time, there is no guarantee that it will be found. Some fossils and archaeology will be buried deep under metres and metres of sediment, or are in very remote places. Most ancient fossils or archaeological sites are discovered by chance. Often, it is human intervention that lays bare the evidence, through mining, quarrying or road-building, for example. At other times, natural erosion might reveal long-buried fossils. And when that evidence becomes accessible, it still may not be recognised.

Although archaeologists and palaeontologists do try to direct their attention to areas where they think there is a likelihood of finding evidence, sometimes the first clues that lead to significant discoveries are entirely serendipitous.

With all this chance and uncertainty involved in fossil-hunting, it is astounding that fossils have been found, in Africa, that seem to go right back to the dawn of our species.

The Earliest Remains of Our Species: Omo, Ethiopia

For several decades two sites have vied for primacy as containing the most ancient fossils of anatomically modern humans. Both these sites are in Ethiopia: Herto and Omo Kibish.

In 1967, a team led by Richard Leakey had unearthed modern human fossils – two skulls and one partial skeleton – from the Kibish formation in the Omo Basin in Ethiopia. Uranium series dating of mollusc shells in the sediments suggested the fossils dated to around 130,000 years ago. Thirty years later, the fossilised skulls of one juvenile and two adult modern humans were discovered in Herto, in the Middle Awash area of the Afar depression. The skulls looked modern in many ways, but were quite robust. The authors, presenting their finds in *Nature*, described them as being 'on the verge of anatomical modernity but not yet fully modern'. The Herto fossils were dated to 160,000 years ago by argon-argon dating.[1]

But in 2005, new – much earlier – dates were published for the Omo skeletons. A team of geologists and anthropologists, led by Ian McDougall from Australian National University (ANU), revisited the site where the Omo fossils had been discovered. Using site records and photographs, they were able to pinpoint the exact locations where the fossils had been found. At the Omo I site, confirmation that they had indeed located the precise findspot came when they discovered additional fragments of fossilised skull that fitted into gaps in the original finds.[2]

Both sets of human remains, although found on opposite sides of the Omo River, lay in the same stratigraphic layer. And this layer was sandwiched neatly between two strata formed of volcanic tuff, from the eruptions of ancient volcanoes. This was highly convenient, because tuff is amenable to dating using argon isotopes.

It turned out that the deeper layer was laid down some time after 196,000 years ago, while the higher layer dated to, at most, 104,000 years ago. The layer in which the human fossils were found lay *just* on top of the deeper tuff. McDougall and his team argued that the fossils were therefore almost as old as this deeper layer of volcanic tuff, at around 195,000 years old. This new date made the Omo fossils the oldest anatomically modern human remains – in the world.

The fossils themselves resided in Addis Ababa Museum, but I wanted to see the place where they had been discovered. It felt like a pilgrimage, and in a way it was: I was going to visit our ancestral home. I had read Richard Leakey's books as a teenager, and I couldn't quite believe I was about to get the chance to visit the places I had read about. The names attached to this landscape seemed epic, even mythical, to me: the Rift Valley, the Omo River, Lake Turkana. But they are real places. However, I was travelling to one of the most remote parts of Africa, and finding the precise place where the fossils were discovered wasn't going to be easy.

On a Monday morning, I set off from Addis Ababa in a small Cessna Caravan, headed for Murule camp (and the closest airstrip to the Omo findspots), piloted by Solomon Gizaw. It was cloudy when we took off, but it soon cleared, and we were flying south-west over green countryside with irregular fields and round, thatched houses of small farming communities. From the air the thatched houses looked like clusters of brown mushrooms. It was clear that agriculture was dominated by small-scale farms, and Solomon was outspoken about the inefficiency of food production in Ethiopia. 'We have so much good land. There should be more than enough to feed everyone,' he said. 'But it is not managed properly.' Certainly the land beneath us looked green and fertile. It was a very different view of Ethiopia from the terrible pictures of the famine in the 1980s. But there was also little in the way of infrastructure; I couldn't see many roads. It must have been extremely difficult to reach people out in rural areas when there was a food shortage.

As we flew over the mountains, Solomon explained that we were into coffee country. So when we stopped off to refuel at Jima, I refuelled with some local coffee. The café at the airport was a wooden shack with a corrugated iron roof. Inside, a group of men were playing draughts using a home-made board and bottle tops for pieces. A woman with a beautiful, serene face came over, holding high a tin kettle, and poured a stream of strong, sweet coffee into my china cup.

Then we were back in the air again. The second leg of the journey took us over more wooded hills and valleys. Suddenly, Solomon pointed down at a narrow, shining ribbon: 'There it is: the Omo River.'

And there it was, meandering southwards through a wide, wooded valley. Then I lost sight of it as we flew up over a ridge of mountains. Solomon handed over the controls of the plane to me, and I followed his directions. 'Continue straight on down this valley, and the Omo will loop round and meet us again on the other side of those mountains,' he said, pointing to the edge of a ridge in the distance. I flew for about another half an hour, taking the plane down from 5000 to 3500ft, then Solomon took control again as the river reappeared and we approached our destination. We had passed the mountains, and were now flying over a wide, flat flood plain, with the Omo thrown in wide, brown coils across the landscape. Next to the river, the land was densely forested, breaking up into scrubby bush away from its banks. But it was still greener than I had expected. The flood plain was huge. I wondered *how* Richard Leakey had ever found those fossils.

As we flew lower, I saw a village on a hill looking over a wide arc of the Omo. It was Kolcho, the nearest village to Murule camp, where I would be staying. We circled round and landed at a dusty airstrip where Enku Mulugeta met me in a Land Cruiser. After unloading the plane, Solomon took off again, bound for Addis Ababa; he would return on Friday to pick me up. And I was in the middle of nowhere.

Murule Lodge was situated right on the banks of the Omo, which was wide and greasy-brown here. It was high on its banks and fast-flowing, after recent heavy rains in the mountains. The small houses of the lodge were surrounded by tall trees. Almost as soon as I arrived I heard branches moving above me, then there was a shower of small twigs and leaves. I looked up to see a black and white colobus monkey staring down at me through a forked branch. We made eye contact and then he was away, throwing himself through the trees with amazing speed and agility. More colobus followed in hot pursuit.

Black and white colobus monkey Murule camp.

I had been expecting a very basic camp, so my little house at Murule came as a very pleasant surprise. The windows were covered in mosquito screens, with curtains inside and raffia blinds outside. There was a large room in the house, with a double bed, a crude wooden chair and a sill wide enough to hold my bag. A single lightbulb hung in the corner of the room, but there were also pots of sand on both sides of the bed, with candles and matches. (Electricity was an occasional luxury here, powered by a generator for just a couple of hours after sunset.) I even had a bathroom, square and concrete, with a flushing toilet, a sink and a shower fitting in the ceiling. Admittedly, the shower was unheated, untreated water straight out of the Omo, but it was all so much more luxurious than I'd anticipated for this remote place in the middle of the Rift Valley – even if I was sharing the place with a few geckos and a strange-looking spider with a penchant for hiding inside toilet rolls, a ploy destined only to scare the life out of both of us.

I fixed up my mosquito net. It was late afternoon and the mosquitoes were just starting to bite. As night fell, I sat out on a wooden chair looking over the Omo, with a packet meal and a bottle of locally brewed St George beer. I was being extraordinarily careful with food and water here: consuming nothing unless it came out of a packet or a bottle. We were too far away from anywhere to take any risks, and I didn't want to jeopardise my one chance of visiting the site where the Omo fossils had been found. I turned in around ten; I had an early start the next day, and I knew it was going to be tough.

I slept well that night, despite the sticky heat and the noisy wildlife. I woke up a couple of times to hear things moving about outside and colobus monkeys barking in the trees above. By 5.30 a.m. I was up and about, as were a vast range of different birds, judging by the Omo dawn chorus. I packed a rucksack with essentials: medical kit, some cereal bars, camera, Ventolin inhalers, notebook, GPS and map, as well as several boxes of crayons and a selection of small toy cars. I filled my Camelback reservoir to the brim, and I took an extra two-litre bottle of water as well. Then Enku and I set off in the Land Cruiser to catch a boat across the Omo.

After a bumpy ride along a dusty dirt track, we pulled up next to the riverbank, opposite the village of Kangaten. We waved at the boatman on the other side, and he brought his boat across. It was small, with a modest outboard motor, but its skipper knew how to beat the swift current by taking it in a wide arc, heading upriver first then falling back alongside us as he approached the near bank. I scrambled into the boat and off we went again, over to Kangaten. I had seen crocodiles in the Omo as I'd flown in, and when I asked the boatman if there were any around he smiled and nodded. I didn't see any, although the odd log drifting down the river made me look twice.

On the other side, it seemed as if most of the village had turned up to greet us. *Lots* of children swarmed around. I took photographs and they clustered around me, squealing in delight as I showed them the photos on the camera. They pointed and laughed as they recognised each other in the tiny screen. Before Enku and I embarked into the four-wheel-drive that was waiting for us on this side of the river, I handed out coloured wax crayons and toy cars. Some of the kids looked healthy, but others displayed the swollen bellies and stick-thin legs of malnutrition. I also spotted patches of skin infection and ringworm on their bare bodies and faces. They were much less healthy than the children in the Bushmen village. I felt as though I should have brought medical supplies and food rather than crayons and toys. I also felt that familiar pang of guilt that I was an academic, not a practising medical doctor any more. At such times I have to remind myself that teaching and research are worthwhile. And I know that Ethiopia's problems can't be solved by aid workers alone. I handed out my cereal bars and got into the car.

Enku introduced me to Soya, who would act as my translator when we reached the village of Kibish. I knew that people from that village had been involved with the dig and discoveries at Omo, and that, more recently, they had helped Ian McDougall and his team when they visited the site. Although I had the map and the coordinates published by McDougall, I was aware that I might end up going hugely out of my way in the bush if I tried to find the site on my own, without help from locals who knew the landscape and the paths through it. And I really didn't want to get lost in the bush. Another issue was security. The tribes around the Omo River – the Mursi, Bumi, Hamer, Karo, Surma and Turkana – seemed constantly to be fighting each other, and there were a lot of men wandering about with guns.

After a further drive along dusty tracks through the bush, having stopped to talk to a group of men with guns who were apparently local police, we reached Kibish. The huts of the village were surrounded by a dense thorny fence, with a slit-like entrance that was hard to discern but presumably excellent for keeping out hyenas or other tribesmen. Soya led the way and took me to the chief, Ejem. Most people in the village seemed to be dressed fairly traditionally. The women wore knee-length apron-like skirts, and masses of bead necklaces on their bare chests. Many had red ochre painted on their chests, necks, faces and braided hair. Smaller children ran around naked. Older ones had painted faces and cloths tied loosely round their waists. One boy was wearing a faded, red David Beckham T-shirt. Some men were dressed traditionally, with small skirts and collars of beads, but the chief, Ejem, was wearing flamboyant basketball shorts, a plastic leopard-print cowboy hat and a necklace of red and yellow beads. His status was made obvious by the wearing of these exotic items.

With Soya translating, I introduced myself to the chief and asked him if anyone knew the place where the fossils had been found. Ejem called over one man and pointed to him. Soya translated. 'Here is one man, the other is coming.' This first man, Kapuwa, was wearing a T-shirt and a brown cloth hat with a turned-up red brim, and carrying a gun. He talked to Soya and I grew excited as he pointed off into the distance and made digging motions with his hands.

'Soya, what did he say?'

'He said, "There was someone with a camera and someone was digging. He found something like a bone, which had stayed there for a long time. I know exactly the place and I can show you now."'

The second guide, Logela, had now come over as well. He was wearing a yellow cloth hat, set at a jaunty angle on his head.

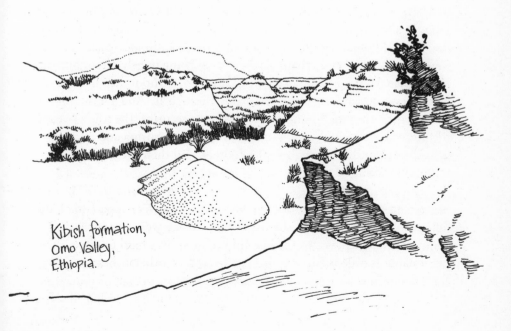

Kibish formation, Omo Valley, Ethiopia.

So Kapuwa, Logela, Soya and I headed off into the bush. We drove for about 4km away from Kibish, through flat, scrubby bush, until we reached an area with dune-like hills and deep channels: the Kibish formation. It was impossible to take the car any further. I waymarked the car on the GPS before striking out into the wilderness. Even though I had started as early as possible that morning, I was still having to walk in the heat of the midday sun. And Logela and Kapuwa set a fairly intense pace. Logela was carrying a gun, and I asked Soya how often violence broke out between tribes.

Apparently it was a fairly common occurrence, but, as we weren't after anyone's cattle, we ought to be safe. It was still just as well to have a gun, though.

'Did you see the scarification on the chief's chest?' Soya asked me. I had. 'It means he is a hero: he has killed a man.'

As we walked, we did meet a couple of other lone men with guns, and I was very grateful for the security of travelling with Soya and the guides from Kibish. And they did seem to know where they were heading. We walked down through the barren valleys of the Kibish formation, which was like a lunar landscape. Then we were on the flood plain again and close to the Omo. As we tracked up the west bank of the river, I started to feel heady and overheated, and so we stopped for a while. We had been going for about an hour, and the sun was now at its highest point, directly overhead and beating down. I drank more water and wrapped a silk scarf around my head, then we carried on. We passed a ridge, perpendicular to the river, and the guides pointed to another, similar ridge. Soya told me that they thought we were almost at the spot. I was very glad; I was starting to feel as if I might have to turn back without reaching the site, but now I knew we were close I could summon the determination to carry on.

We marched on, along dusty paths through the bush that looked as though they were used more often by animals than people. Umbrella thorns (*Acacia tortillis*) raised their branches above the other shrubs, and every now and then, among the scrubby bushes, there would be a beautiful little Elemu tree, with five-petalled pink flowers. It seemed strange in this dry and dusty landscape to see something so exuberantly colourful. These shrubs were between one and two metres high, with bottle-bottomed trunks and very bendy branches, I discovered, by doing just that – bending them. And then, as we walked up to the second ridge, Logela and Kapuwa came to a halt and gestured round at the landscape. 'This is the place,' said Soya.

We were standing at the foot of the ridge, with smooth, light-brown slopes formed of silty sediments, shaped by the biannual flooding. There were also a couple of darker layers, one close to ground level, the other higher up the slopes. On closer inspection, these layers were harder rock, dark brown and almost black in places. The bottom of the slope was covered with fragments of this tuff.

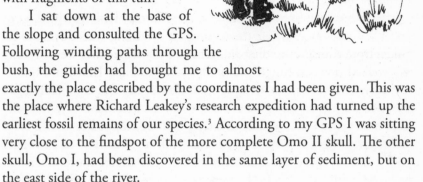

herdsman near Omo River

I sat down at the base of the slope and consulted the GPS. Following winding paths through the bush, the guides had brought me to almost exactly the place described by the coordinates I had been given. This was the place where Richard Leakey's research expedition had turned up the earliest fossil remains of our species.[3] According to my GPS I was sitting very close to the findspot of the more complete Omo II skull. The other skull, Omo I, had been discovered in the same layer of sediment, but on the east side of the river.

The revisit to the site by Ian McDougall's team was extremely important in producing the new, and very ancient, dates for the Omo fossils, but also in confirming that the two skulls came from the same layer. This was interesting, because the skulls are quite different shapes. Omo I is anatomically modern, through and through, if a little more robust than many modern skulls.[4] The braincase is globular, with the broadest point high up on the parietal bones, rather than low down, around the ears, as on earlier, archaic human skulls. It has a prominent browridge, but this thins out at the sides. The shape and size of the teeth and the prominence of the chin also mark it as modern.[5] But Omo II, was a bit odd-looking.

Sitting on that silty ridge near the Omo River, I pulled a cast of the Omo II cranium out of my bag. The face was missing, but most of the braincase was there. The sutures (joints between the plates of the skull) were practically closed, which suggested that its owner was adult, and getting on a bit. It was modern in some ways: in its round head, slight browridge and steep forehead. The volume of the braincase has been estimated as a very respectable 1435ml. But it also had an angled occiput (back of head) with a very marked horizontal ridge, or 'occipital torus', for the attachment of neck muscles, and a very slight ridge along the midline, called 'sagittal keeling'.[4] These were archaic features: they are things that *Homo erectus* and *heidelbergensis* had as a matter of course, but they are features that have disappeared from modern populations today. Michael Day, the physical anthropologist who wrote the first report on the Omo skulls in 1969, noted that Omo II was more 'archaic' but classified both skulls as modern human, *Homo sapiens*.[4] When Michael Day and Chris Stringer reassessed the skulls in 1991, they again drew attention to the more archaic-looking features of Omo II.[6] It's a skull that may be best described as 'on its way' to being modern human.

That the earliest fossils of modern humans we have show some archaic features is not really surprising. It would be more remarkable if there was a sudden, global change in skull shape with the arrival of modern humans. Evolution proceeds gradually, step by step, and species change slowly over time (although, of course, each change relates to a mutation in a gene, which may have quite widespread effects on body shape and size). Looking at living species today, there are usually very clear morphological species distinctions. But when you are looking at species over time, it's quite difficult to pin down when speciation 'happens', in other words when enough changes have occurred for the descendant population to be called a new species. Pinning a specific date on the emergence of a new species may be futile, because those gradual changes accumulate over time.

Between 600,000 and 300,000 years ago there was an earlier human species, *Homo heidelbergensis*, in Africa, represented by fossils such as the Bodo cranium from Ethiopia, and Kabwe (Broken Hill) from Zambia. This species seemed to combine some archaic features (similar to those seen in the even earlier species, *Homo erectus*) with some more anatomically modern features.

By 195,000 years ago we had the anatomically modern Omo skulls, the first of many: *Homo sapiens* was firmly on the map, and *Homo heidelbergensis* was no more. It is wrong to see this as an extinction, though. The descendants of *Homo heidelbergensis* were still alive: they were modern humans (and Neanderthals in Europe – but more of that in a later chapter). So, somewhere between Kabwe at 300,000 years ago and Omo at 195,000, there was speciation. But presumably it was gradual,

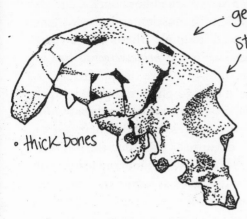

gently sloping forehead

strong brow ridge

Bodo
cranial capacity 1,250 mL

• thick bones

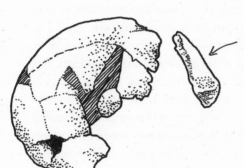

steep forehead, non-protruding brow

Omo I

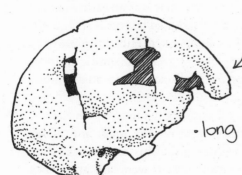

non-protruding brow

Omo II
cranial capacity 1,435 mL

• long but domed braincase

Bodo & Omo crania

for otherwise we have to imagine a pair of *Homo heidelbergensis* parents producing a baby that looked most unlike them: a little anatomically modern human. So it is to be expected that the earliest anatomical modern humans would retain some archaic features; some anthropologists have called them 'archaic *Homo sapiens*' to distinguish them from the later, more gracile, more 'modern' forms. They were certainly more chunky than most of us today.

Of course, being anatomically modern is one thing. Those early Omo people may have looked like us (albeit with a slightly angular occiput, or back of the head) but did they think and behave like us as well? The only way we can even begin to tackle this question is by looking for clues as to how these people lived, what they made and, from that, try to deduce behaviour and patterns of thought. But the Omo site contains nothing of these sorts of clues: it is purely palaeontological, rather than archaeological. The fossilised bones of our ancestors were preserved there, which is wonderful, but it leaves us wondering what they were *like*.

I stayed for a few days around Omo, and spent some time in the village of Kolcho, near Murele camp. This was a Karo village, and body-painting was a strong tradition for this tribe. The first day I visited I met a young man called Muda whose torso was completely covered in white, finger-painted spirals. He spoke some English, and I asked if the painting symbolised anything. He was doubtful about any messages in the design itself, but said that men and boys had their bodies painted, while women and girls had their faces painted.

In the village a group of women sitting under a low, loosely thatched shelter invited me to sit with them. A couple of children were too, watching the women sewing. One girl was painting another's face, and, when she was done, she started on mine. She had a small tin pot with white clay in it, and she used the round, blunt end of a nail to apply white spots on my face. Each one was like a little spot of coolness on my hot face. I found out that the little girl was called Buna. She introduced me to the other women and the growing crowd of children who were gathering to see the white-faced woman getting Karo spots painted on her face.

Buna was applying the spots in a careful pattern, leaving a wide circle around my eyes clear. Another little girl came to help, and Buna directed her. Eventually, Buna put the nail and the pot down and looked at my face, very seriously, then pronounced me done.

Muda reappeared and took me off to meet one of his friends. He stooped down at the doorway of a low, thatched hut, and I could see a woman inside. She invited us in, and we crouched down and entered her house. She was kneeling on the floor, roasting coffee, and Muda and I sat down opposite her. 'My friend: Chowli,' said Muda, in slow and careful English. Chowli was dressed the same way as all the other women in the village, in an apron-like soft leather skirt, which hung down low to the knees in front and behind, tied at the sides. She wore masses of bead necklaces, and her forearms were stacked with brass bangles. Buna followed us in and sat by me: I discovered that Chowli was her mother. Through Muda's little bit of English, we managed to have a weird kind of conversation; I don't think any of us really knew what it was about. But Chowli did indicate that she was impressed by Buna's work on my face.

The smell of the roasting coffee was fragrant in the air; Chowli lifted the pan off the fire and emptied the coffee into a half-gourd and poured in hot water. Then she handed it to me to taste. I lifted it to my lips with great reservation: I had been *so* careful about what I ate and drank in Omo, and now it was all about to go to pot. I knew, as I took a sip, that I was risking vomiting and diarrhoea later that evening. (Thankfully, I escaped unscathed. And it was good coffee.) Before I left, I gave them some cereal bars, and Chowli passed over one of her brass bracelets, which Muda opened into a wide 'C' and then closed around my wrist. 'Friends,' he said, gesturing between us. Not to be left out, Buna gave me a yellow and blue bead bracelet.

As well as visiting the Omo fossil site, and feeling utterly in awe at just being in a place that we can truly call the birthplace of humanity, I would also treasure experiences like meeting Muda, Buna and Chowli. And writing about it makes me think again about how those ancient Omo people might have been. If I was able to travel back in time, would the original tribespeople invite me in for coffee? Would they understand friendship? These are philosophically interesting, but ultimately unanswerable questions (although I suspect coffee drinking may be a somewhat later cultural development). But there are other aspects of humanity that archaeologists *can* expect to find traces of, and one of them is the urge to decorate: to paint and adorn ourselves and our environment. There was an explosion of art – including cave art, figurines and beads – from around 30,000 to 35,000 years ago in Europe, but there are traces of art and adornment going back much, much further than that. And I think that art, like music and language, implies a level and quality of consciousness that we can call 'modern human'.

Modern Human Behaviour: Pinnacle Point, South Africa

I left Omo and headed south. There are a number of quite famous Middle Palaeolithic sites in South Africa, where evidence of what looks like modern human behaviour has been found, including Blombos Cave, Klasies River Mouth, Boomplaas and Diepkloof.

Although these sites are broadly classified as Middle Stone Age, there are particular technological and cultural features that make them stand out from older MSA sites. In other words, the archaeologists argue that while the sites may be MSA, they are MSA with signs of *modern human* behaviour. (This is an important distinction to make because earlier species like *Homo heidelbergensis* also made MSA tools, and in the last decade of the twentieth century many archaeologists still believed that humans didn't become 'fully modern' until 45,000 years ago.)[1]

The 'modern' features from these South African sites, dating to between 55,000 and 75,000 years ago (well before the colonisation of Europe and the Upper Palaeolithic), include a new way of flaking bone, using a soft hammer (perhaps bone or antler), specialised end-scraper tools that archaeologists believe were used for cleaning hides, and pointy burins – presumably for making holes in leather or wood. The assemblages also included the first bone tools, including things that may be spear points and awls. There are also tiny stone flakes, which may have been used as inserts in spears to make them harpoon-like, or perhaps even as arrowheads (though there is no definite evidence of bow and arrow technology until much, much later – about 11,000 years ago). Small pieces of stone may not sound impressive or important, but they suggest that people were making composite weapons: similar to the type of technology that is seen, much later, in the Upper Palaeolithic of Europe and Asia. And even if the tiny blades aren't evidence of archery, they do at least seem to suggest that new, more effective hunting tools were being developed. From the original source of the stone material used, it seems that there was some fairly long-distance trading going on, perhaps indicating widening and increasing complexity of social networks. The presence of tiny bladelets within an African MSA assemblage was first noticed in a site in South Africa called Howiesons Poort.[2] Similar toolkits have also been found in East Africa, at the site of Mumba in Tanzania, and at Norikiushin and Enkapune ya Muto in Kenya.[3]

But as well as these technological advances, sites in South Africa have also produced tantalising suggestions of early art and ornamentation. In Blombos Cave, archaeologists found pierced sea snail shells; careful inspection of the holes in them showed that these perforations were not natural, and experiments on shell with a sharp bone point produced similar results. The edges of the holes and the lip of the shells were slightly worn down, showing they had been strung and worn.[4] Numerous pieces of ochre had been transported into the cave at Blombos, including some lumps with geometric patterns scratched on to them, dating to around 75,000 years ago. These engraved ochre pieces have been held up as the earliest examples of 'abstract art'.[5, 6]

Archaeologists have linked these technological and cultural developments to environmental changes happening between 80,000 and 70,000 years ago. This was the time of the transition between the warm interglacial known as OIS 5 to the chilly glacial OIS 4, and that change was marked by dramatic swings between wetter and drier conditions. The eruption of the Toba super-volcano 74,000 years ago may have destabilised the global climate as well. So perhaps it was these environmental challenges that drove the invention of new technologies, the expansion of social networks and somehow produced a need to mark identity, and even communicate, through the use of art and ornamentation.[6]

Good. That all seems to stack up quite nicely. Things get a bit tough in Africa about 80,000 years ago, with unpredictable weather patterns setting in. And then, from about 75,000 years ago, we see humans rising to the challenge, engaging their anatomically modern brains to invent new ways of life, better ways of hunting, and discovering their artistic talents.

Then, in the late nineties, someone came up with the idea of building a golf course at Pinnacle Point, close to Mossel Bay on the west coast of South Africa. The setting was stunning: a dramatic rocky coastline with native fynbos covering the tops of the cliffs. Fynbos is South African coastal heathland: a hugely diverse mix of proteas, heathers and reeds. But as well as its natural beauty, Pinnacle Point held archaeological treasures. For a long time, archaeologists had known that the caves there were used during the Stone Age, but it was not until a detailed survey of the caves was carried out, in preparation for the golf course development, that the wealth of material preserved in the caves was fully recognised.

I drove up the garden route from Cape Town to Mossel Bay to meet archaeologist Kyle Brown, who had worked – and was still working – in the caves along this spectacular coastline. We met at the golf course clubhouse and headed for the cliff edge, descending a steep set of wooden steps to the rocky foot of the cliffs. Golfers practising their swings watched us with an air of curiosity as we disappeared over the edge. At the bottom of the steps, part of the wooden walkway had been smashed up in recent storms, and we had to jump down and scramble over the rocks. Kyle explained that there were twenty-nine archaeological sites along this short stretch of coast, at least eighteen of which were caves. Climbing up more steep steps, we eventually arrived at a large, teardrop-shaped cave entrance. 'This is Cave 13B,' said Kyle. 'The very first place we excavated.'

I stood in the cave and looked out at a beautiful view, rolling waves crashing on to yellow-brown rocks. It was September and two whales were swimming very close to the coast. I could see their flippers raised in the air as they rolled over, and the jets from their blowholes. The cave seemed welcoming and homely; I could imagine camping out there. Kyle said it was a great location for protection from the elements on this stormy coast. He had worked here in all weathers and found that it provided excellent shelter from rain and prevailing winds.

Once the potential of this coast had been recognised, archaeologists started a long-term project, aimed at investigating the MSA archaeological record in the caves and tying this in with palaeoclimate data. The project was directed by Curtis Marean, of Arizona State University, and Peter Nilssen, of Iziko South African Museums.[7]

The caves had originally formed about a million years ago, in quartzite cliffs. But there were also layers of limestone higher up in the cliffs, which dissolved and then percolated down through other layers, cementing them and forming breccia. 'At different times in the past, some of these caves were open, whereas others were sealed off by huge sand dunes blown up the cliffs,' Kyle explained. Breccia would form inside the sealed caves, and Kyle told me that the climate record contained in these layers amounted to an almost unbroken sequence for the last 400,000 years (apart from what he called a 'short' 5000-year gap). But at any one time, *some* caves would be open and available for human occupation. The limestone-impregnated breccia layers in the caves were important not only for dating and climate

·Pinnacle Point·

reconstruction, but also because they sealed in earlier archaeological remains and stopped them being washed away. This meant that Pinnacle Point provided a unique, long view: a record of both climatic and archaeological evidence, in one place.

The floor of the cave was covered in sandbags, to protect the archaeology. On one side, Kyle removed some sandbags to show me a small, excavated area. I touched the section: it looked like silt but it was rock-hard breccia. 'This stuff is part of the reason we have the archaeology so well preserved,' said Kyle. 'It's been cemented by flowstone that has come down the cave wall.'

'Presumably you can't dig this with a trowel?' I asked.

'No – it's all dental picks and hand drills; it's very, very difficult stuff to dig. This small pit took about four excavation seasons to dig. But it was well worth the effort.'

The lowest layer was the least cemented, and still quite sandy. It contained streaks of burnt material, possibly ancient hearths, as well as lithics and animal bones. Optically stimulated luminescence (or OSL) dating of this layer placed it at around 164,000 years ago. At that time, during OIS 6, the sea level was somewhat lower than today: the cliffs would not have been on the coast, but set back some 5 to 10km from it. Above those very ancient sediments was a layer containing hearths, but with fewer artefacts, dating to about 132,000 years ago. Sitting on top of that was a heavily cemented layer containing another layer of shells dating to about 120,000 years, and topped off with cemented dune sand and flowstone that had sealed the cave, and kept all that archaeology intact, building up from between 90,000 and 40,000 years ago. 'These are some of the oldest archaeological deposits associated with early modern humans,' said Kyle, quite proudly.

Kyle brought out a selection of stone tools that had been discovered in the excavations. 'These are very typical of the stone tools we find in this cave: blades and points made of quartzite, locally available on the beach down here. And alongside those larger tools, we find these very small bladelets.' They were indeed tiny, less than 1cm in width and perhaps 2cm long.

The stone tools from Pinnacle Point are a mixed bag of MSA and bladelets – like Howiesons Poort. In fact, more than half of the stone tools from Pinnacle Point are bladelets.[7] It looked like these people were using composite tools. 'It's hard to see how these tiny blades would have been used without hafting them, setting them in a handle. So these suggest advanced tool-making techniques,' said Kyle.

The shells found in the upper cemented layer are also interesting. They include all sorts of edible shellfish, types that still thrive along this rocky shore, including many brown mussels (*Perna perna*), limpets (*Patella spp.*) and giant periwinkles (*Turbo sarmaticus*). There was also a fragment of whale barnacle, which may have been scavenged from the skin of a beached whale.

Kyle told me that the giant periwinkles provided extra information about past climate. While the shells themselves were often quite smashed up, the operculum, or trap door of the periwinkles was usually very well preserved. He showed me some: they were like small, white, slightly domed buttons, with a spiral on the flat side, and easily visible growth layers. Sampling oxygen isotopes from these opercula gave the archaeologists an idea of the ocean temperature and more general climate at the time the sea snail had been alive. Before they started on the ancient shells, the archaeologists wanted to check their calculations comparing modern shells against recent climate records. For two years, they had collected giant periwinkles from the shore to study their opercula. And Kyle said that they were delicious.

It might seem unremarkable that these hunting and gathering ancestors of ours were eating shellfish. But this is the first example of *any* human species exploiting marine resources. For millions of years, australopithecines and earlier *Homo* species had been restricted to eating land plants and animals. But it seemed that *Homo sapiens* developed a taste for fish and shellfish: exploitation of coastal resources is seen as another type of definitively *modern human* behaviour. From the evidence from other South African sites, it had been thought that this adaptation to coastal living came along perhaps 70,000 years ago. Archaeologists had argued that this development set the stage for the coastal expansion of modern humans out of Africa and into Asia. But once again, Pinnacle Point pushed the dates back – this time to about 120,000 years ago. Marean and his team suggested, based on the Pinnacle Point evidence, that shellfish may have become an important food source during OIS 6. During this glacial period, between 190,000 and 130,000 years ago, the environment became terribly dry, and humans would have struggled to find food. Turning to the resources on the coast may have been crucial to the survival of those early hunter-gatherers.

But the evidence for modern behaviour at Pinnacle Point didn't stop there. In the lowest layer of the excavation, the archaeologists also found lots of pieces of red ochre: fifty-seven in total. And these weren't just natural pebbles. They had been scratched and scraped. Kyle handed me

one of the pieces of ochre that had been found in
the cave. It was clearly faceted where it had been
ground down, and scratched on one side. I had
seen photographs of the ochre pieces, but it was
so much more convincing when I held one and
looked at it. There was no way that the marks
and shaping of this lump of pigment could have
happened naturally. So there it was, sitting in my
hand, the earliest evidence for pigment use in the
world: 164,000 years ago, the people of Pinnacle
Point had been painting *something*.

Scratched
and ground-down
ochre from Pinnacle
Point

'We really have very little to go on,' said Kyle.
'But this ochre is our best evidence that these
people were practising some kind of symbolic
communication.' The red ochre from Pinnacle
Point is certainly *suitable* for body painting, but we'll never know
for sure what it was – the cave walls, or some object, or themselves –
that they were covering in this colour, or what it meant to them. But
I couldn't help thinking of the Kibish women with their braided hair,
faces, necklaces and breasts painted in deep, rich, red ochre.

Mellars[6] cautions against assuming that South Africa was where
modern human behaviour appeared, because the preponderance of sites
may simply reflect the fact that extensive investigation has been carried
out there. And in fact there are similar sites in Tanzania and Kenya,
although the dating of these has proved problematic. As we have seen,
it is difficult to know exactly when our ancestors became, anatomically,
modern human – but we know they had pretty much 'got there' by the
time of the Omo fossils, 195,000 years ago. The genetic evidence also
suggests an origin of our species around 200,000 years ago.

The evidence emerging from Pinnacle Point pushes back the date for
the emergence of modern behaviour much closer to the earliest dates we
have for the emergence of modern human anatomy.[7] Like the anatomical
features, it is likely that behavioural traits we consider as modern appeared
one by one, gradually coming together to form a mosaic, a 'package' of
modern features. But what Pinnacle Point shows is that by 160,000 to
120,000 years ago the people living there had a fairly comprehensive
portfolio of dietary (shellfish), technological (bladelets) and cultural
(pigment use) behaviours that all shout out 'modern!'.

The First Exodus: Skhul, Israel

It is difficult to trace population expansions within Africa: there has been a great deal of time for populations to move around, over several glacial cycles. Both archaeological and genetic investigation have been biased towards more developed and politically stable countries, meaning that there is very little evidence about early modern humans across great swathes of Africa. Nevertheless, genetic studies provide clues about the geographic origins of modern human populations in Africa. The most ancient mitochondrial lineage, L1, is found in the Bushmen of South Africa and the Biaka pygmies from the Central African Republic. The most ancient Y chromosome haplogroup is found in East Africa, among Sudanese and Ethiopians, as well as among the Bushmen and other Khoisan populations. Genetic lineages appear to have expanded from East Africa to the south and north, as well as out of Africa. African genes also record a much later expansion, of Bantu speakers, from a homeland in West Africa, towards the east and the south, some 3000 years ago.[1]

So when – and from where – did humans expand out of Africa? Migrations out of Africa may have depended on our ancestors being able to usefully exploit marine resources and spread along coastlines. But those migrations were also constrained and determined by geographic and climatic factors, factors that vacillated with the changing environment of the Pleistocene.[2]

Geographically, at least four routes from Africa to Eurasia seem possible: from Morocco, across the Strait of Gibraltar; from Tunisia to Sicily to Italy; a northern route from Egypt into the Sinai Peninsula, and up into the Levant; and a southern route from Eritrea, across Bab al Mandab (the 'Gate of Tears') at the southern end of the Red Sea. All would have involved sea crossings apart from the Sinai route, but, as we have seen, the occupation of Australia, by perhaps 60,000 years ago, would have required a sea crossing.[3] So which of these routes may have been used, given the genetic and archaeological evidence?

As many of these routes suppose a spread out of North Africa, what is the evidence for the earliest modern human presence in that area? In the 1960s the fossil remains of four hominins were discovered in Jebel Irhoud cave, along with Middle Palaeolithic Mousterian tools, in a quarry in Morocco. Animal fossils suggested that these might date to the end of the Pleistocene. Recent uranium series and ESR (electron spin resonance) dates on a juvenile mandible from Jebel Irhoud place the fossil around 160,000 years old.[4] Some researchers claimed that the skulls were

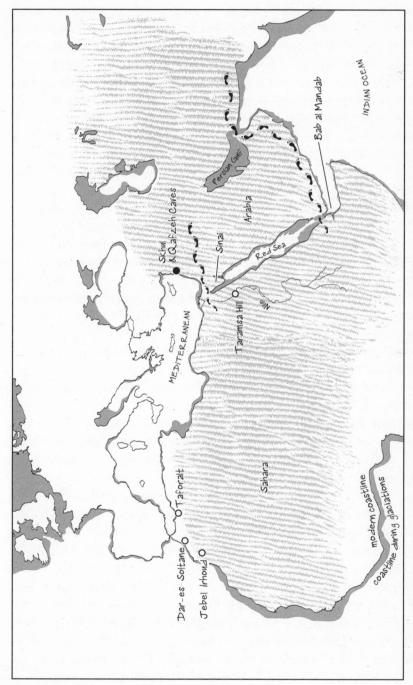

Routes out of Africa. The trails of footprints indicate the northern and southern routes at either end of the Red Sea.
The dune texture shows the maximum extent of deserts – during glaciations.

Neanderthals, but recent analyses have concluded that, though certainly robust, these people were early modern humans.[5] There are some sites, like Dar-es-Soltane in Morocco, where early modern human fossils were found associated with Aterian tools, and there's other evidence of modern behaviour, with pierced shell beads found alongside Aterian points at the site of Taforalt in eastern Morocco, dating to around 82,000 years ago.[5,6]

There is, however, no actual evidence of any spread of modern humans from North Africa into Europe: it seems that the Mediterranean was a major barrier to expansion and colonisation. The dates of archaeological sites in Europe, as well as genetic studies of modern Europeans, suggest an east-to-west spread, making an exit from East Africa most likely.[7] So that leaves the two routes out of East Africa: a northern route via Sinai, and a southern route across Bab al Mandab. The navigability of these routes would have varied during glacial cycles.

In his book *Out of Eden*, Stephen Oppenheimer discussed the likelihood of each of these routes as an exit point for modern humans from Africa, taking into account the climate and environment at the time.[8] For most of the Pleistocene the northern route out of Africa would have been 'shut': the climate would have been cold and dry, and both the Sahara and Sinai deserts would have been impassable. But the Ice Age was punctuated by interglacials, about one every 100,000 years, when the climate temporarily warmed up and the monsoon returned. During these periods, some of what had previously been desert would have turned green. Oppenheimer vividly describes this event as being like a 'science-fiction stargate'. Sub-Saharan animals would have been able to move into areas that had previously been desert, expanding their range from equatorial regions into temperate zones. African fauna could move up into the Levant through a green 'environmental corridor' right through the Sinai Peninsula.[9]

We are enjoying an interglacial at this moment: a nicely warm phase that started 13,000 years ago. The last interglacial before this one, the Eemian or Ipswichian, corresponding with OIS 5, started about 130,000 years ago: it seems that the climate was particularly warm and wet between 130,000 and 120,000 years ago.[10] It is also around this time that the first traces of modern humans appear outside Africa, in the form of fossils from the caves of Skhul and Qafzeh, in Israel. It seems likely that the ancestors of these people left Africa by that newly green northern route. We are somewhat trapped here by our own appreciation of geography, our ideas of where continents begin and end. In fact, rather than thinking of these humans as having 'left Africa', it is probably better to think of the

Levant of 125,000 years ago as an extension of north-east Africa: it was essentially part of the same environment, with the same range of animals, and modern humans were part of that African fauna.[7, 11]

———————

So I made my way to Israel, to visit Skhul Cave. From Tel Aviv, I drove north and followed directions to the place, leaving the main road to enter a canyon called Nahal Mearot (or Wadi el-Mughareh), near Mount Carmel. The valley was framed by great limestone ridges on each side, and there was a series of caves on the southern escarpment. Still following directions, I headed up the valley a little way on foot, and then up the hillside to Skhul Cave. It was an unprepossessing place: small and short, opening on to a flat terrace in front. I could see where spoil from excavations had been piled in mounds, now covered in thorn bushes, just in front of it. I noticed lots of tiny pieces of flint debitage (debris from making stone tools), scattered on the ground in front of the cave.

I sat on the rocks outside the cave and waited for Professor Yoel Rak to arrive. Yoel Rak was an anatomist at Tel Aviv University and a distinguished palaeoanthropologist. As well as working on Pleistocene sites in Israel, he had also worked with Don Johanson and William Kimbel in Ethiopia, finding the first, almost complete, skull of the early hominin *Australopithecus afarensis*.

Yoel told me about the discoveries at the Mount Carmel caves. As with so many sites, the Mount Carmel caves had been found by accident, by British geologists planning a new road and port at Haifa in what was then British Mandate of Palestine. Archaeologist Dorothy Garrod, who later became the first female professor at Cambridge University,[12] arrived to direct excavations, accompanied by palaeontologist Dorothea Bate, from the Natural History Museum in London. The excavations continued from 1924 to 1934. Yoel said that Dorothy was quite a feminist, and that the excavation team had consisted almost entirely of women from the nearby Arab village, but apparently some men had been drafted in when it came to the strenuous job of hand-drilling and lifting slabs of limestone breccia. Yoel pointed out the marks left by the drills on the terrace in front of Skhul Cave. The archaeologists found thousands of Mousterian stone tools in the cemented sediments, and in the lowest layer they discovered ten burials of modern humans.

Just around the corner of the canyon from Skhul, the archaeologists discovered Neanderthal remains in Tabun Cave. Garrod thought that the modern human remains probably dated to around 40,000 years ago, while the Neanderthal bones were more than 50,000 years old. This fitted with the idea at the time that Neanderthals were predecessors to modern humans.

But when absolute dating techniques were applied to the site in the 1980s the modern human burials were found to be much older. ESR dating of bovine teeth from the same layer as the burials produced a date of around 90,000 years.[13] Even more recent dating, using uranium series and ESR on fossil human bone and teeth, and on two animal teeth associated with the burials, indicate that the Skhul burials date to some time between 100 and 130 thousand years ago.[14]

After visiting Skhul Cave, I went to see some of the human remains themselves, in the Rockefeller Museum in Jerusalem. To reach them I walked through echoing galleries full of classical sarcophagi, Bronze Age burials and ossuaries, but the bones I was going to see were much more ancient.

There were two of the Skhul individuals in the museum: the bones of a four-year-old child (Skhul I) and those of an adult male (Skhul IV). Both skeletons were incredibly well preserved: more complete and in better condition than many medieval skeletons I have looked at in the bone lab at Bristol – and so much older. Soils on limestone generally make for good preservation of bone, but these bones wouldn't have survived had they not been buried.

The bones of Skhul IV were arranged just as they had been found by the archaeologists. The body was not carefully laid out as in later burials, straight or bound into a crouched position. The skeleton lay awkwardly, with the legs bent at the hips and knees, and the torso twisted on to its front, with the skull resting higher up, and the arms bent to bring the

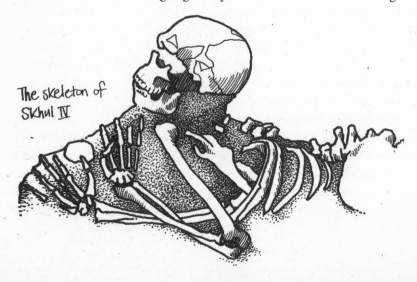

The skeleton of
Skhul IV

hands close to the face. There was a flint scraper between the hands. It's very difficult to know if this was placed in the grave deliberately or whether it was just in the soil that had been heaped on top of the body at burial.

But there were other objects with the Skhul burials that definitely appear to have been associated with the burials, to have been placed in the ground at the same time as the bodies. Skhul V was buried with a boar's mandible clasped in its arms, and there were also two shell beads in the same layer as the burials.[15]

Later in the 1930s, more modern human remains turned up in a cave near Nazareth, called Qafzeh. Initially, seven individuals were found, but when the cave was reinvestigated in the sixties and seventies the remains of a further fourteen individuals were uncovered.[16] One of these, an adolescent, was buried holding deer antlers in its arms, and there were also pieces of worked ochre in the ground with the burials.[14]

The two sites at Skhul and Qafzeh represent the earliest evidence anywhere of burial. Placing grave goods in the ground with the body is further evidence of ritual, and a spiritual dimension to these people's lives. Like the traces of art and ornamentation from South Africa, we seem to be seeing something here which implies modern ways of thinking and behaving: an approach to life and death that seems somehow familiar to us. Although we can't know what these objects signified, we can assume they meant *something* to the people who placed them in those graves around Mount Carmel all that time ago. It is tempting to imagine that the inclusion of personal ornaments and animal remains might even imply some kind of belief in an afterlife.

But after those burials at Skhul and Qafzeh, traces of modern humans in the Levant disappear, for around 50,000 years. I asked Yoel Rak what he thought had happened. He said it was difficult to prove that people had actually disappeared from the region. They might have stopped burying their dead. There are no more anatomically modern bones, and no more pierced shells, for a long time. And as the stone tools at Skhul were very basic, Mousterian tools, in fact exactly the same as Neanderthal technology, the presence of modern humans could not be demonstrated through tools alone, that is, until more sophisticated technology appeared in the Levant, around 45,000 years ago. Perhaps the evidence is there but hasn't been found yet. But maybe that absence is real, and Yoel believed that the disappearance of modern humans from the Levant, perhaps 90,000 years ago, could be explained by looking at what was happening to the climate and the environment at the time.

With the return of cold, dry conditions, the Sinai and Sahara deserts would have spread out again, blocking the northern route out of Africa, and cutting off North Africa from the south. 'The Middle East is the borderline between Africa and Europe, and it shifted back and forth,' Yoel said. During wet periods, the border effectively moved north and African fauna (including modern humans) colonised the Levant. In dry periods, the border moved south: the African fauna shrank bank, and European fauna moved in. And that fauna included Neanderthals. 'They felt comfortable living in the shadow of the glaciers,' said Yoel. I asked him what he thought had happened to the modern humans in Israel – had Neanderthals pushed them out? He thought not; 'There's no evidence for a dramatic scenario.'

Between 90,000 and 85,000 years ago there was a vicious, cold, dry period, known as the 'Heinrich 7 event', or H7. Heinrich events are characterised by great icebergs breaking free from ice sheets and floating in the North Atlantic, bringing down the surface temperature of the sea. In southern Asia, these events spell a reduction of the monsoon rains and very dry conditions. Perhaps this is what drove those pioneers out of the Middle East.[10] Yoel imagined the people living around Mount Carmel migrating, following herd animals heading south as their northern pastures disappeared. 'We could see them as a population who lived here happily, buried their dead, and then moved on.'

Neanderthal remains from Kebara and Amud in Israel date to around 60,000 years ago. But redating of the Neanderthal fossils from Tabun Cave, just around the corner from Skhul, in the same valley, has shown them to be 120,000 years old, roughly contemporaneous with the modern human burials at Skhul and Qafzeh.[14] It looks as if, for a while, between 100,000 and 130,000 years ago, the ranges of the 'African humans' (*Homo sapiens*) and 'European humans' (Neanderthals) may have overlapped in Israel.[17] But we're working with very broad estimates of age here, that don't allow us to resolve whether the modern humans and Neanderthals were actually contemporaries – they could have missed each other by hundreds or thousands of years. As Yoel Rak put it, 'Now this is not to say they were sitting in the cave and playing cards. We cannot know even if they saw each other.' And, in fact, we don't even know who got there first.

But what is very clear is that the modern human graves at Skhul and Qafzeh represent the earliest known symbolic burials, and that these people were therefore modern in their behavior as well as in their anatomy.[14] And it is a long time until we see evidence of modern humans in the Levant again.

Some anthropologists talk about the modern human people repres-ented by Skhul and Qafzeh as part of a 'failed exodus' from Africa. But although their descendants didn't spread into Asia and Europe, I think it's disingenuous to call this a 'failure', as there could never have been an *aim* to expand into other continents. We can only see it as a failed exodus with the benefit of hindsight. However, having said that, it seems that this particular expansion did not lead on directly to the colonisation of Asia and Europe.

So that leaves us looking for a later expansion out of Africa. It is possible that the northern route, through the Levantine Corridor, remained a viable option through the cold, dry period. Indeed, it seems that some modern human populations did survive in pockets of habitable environment, or refugia, in North Africa. In 1994, a child's skeleton was found at Taramsa Hill, on the west bank of the Nile in Egypt. OSL dating of the sands in which the skeleton was found suggested the burial was between 50,000 and 80,000 years old. The skeleton was very fragile, but enough of it – in particular the skull – was present for the anthropologists to be sure that was an anatomically modern human,[18] and some geneticists have suggested that there is Y chromosome evidence that migrations via this northern route contributed to living populations.[19]

A re-emergence of modern humans from Africa, perhaps around 50,000 years ago, via the northern route, seems to fit well with archaeo-logical and fossil evidence in Europe, but it is too late for the most recent estimates of the date of colonisation of southern Asia and Australia. Marta Lahr and Rob Foley of Cambridge University have suggested that there were at least two dispersals out of Africa: one via the southern route, from the Horn of Africa to the Arabian Peninsula, and along the Indian Ocean coastline, some 70,000 years ago, and then another exodus via the northern, Levantine Corridor up into Europe around 50,000 years ago. They suggested that each of these dispersals carried a different archaeological 'signature': the first migration from the Horn of Africa was associated with Middle Palaeolithic stone tools, while the later expansion was characterised by the appearance of more advanced, Upper Palaeolithic tools. They also argued that this model of multiple dispersals explained the range of anatomical variation, especially in skull shape, among both fossil and living modern humans, and that genetic studies also supported this theory.[7, 9]

Julie Field and Marta Lahr produced an ingenious GIS-based computer model to examine potential routes of expansion out of Africa. The model was based on studies of the palaeoenvironment during the cold, dry period of OIS 4, between about 74,000 and 59,000 years ago. Growing glaciers in northern and southern latitudes would have trapped an enormous amount of water as ice: sea levels would have dropped around the globe, to around 80m below modern levels. While the Persian Gulf would have been drained dry, the Red Sea would still have existed, though its coast would have been further out than today. North Africa and Arabia would have become increasingly arid, and the deserts would have expanded.

Field and Lahr's programme worked on the principle of finding the 'route of least resistance', taking into account obstacles like mountains, and wide lakes and rivers, as well as the availability of crucial water sources along the way. As colonisers would not have been heading in any particular direction or with any specific destination in mind, the programme was designed to 'wander', and to explore routes within a 60km radius of an origin. That starting point was set as Omo Kibish, then the computer was set free to wander. The 'route of least cost' across the ancient landscape, with sea levels much lower than today, took the virtual colonisers to the coast of the Red Sea, and (in a version 'without boats') up the west coast. Then a range of hills, level with modern-day Aswan, forced the virtual colonisers west, into the Nile Valley, and up to the Mediterranean coast. The route continued north, close to Mount Carmel, then headed east, to link up with the Euphrates, and followed that river down through the vast plain that is the Persian Gulf today. However, as Oppenheimer pointed out in a more recent paper, this northern route involved the virtual colonisers making three journeys of over 300km each, through deserts: from the Red Sea to the Nile, from the Nile to the Dead Sea, and across the Syrian desert to the Euphrates, not a mean feat for a non-desert-adapted animal that relies on plentiful supplies of water.[2] Even the modern Bushmen of the Kalahari need water.

In a second version, Field and Lahr allowed their colonisers the luxury of a boat, to cross Bab al Mandab. Then the route split, going north and coming to a stop near the Gulf of Aqaba (though presumably, if virtual boats were permitted here as well, they could have have continued around to the west coast, and there could have been a thriving virtual community all around the Red Sea coastline), or going east, along the shores of modern-day Yemen and Oman.[9]

The computer model depends on assigning values to all sorts of variables: deciding how readily people would cross rivers, lakes and mountain ranges, for instance. It's important to recognise that its authors didn't suggest that it would be able to predict actual, ancient routes of migration. It was really designed as a tool that could provide a different way of looking at potential pathways through ancient landscapes. But it is slightly bizarre in that it tends to show people wandering about in the desert, rather than sticking to areas with better water supplies. However, even with its limitations, the model does suggest that, between 60,000 and 70,000 years ago, both a northern and a southern route out of Africa would have been possible.

But now, genetics comes to the table to force the palaeoanthropologists' hand. Geneticists studying mtDNA trees have suggested that a single migration is most likely.[8, 11, 20, 21, 22] All non-Africans derive from a particular lineage called L3, which originated in Africa around 84,000 years ago.[22] The two 'daughters' of L3, haplogroups labelled 'M' and 'N', arose about 70,000 years ago.[2] The greatest diversity of M lineages is found in South Asia, suggesting that this is where the haplogroup originated; a branch of M, haplogroup M1, is also found in East Africa, but this may represent a late back-migration, after the Last Glacial Maximum (or 'LGM').[20] The N lineage is almost exclusively non-African. The simplest, most parsimonious explanation for this pattern is that a branch of L3 emerged out of Africa as a single migration, some time between 85,000 and 65,000 years ago, and that M and N then sprang forth, somewhere around the Indian subcontinent. Later, the first modern human Europeans would have come, not through the Levantine Corridor from North Africa, but from populations that had established themselves in the Indian subcontinent.[2]

Proponents of this model argue that the Y chromosome evidence for the northern route has been misread, and the relevant genetic markers represent a much later expansion from North Africa. It looks like there are two or three Y chromosome lineages that contribute to non-African populations (compared with the single mitochondrial L3) but this does not necessarily mean there were multiple migrations: the two or three ancestral lineages could have been carried out in a single exodus. The distribution of Y chromosome haplotypes outside Africa seems to support this proposition, and studies of genes in the other chromosomes also fit with a single exit.[11]

A single exit out of Africa means choosing between those northern and southern routes,[23] but it is hard to pin down the route taken, whether via Sinai or Bab al Mandab, on the basis of the genetics alone.[21]

For Oppenheimer, though, the choice is an easy one, if you take a look at the palaeoclimatic evidence. He argues that this suggests that, while the northern route may have been shut off during the cold of glacial stages, the southern route would have been open. Sea levels dropped significantly 65,000 years ago, at the Heinrich 6 event in the middle of OIS 4;[19] this was in fact the coldest, driest episode of the last 200,000 years. There was also a huge drop in sea levels 85,000 years ago, corresponding with the Heinrich 7 event (that Yoel thought may have forced the people of Skhul southwards). Although most of the Arabian Peninsula would have been dry, unwelcoming desert, the coast might just have still received enough monsoon rain to form a route for the beachcombers to follow. So a southern route across Bab al Mandab would have kept the colonisers close to sources of fresh water.[2, 8, 11]

There is evidence of people living on the East African coast by 125,000 years ago, with shell middens and MSA stone tools found at a site in Eritrea (although, in the absence of fossils, it is not definite that these beachcombers were modern humans).[24] For Oppenheimer, the onset of dry conditions combined with reduced sea levels would have provided people living around the Horn of Africa (modern-day Djibouti) with both the impetus and the means to leave Africa. Increasing aridity, leading to scarcity of food and hunger, may have prompted the migration, while at its lowest the reduction in sea level meant that there would have been just 11km of sea to cross at Bab al Mandab, at the southern end of the Red Sea.[25] Other climatologists suggest that the most likely time for the migration, from a climatic perspective, would have been slightly later, after the Heinrich 7 event, when the climate was warm and wet again, with the monsoon in full operation. And the last major wet phase in the Arabian Peninsula was between 78,000 and 82,000 years ago. This would still fit between the suggested dates for the emergence of L3 and her daughter haplogroups, M and N.

There is no evidence for boats from this long ago, but it seems reasonable to me to assume that modern humans living on the coast would have had the ingenuity to invent them. Small watercraft would have meant that people could have crossed the mouths of rivers and better exploited coastal resources. If they could see Arabia across the water, and if their families were struggling to survive on the African coast, as Oppenheimer suggests, that voyage seems like a very sensible option. In fact, there might even have been a maritime community all around the shores of the Red Sea, providing a base from which to spread along the southern coast of Arabia. Is it possible that we're getting too mired in this debate about the northern versus southern migration routes out of Africa? During warmer, wetter phases modern human could have been spreading out, along coasts and rivers, and around edges of deserts. They could have made their way out of Africa via both northern and southern routes, perhaps meeting up again somewhere around the Gulf. During arid phases, populations may have become restricted to green refugia.

It's very hard to trace movements of modern humans by archaeology alone. Before the Upper Palaeolithic or Later Stone Age, it's difficult to tell apart tools made by modern humans and by other archaic species like Neanderthals. It would be great to have some fossils of modern humans from this important period in the Middle East, dating to between 80,000 and 50,000 years ago, but at the moment there are none. Nevertheless, some archaeologists think that they can discern the presence of *Homo sapiens* from the tools they made, or other signals of modern behaviour, and claims have been made for modern humans being in India nearly 80,000 years ago, and in Australia by 60,000 years ago, whereas the earliest dates for the Levant (after Skhul and Qafzeh) and Europe are around 50,000 years ago. If those archaeological claims are to be trusted – and we must treat them with some caution in the absence of any accompanying modern human skeletons – then Oppenheimer's hunch that modern humans made it out of Africa via an early, southern route of dispersal may turn out to be true.

When you've got information coming in from so many different areas of science, from archaeology, from studies of skull shape, from reconstruction of ancient environments and from genetics, perhaps it's not surprising that there are different views, and it may take a long time for a consensus to be reached. But each new bit of evidence adds to the picture and helps to make it clearer. There are still missing pieces in this puzzle, and it seems that the debate about routes out of Africa won't be settled until more hard, archaeological and fossil evidence emerges from North Africa and the Levant, and from what, at the moment, is a mysterious, big black hole: Arabia.

An Arabian Mystery: Oman

Political unrest in South Arabia has meant that Palaeolithic archaeology has been very low on the agenda in the countries that form a potential 'Arabian Corridor', from Bab al Mandab, along the coast between the Red Sea and the Persian Gulf.[1] Yemen, in the west, was embroiled in civil war until the mid-nineties, and Oman, to the west, only opened its doors to outsiders in 1970.

One of those outsiders who had recently been tramping around the dry, desert landscape of Oman looking for traces of early human occupation was archaeologist Jeff Rose. I flew into Muscat and then on, across a vast, rocky, desert plain, to Salalah, in the Dhofar region, and there I met Jeff. He was fascinated by Arabia and the Gulf region. His obsession with Sumerian legends was evident, not only from his conversation, but by the fact that he appeared to be liberally covered in them, in the form of colourful tattoos. But it was prehistory that had drawn him to Oman. The scarcity of well-dated evidence from the Middle Palaeolithic of South Arabia, and its crucial position and relevance to the debate about routes out of Africa, made it irresistible to Jeff.

As far as modern human fossils are concerned, there are none at all in Arabia, and the earliest ones from the Indian subcontinent date to later than the colonisation of Australia. There *are* collections of Middle Palaeolithic tools from Arabia, but the dating of these is uncertain and there's nothing to mark them out as modern human. These tools could have been made by *Homo sapiens*, but, equally, they might have been manufactured by previous, archaic species.[2] To make things even more difficult, many of these collections of stone tools have been discovered lying on the surface of the ground, and so they are practically impossible to date.

Jeff and I drove north from Salalah, up into the rocky desert of Djebel al-Qara. Turning off the road, we headed into the desert landscape, dotted with rocky outcrops and frankincense trees. Jeff spotted a darkish area on one raised area of rock and we alighted and climbed up it to investigate. The ground was littered with angular, dark brown stones. And they weren't just any old stones. Most of those I stooped to examine had been flaked and shaped, not by natural processes but by humans. It seemed amazing to me to have driven into what seemed like the middle of nowhere, into a dry and unwelcoming landscape, and find this evidence of early human activity just lying there, on the surface of the ground. 'Pretty much anything you pick up here has been created by humans in the past,' said Jeff, reaching for what looked just like any old stone from a distance, but, on closer inspection, proved

a sand lizard
in the Omani desert

to be yet another lithic. 'Look at this, for instance. You can see where it's been flaked. It's a burin – like a chisel – for working softer materials like wood, hide or bone.'

Then he spotted a larger, more angular pebble. 'This one's a core – and they would have struck it like this to remove slivers of stone.' He held the core and imitated the action, without actually hitting the stone. 'Cores are important because they show us how they made their stone tools. And this technique is very distinctive. This is a blade core: it's been used to make long, thin blades.' We went on, scanning the ground for stones. It seemed every other stone was worked. I found one of the thin blades that had been made from a core like Jeff had showed me.

I asked Jeff how old he thought these tools might be. 'It's hard to say. It's a surface site so it's impossible to be specific about dates. But based on the type of technology, it could be anywhere from 12,000 to 70,000 years old, maybe even older. There was a site found recently on the coast of Yemen, with similar technology, that was dated to 70,000 years ago.'

It is very difficult to argue for the presence of modern humans based on the style of undated stone tools, but archaeological deposits in caves and filled-in basins have been reported in Arabia, if not in detail, and this made Jeff hope that there was good, datable archaeology to be found in Oman.[3]

It was hard to believe that people could have survived in the desert, though. Today, most of southern Arabia is desert, with the vast Rub' al-Khali sand sea filling the interior. But Jeff was very keen to show me that not all of Oman was dry, all of the time, even today. We drove east from Salalah and headed inland to the Wadi Darbat. As we drove, the landscape suddenly changed, from rocky desert … to a verdant oasis. We were there at the end of the rainy season, and the monsoon had filled the wadi with a turquoise river, fringed by lush grassy meadows, framed by green, wooded slopes. Herds of cows and camels shared the pasture and the cool waters. But this was temporary; during the dry season, the greenness would fade back to desert, and plants would lie dormant, waiting for the next summer's rains.

The verdant green valley of Wadi Darbat reminded me of south-west France. It seemed as though I'd been magically transported away from the hot and unforgiving desert that characterises most of Oman, even during the rainy season, to an idyllic valley full of plant and animal life. The desert had been almost silent. But here the air was full of the sound of birdsong. Life abounded. And it all came down to the presence of water.

It wasn't just the greenery that reminded me of the Dordogne. Up on those wooded slopes were huge rockshelters. They looked like good

· a camel in the Wadi Darbat ·

places to start looking for traces of human habitation. Jeff knew that stone tools had been found in some of them, and they were exactly the type of site where 'good archaeology', sealed in layers of sediment and datable, might be discovered. I could see why Jeff found Oman so exciting: there was so much promise here, and the chance to find the answers to so many open questions.

The transmutation of desert to a green Eden is a seasonal occurrence in Wadi Darbat. But just such a transformation, across vast swathes of South Arabia, has happened at various times in the deep past. Looking back into the Pleistocene, the environmental state of the region has fluctuated, depending on the intensity of the monsoon rains. Marine cores from the ocean around Arabia show that monsoons became heavier at the onset of interglacials, as ice sheets retreated and ocean surface temperatures increased. At the beginning of the last interglacial, about 130,000 years ago, there was a dramatic increase in the amount of rain falling on South Arabia, lasting for about 10,000 years. There was another peak in rainfall at the end of OIS 5, between 78,000 and 82,000 years ago.[4,5] And at those times, South Arabia would have become a very agreeable place in which to live.

Sites like the one we had visited in the desert were difficult to understand until I started to think about how much the environment had changed over time. In Oman, there are a number of sites where scatters of stone tools have been found on what is now a bone-dry landscape, but which would have been well watered during interglacials. Many sites are close to ancient, but now dry, wadis or river channels, and relict lakes, which would have been full of water during interglacials.[3, 6]

Many of the Arabian Middle Palaeolithic stone tools are small oval or leaf-shaped bifaces, which look as though they have been flaked using a soft hammer. Oxygen isotope stage 6 would have been too dry for any humans, archaic or modern, to survive in Arabia, so when conditions improved in OIS 5, the biface-makers must have moved into the region from a neighbouring area: from refugia in the Levant, in East Africa, or in the Zagros Mountains. The closest technological link is to East Africa; there are no tools anything like the Omani ones in collections of Mousterian tools from the Levant or Zagros Mountains.[6]

The fauna of southern Arabia shows links with Iran and Pakistan to the east, and with Africa to the west. Baboons, indigenous to the arid uplands of Ethiopia and Somalia, are also found in Yemen. Baboon mtDNA lineages suggest that they originated in East Africa between 50,000 and 150,000 years ago, and later migrated into Arabia.[1, 4] (So baboons appear to have migrated via the southern route, without boats, presumably making their way around the coast of the Red Sea.)

Many sites around South Arabia may lie beneath seas: either under the Rub' al-Khali sand sea inland, or submerged under the Arabian Sea as the oceans have risen. Jeff was particularly interested in the idea of a submerged coastal plain where humans could have lived. Yemen and Oman may have been liberally doused with monsoon rain when interglacials arrived, but even during glacial periods it seems that there may have been enough water for humans (alongside other flora and fauna) to exist, along the coast. Because underwater, but close to the coast of the Arabian Sea and the Persian Gulf, there are submerged freshwater springs. 'One of the strangest things about Arabia is that we have this completely arid landscape, and yet beneath the surface there's heaps of fresh water, running towards the coast and coming up under the sea,' Jeff explained. 'If you were to dive down with a canteen you could fill it up with fresh water and have a drink.'

When sea levels were lower during glacial periods, these springs would have watered the exposed coastal plains, creating a long oasis stretching along the southern coast of Arabia from Yemen to the Gulf.[4] This means that when sea levels dropped, reducing the Bab al Mandab gap to some 11km, there would have been this coastal oasis on the other side. So, for Jeff, the Red Sea was not a barrier to human migrations but a conduit between Africa and Arabia. And, indeed, Middle Palaeolithic tools have been found at sites along the eastern coast of the Red Sea.[3]

The Arabian oasis ran right along the southern coast and up into the large plain that is now the Persian Gulf. 'It's the shallowest inland sea in the world, just 40 metres deep, so when the sea level was lower the entire area was an exposed flood plain – a beautiful verdant paradise,' Jeff enthused. 'It sounds idyllic,' I said. 'Well – the Sumerians called it "Eden",' he replied.

The Gulf Basin would have received water from underground aquifers, welling up as springs, from the Tigris and Euphrates rivers, and from rivers flowing down from the Zagros Mountains to the east. All this water would have flowed into a great river which ran down the length of the Gulf Basin; topographical studies of the seabed, or bathymetry, have shown the wide, deep groove created by this ancient waterway. So Jeff argued that, between 115,000 and 6000 years ago, the Gulf Plain would have formed a refugium where humans and other animals could have survived, even while harsh and arid conditions prevailed elsewhere.[4]

Studies of the palaeoenvironment of Arabia are fascinating. It seems that there would have been favourable habitats for humans to exploit there during both interglacials, with the monsoon rains greening the Arabian deserts, and, at the height of glacials, where the interior would have been incredibly arid, the exposed coastal plains would have been kept oasis-like by freshwater springs.

But none of this means that the southern route and the Arabian Corridor *were* used as the main or only route out of Africa for modern humans. The Middle Palaeolithic archaeology of Arabia shows someone was there, but it really is impossible to pin down whether it was *Homo sapiens, heidelbergensis* or *neanderthalensis* on the basis of the currently available evidence.[5] It seemed that Jeff had his work cut out for him.

So Arabia is still a bit of a mystery. But once the modern humans had expanded out of Africa, they continued expanding, with a wave of colonisation pushing east and perhaps even reaching Australia, before modern humans made their way northwards into Europe. Climate change may have pushed modern humans to the coast of South and East Africa, forcing them to adapt to this habitat and diversify their subsistence, but once they had learned to make use of coastal resources, the coast would provide them with a benign habitat that stayed relatively stable compared with inland environments, through the fluctuating climate of the Pleistocene.[7] The next stage of my journey would be to head east and try to pick up the traces of the migration along the coastline of the Indian Ocean.

2. Footprints of the Ancestors: From India to Australia

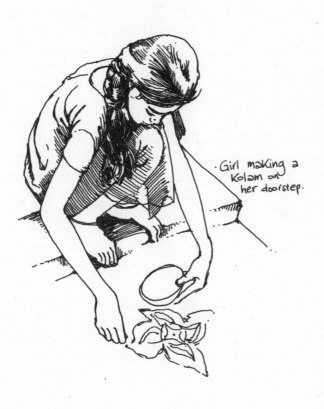

Girl making a Kolam on her doorstep.

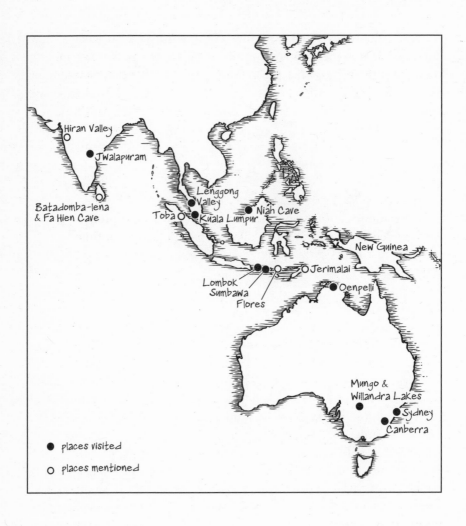

places visited

places mentioned

Archaeology in the Ashes: Jwalapuram, India

Around 74,000 years ago a massive volcano erupted on what is now Sumatra.[1] It was the greatest volcanic eruption of the last two million years,[2] and the biggest that the human species has yet experienced: tens of thousands of times larger than any eruption recorded in history. The huge 100km-wide crater left by the eruption is now filled with a massive lake: Lake Toba.

When Toba volcano erupted, there were humans, although not all of them modern, in Africa, Europe and parts of Central and South-East Asia. There were Neanderthals in Europe, *Homo heidelbergensis* in China, and perhaps there were still some *Homo erectus* in South-East Asia. Anatomically modern humans were in Africa, and probably in the Arabian Peninsula, too. But had they reached India by that time?

The first part of my journey would take me closer to Toba, not to Sumatra but to India, to Jwalapuram in the Kurnool district of Andhra Pradesh, where an archaeological excavation through the Toba ash layer was under way. When Toba erupted, it spewed out massive clouds of heated ash, and this ash has been found deep beneath the seabeds of the South China Sea and the Indian Ocean – as well as on the Indian subcontinent (India, Pakistan, Bangladesh, Sri Lanka, Bhutan and Nepal), more than 3000km away from the volcano. From the pattern of dispersal of the ash it seems likely that the eruption occurred during the summer monsoon season, when southerly winds would have blown the ash northwards towards the continent.[3]

There is still some uncertainty about the effect of the Toba super-eruption on climate and its impact on plants and animals – including humans – as well as controversy over whether or not humans had reached India by that time.[2] These unanswered questions explain why the archaeologists are still there, in that hot and dusty district of central India, digging through the ash for more clues, more pieces of the puzzle. So, having flown into Chennai and on to Bangalore, I caught the train to Nandyal in Andhra Pradesh to meet archaeologists Mike Petraglia (Cambridge University) and Ravi Korisettar (Karnatak University), who head up the international team working on the Toba ash layers at Jwalapuram.

We left Nandyal early the next morning, and I shared a jeep with Ravi Korisettar. Our fearless driver thought nothing of overtaking – or undertaking – on bends or hills, speeding up whenever confronted with people, chickens, dogs or cows on the road. The journey took about an hour and a half, along potholed roads.

As we drew near to the site the landscape changed, and the dry, dusty fields and brickworks around Nandyal gave way to a lusher landscape of paddy fields, where flocks of egrets wheeled then settled in regimental lines. We drove along roads lined with neem trees, then turned off the tarmac road on to a track, which wound its way through the small village of Jwalapuram and further on into the Jurrera River valley, framed by limestone escarpments topped with quartzite boulders. The slopes were covered in scrubby bushes, and some massive limestone boulders had tumbled down, forming ready-made rockshelters. The Jurrera River was penned behind a dam, with some small streams trickling through to irrigate the paddy fields at the edges of the basin.

Ravi told me that the archaeological site near Jwalapuram had revealed evidence of ancient humans living around the shores of a lake. Back then, in the Pleistocene, a high water table would have meant that springs gushed from the limestone hills all year round, and the floor of the basin would have contained a lake, fed by the springs and by monsoon rains. 'As a result of these perennial fresh water and spring water resources, this region would have been covered with lush green vegetation, wooded forests, a diverse variety of plant foods. This would have attracted animals, and then humans would have been drawn to this basin because of fresh water, plant food and animal food resources. And, above all, what we have here are ideal rocks – limestone, quartzite and chert – for manufacturing a variety of stone tools.'

Ravi suggested that the basin would have been such a great place to live that the hunter-gatherers could have moved around within it, finding plenty of resources, and never really needing to move out.

This is a useful thought to keep in mind as it puts a brake on the idea of anatomically modern humans furiously racing across Asia. It may have been a rapid dispersal in the grand frame of geological time, but on a more human time scale the spread would have been gradual, a wave of advancing colonisation edging east. Behind the wave, hunter-gatherers would have been left, populating the landscape, albeit sparsely.

'How many people do you think would have been living around the lake here?' I asked.

'Prehistoric populations were very small,' replied Ravi. 'I suspect there would have been bands of perhaps fifty to a hundred people, living a hunting and gathering, nomadic way of life here.'

The central part of the basin, where we were headed, was very dry and dotted with vicious, thorny acacia bushes.

We saw the clouds of white ash in the air long before we reached the site. The archaeological site was being revealed in part – as well as being destroyed – by mining. And the workers were after not precious metals or gems, but ash. The ancient volcanic ash was being mined, raked and sieved, bagged up, all by hand, and then sent away to be turned into

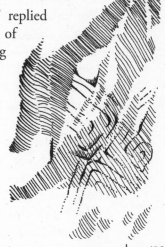

· engraving at Kurnool caves·

brass-cleaning products. Some of this ash was being shipped back 'home' to Indonesia. It was shocking to see people scraping back the hardened ash layer with wooden forks, sending clouds of silicate into the air to be breathed into unprotected lungs: no one was wearing a mask or even covering their nose and mouth. Even more disturbing was that most of the workers were children. The caste system is meant to be a thing of the past, and child labour is illegal, but these low-caste children from the local village did not attend school and were sent out to work instead.

We pulled up just beyond the ash-mining area, where the unmistakable evidence of Palaeolithic archaeologists at work could be seen: a deep, square hole in the ground, with almost impossibly straight edges and perfectly flat and perpendicular sides. We piled out of the jeeps and met Ravi's collaborator from Cambridge, Dr Mike Petraglia. While Ravi was the epitomy of an Indian professor, reserved and thoughtful, Mike was a zealous American archaeologist with a penchant for ancient and exotic investigations, and was even sporting an Indiana Jonesesque hat. He strode around the landscape, taking me on a guided tour of all that Jwalapuram had to offer.

In a square trench, or Jwalapuram 22 as it was more properly known, excavations were ongoing, and a slender whitish line about 57m below the surface marked the Toba ash layer. Villagers-turned-diggers were carefully trowelling around calcrete lumps, which visiting geologists later pronounced to be infilled termite tunnels, marking the basin floor before the ash fall. The trench was sited where the lake shore would have been before the Toba eruption. The ash layer was subtle there, so Mike took me to the edge of the ash-mining area, where a roughly cut section through the sediments revealed a stonking 2m-deep layer of white ash. Here was Toba ash – and lots of it. It was almost as though someone had walked along the section white-washing everything between shoulder and ankle height. Mike's team had tested the tephra deposits and confirmed that they were indeed ash layers from the eruption at 74,000 years, called the Youngest Toba Tuff (YTT), as opposed to ash layers produced by much earlier eruptions. They analysed the microstructure of the tephra and found minute volcanic glass shards that were very similar to those seen in other proven YTT deposits in Sumatra.

Ravi explained to me why there were such wide-ranging differences in the thickness of the ash layer across the valley floor. The initial ash fall probably covered the landscape in a layer some 10–15cm deep, like a heavy snowfall. Then, rain – some immediately precipitated by the dust in the atmosphere – would have washed the ash down into the lake. The

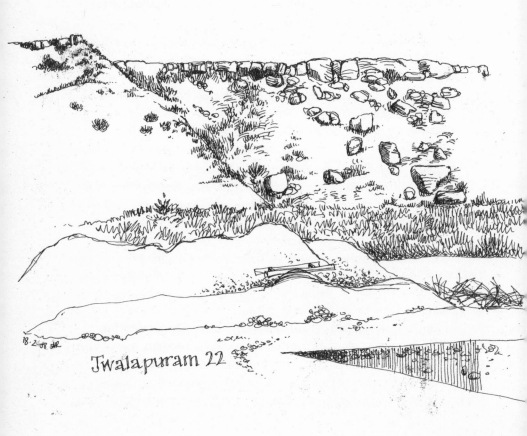

Jwalapuram 22

ash would have gradually settled out on the lake floor, building up into a thick sediment, which explained the thick band in the section where the ash mining was going on. The bottom of the band of ash sat on the red-brown clay of the ancient lake bed. Ravi said that the lake appeared to have dried up soon after the Toba eruption and the ash fall.

What Mike and Ravi were really interested in, though, was not the Toba ash itself, but stone tools that they had found, closely associated with it. But were the toolmakers modern humans or archaic species? Hand axes have been discovered in the vicinity of Jwalapuram (though not associated with the ash), showing that archaic humans had certainly been in the area. And there are many other sites in India with ancient stone tools made by earlier hominins, as well as direct evidence in the form of the 'Narmada cranium' of *Homo heidelbergensis*, from between 250,000 and 300,000 years ago.[4]

But both archaeologists were convinced that the tools associated with the Toba ash, dating to between 70,000 and 80,000 years ago, were indicative of modern humans.

Back in the 'lab' – a hotel room in Nandyal – Mike showed me some of the tools from the excavation. The OSL dates for the layers in which the tools were found ranged from 74,000 years, above the ash, to 78,000, below the ash. There was a Levallois core from below the ash, which had been shaped around the edges prior to flakes being struck off. There was a large selection of flakes that had been knapped from cobbles in just this way, and then carefully chipped or 'retouched' to produce serrated edges. Tools like this were probably used as scrapers, perhaps to clean hides or process plant material. These flakes came from layers both above and below the Toba ash. There were also some longer blades, which looked like useful cutting tools, and some pointy, burin-like pieces, from below the ash.

But Mike was particularly excited by a recent find, a tanged point, perhaps the tip of a spear or dart, found in the layer below the ash. There were also some tiny blades, from above the ash, which Mike called 'microblades', quite similar to the ones I'd seen at Pinnacle Point in South Africa. As far as Mike was concerned, the tanged point and microblades quite clearly indicated advanced, composite tools – and modern humans. 'It suggests an African connection – these tanged points are very characteristic of industries being made by modern humans in Africa between ninety and sixty thousand years ago,' he explained. 'But our findings are controversial in the sense that we're finding stone tools that date right back to seventy-eight thousand years. It's much earlier than people have previously suggested that modern humans had reached India.'

At the moment, the earliest indisputable evidence of modern humans in the Indian subcontinent comes from Sri Lanka, where modern human fossils and microlithic stone tool assemblages have been found at the sites of Fa Hien Cave, dated to about 31,000 years ago, and at Batadomba-lena, at about 29,000 years ago.[4] So Mike and Ravi's claim is understandably controversial: the dates from Jwalapuram are more than twice as old.

JWP 22
#33
tanged point

For Mike, the archaeological findings also suggested that modern humans in the area may have survived the fallout from the super-eruption. 'What's so exciting is that we are finding tools above and below the ash. The tool types and styles don't change dramatically.'

This seemed quite remarkable. Some scientists have suggested that the Toba super-eruption would have plunged the planet into a 'volcanic winter'. Volcanic dust in the atmosphere could have caused an average 5 degree C drop in the temperature of the northern hemisphere, lasting several years. While the earth was already cooling down when Toba blew, the massive eruption may have accelerated the transition into the glacial period of OIS 4, by cooling the world and promoting the growth of northern ice sheets, resulting in more heat being reflected from the earth's surface.[1] Some geneticists have argued that the pattern of variation in genes across human populations today shows that, around this point in human prehistory, there was an 'evolutionary bottleneck': numbers of our species dropped to catastrophically small levels, making us an endangered species for a while.[5, 6]

But other scientists have said this is too extreme a model, and that, while Toba was certainly a massive eruption, spewing out something like 28,000 cubic *kilometres* of lava, its effects on climate may have been more subtle. A more conservative view sees Toba producing a global

temperature reduction of 1 degree C, and unlikely to have triggered the last glaciation.[3] But even if Toba did not create significant climatic disruption, the ash fall would undoubtedly have had an impact on human populations.[7]

Neither Mike nor Ravi wanted to underplay completely the environmental

JWP 23   JWP 23   JWP 23
#37   #136   #67

impact of the eruption, though. 'There's no doubt that the Toba super-eruption had an ecological impact: the water would have been poisoned, plants would have been suffering, animals would have been suffering. And the people subsisting on plants and animals would have been affected, of course,' said Mike. Once the ash had settled, the ancient hunter-gatherers of Jwalapuram would have faced an ecological disaster. With the lake poisoned with volcanic ash, the springs in the limestone escarpments may have provided an essential source of untainted water.

'But you don't think it wiped the population out?' I asked.

'Well, the Toba super-eruption didn't seem to have quite the devastating impact it was once thought to have had. The population here didn't crash, and people continued to live in this area.'

Mike was very careful to say that he couldn't be sure the tools above the ash were made by *precisely* the same individuals as below the ash, but he still believed that the human population in the general area had survived Toba.

In fact, Mike also thought that the ancient people of Jwalapuram had probably coped better than *we* would if something the size of Toba went off today. 'Hunter-gatherers were able to perhaps cope with environmental catastrophes in a much better way than settled societies, because they were flexible. They could shift their dietary strategies; they could move on to places that were more desirable. Today, in a sense, we're trapped. We can't move. If a super-eruption happened now it would make a massive impact on our societies.'

Although there was general continuity in the tool types at Jwalapuram, it was interesting that the microblades seemed to appear only above the ash. Perhaps they represented humans experimenting with different ways of hunting and surviving in the cooler and drier climate that set in after the Toba eruption. The landscape would have become less wooded and more open; more sophisticated projectile technology may have been a very useful adaptation in this changed environment.

It's tempting to turn the Toba event into a story of humans triumphing against adversity, using their superior skills and cunning to survive when other, lesser species would have curled up their toes and died. Of course, humans weren't the only species to suffer the effects of the Toba super-eruption. A recent study of other animals in South-East Asia around the time of the eruption revealed some fairly predictable, but also some quite surprising results. Some species certainly appeared to become extinct following the eruption – but actually relatively few. Most survived it, retreating into refuges that allowed them to bide their time until the worst was over, and then sprang back quite quickly, spreading over large areas of the landscape within a century of the environmental devastation caused by the volcano. It shows how robust mammals can be, and that it wasn't just humans that bounced back.[8]

There were still many questions left to be answered at Jwalapuram, and the Indian stage of the human journey is far from clear. Many palaeoanthropologists remain unconvinced that the tools are evidence of modern human presence at the time of the Toba eruption. But the wider context seems to provide some support for the idea: the ability of the population to survive the Toba devastation and continue producing similar tools; the fact that the Jwalapuram toolkit resembles sub-Saharan MSA assemblages; and the dates also seem to fit the genetic suggestion of an African exodus around 80,000 years ago.[9]

There's no doubt that both Mike and Ravi would love to find some indisputable evidence: remains of modern humans themselves. As Mike put it: 'That would be like the Eureka moment – a human fossil from this period, in India, would be a major find.'

Ravi had an idea that early modern humans had made their way across India following a chain of lake basins – like Jwalapuram. But there are other possibilities. If indeed the Jwalapuram tools were made by modern humans, this only indicates their existence in that place, and does not negate their presence elsewhere. And Ravi's transcontinental, basin-hopping route across India flies in the face of the more traditional explanation of the eastwards spread of modern humans out of Africa: along the coasts. Paul Mellars[10] talks about a 'coastal express route', with modern human populations fairly nipping along the shores of the Indian Ocean. The genetic evidence certainly suggests modern humans moved quickly eastwards, reaching Malaysia and the Andaman Islands by 55,000 and possibly even 65,000 years ago. But Mellars recognises that a major challenge to this rapid, coastal dispersal is the lack of hard archaeological evidence from India and Arabia. From Mellars' point of view, Jwalapuram is one of a very few sites of sufficient antiquity that could represent modern humans in India at or close to the 'colonisation front'. But he does not think that Jwalapuram stands as a challenge to the coastal route – why shouldn't modern humans be penetrating the interior of the continent as well as moving along the coast? On its own, the site certainly doesn't prove that there was an east–west, transcontinental route across India, as opposed to a coastal route, and it could even just be a cul-de-sac.

The attraction of the beachcombing, coastal route, in spite of the lack of hard evidence, is that it makes a lot of sense from an ecological standpoint. Attempts to identify the easiest and most likely routes of colonisation, based on the environment of South Asia between 70,000 and 45,000 years ago, have suggested that coastlines would have been ideal (with a secondary route through the Western Ghats, along tributaries of the Krishna River and along the northern border of the Deccan Plateau). There are a few Middle Palaeolithic sites close to the coast of India, but most of them are far too old to be modern human, except for those in the Hiran Valley, with Middle Palaeolithic tools dating to around 60,000 years ago.[11, 12] It's very important to remember that sea levels have changed and are much higher today than they were for most of the Pleistocene: the ancient coastline is now beneath the waves. Perhaps the archaeological evidence for these earliest modern human migrations is down there, too.

roadside shop just south of Cochin, Kerala.

Hunter-Gatherers and Genes in the Rainforest: Lenggong, Perak, Malaysia

Moving eastwards, the archaeological record is still very patchy. The same problems that beset my search for evidence in India affect the entire putative route along the northern rim of the Indian Ocean: sea levels have risen, hard fossil evidence is missing, and toolkits are hard to assign to modern humans with any degree of certainty. The next leg of my journey would take me to Malaysia, to rendezvous with Stephen Oppenheimer and the Lanoh people of the Lenggong Valley.

Stephen is a geneticist and a force to be reckoned with in palaeo-anthropology. He has published widely in the scientific literature, but also has a strong interest in letting people outside of the illuminati know what's going on, and has written several popular books on the subject, one of which, *Out of Eden*, I had in my bag as I went to meet him. I knew that Stephen was a medical doctor, and I wanted to know what had got him interested in genetics and, in particular, in early human migrations. I met Stephen in a hotel in Kuala Lumpur, and the following day, as we drove to the Lenggong Valley in the north of peninsular Malaysia, his story started to emerge.

It turned out to be a mixture of curiosity, a wandering spirit and family ties that had drawn Stephen to the Pacific, and genetics. He had studied medicine at Oxford University, spending his clinical years in the Royal London (teaching) Hospital, where he had stayed on for his first job as a junior doctor. But a year later – as soon as he was a fully registered doctor, wanderlust had got the better of him. 'Within a week of finishing those jobs and getting registration, I headed off out east – to Hong Kong in the first place, where I worked in a mission hospital for a few months. From Hong Kong I went to Bangkok, and then I worked as a flying doctor in Borneo.'

After a year in the Far East, Stephen returned to England for three years, to specialise in paediatrics. Then he hurried back east, this time to work as a paediatrician in Papua New Guinea. I asked him how he became interested in genetics. 'After getting more clinical experience, I started doing research in New Guinea,' recounted Stephen. 'I was interested in iron deficiency anaemia – its causes and prevention. And at that time I noticed that there was a very high rate of a genetic blood disorder which also causes anaemia: α-thalassaemia.'

In fact, this disorder was *so* common in these South-East Asian populations that Stephen had actually found the highest rate of any genetic defect, anywhere in the world. He also knew that another genetic form of anaemia, sickle cell anaemia, was very high in African populations. And the reason for this was well known: although the genetic 'defect' caused anaemia, it was also protective against malaria.

Stephen wondered if α-thalassaemia in his New Guinean patients was doing the same thing. It turned out that he'd hit the nail on the head.[1, 2, 3] So that explained how Stephen had become interested in genes, and medical genetics was a thoroughly respectable area of research for a clinician to engage in. But Stephen also realised that the genes he was looking at didn't just represent the adaptations of a population to the tropical environment they now inhabited, but also acted as some sort of record of where those populations had come from: they could be used as markers of human migration.

'What was interesting was that the various mutations causing α-thalassaemia were present in different frequencies in groups of people speaking different languages,' explained Stephen. 'There was a particular α-thalassaemia mutation which was specific to Austronesian-speaking coastal and island populations, and different from the mainland New Guinea variant. There seemed to be a genetic trail out to the Pacific – probably very ancient – and you could still see it in the population along the north coast of New Guinea.'

He was entering a different world – the world of archaeology and anthropology, with theories usually built on fragments of fossils and stone tools – but he was confident that the genes within modern populations held important clues that could help unravel the riddle of human origins. 'I found it fantastically exciting that very specific genetic mutations could act as trailmarkers for ancient migrations. I haven't really stopped thinking about it ever since, and that was twenty-five years ago.' Although he continued to work as a clinician, Stephen had also published a huge amount of research into genetics and the dispersal of modern humans, looking in particular at mitochondrial DNA. And that is what had brought him to the Lenggong Valley rainforest. He was there to gather samples of mtDNA from a group of Orang Asli ('original people'), who have long been presumed to be Aboriginal Malaysians.

There are many tribes of Orang Asli, divided into three main groups, of which the smallest is the Semang; this group is also thought to be the most ancient. Some sources have described them as African-looking, with very dark skin and thick black hair. The Semang, of all the Orang Asli, have also held on to their traditional way of life as hunter-gatherers the longest. The people we were going to see were one of those Semang tribes, the Lanoh. They live in the Upper Perak district, and they have another name for themselves, *Semark Belum*, which means 'the original people of the Perak River'.[4]

Stephen had already investigated mitochondrial lineages among the majority Malay population, and within other Orang Asli groups, but the DNA of the Lanoh people had never been sampled before. I was interested in the story held in their genes, but I was also looking forward to meeting some people who were still living as hunter-gatherers. Sadly, though, they are very much hunter-gatherers under threat. Writing in 1976, anthropologist Iskander Carey was able to state that the Semang were just about the only Orang Asli group to 'practise little or no cultivation' and that they were 'the only true nomads' in Malaysia. But the Lanoh tribe no longer fitted that description. Since the 1970s, the Malaysian government had taken measures to 'resettle' the Orang Asli, building villages where the once-nomadic tribes could be brought together to settle down and make a useful contribution to the economy. We were to meet the Lanoh in one such village at Kampong Air Bah ('Flooding River Settlement'). For these people, who had lost not only most of their jungle, but also any rights to their former territory, hunting and gathering was now little more than a pastime. Real labour meant working on rubber plantations, farms and logging camps. Like many groups of people subsisting as hunter-gatherers on the fringes of agricultural and industrial populations, the Lanoh are in the process of being swept up and forced to leave the old ways behind. It is a story that must have happened thousands of times, over thousands of years, since the time that some people started to farm and build civilisations.

The commercial pressures that were changing the lifestyles of the ancient populations of Malaysia were also changing the ancient landscape: logging was stripping the hillsides bare, and palm oil monoculture followed in its wake. The effect was quite astounding from the air: on the flight into Kuala Lumpur I'd seen hills and valleys completely divested of green, with the bulldozer tracks forming strange ridged patterns like pink thumbprints on the land. Some were starting to grow green again, but with palm oil seedlings rather than ancient rainforest. The palm oil plantations covered vast areas, in regular patterns and standard green. On the ground, we drove through acres and acres of oil palms, durian orchards and rubber tree plantations, before we reached Kampong Air Bah, tucked inside a remaining bit of forest.

Individual houses occupied the lower ground. These were modern houses, still on stilts, but many had more traditional extensions of bamboo

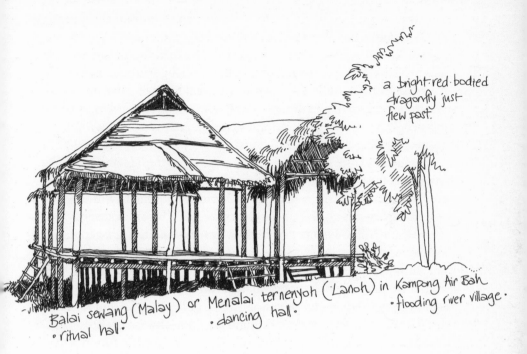

a bright·red·bodied dragonfly just flew past.

Balai sewang (Malay) or Menalai ternenyoh (Lanoh) in Kampong Air Bah
·ritual hall·   ·dancing hall·   ·flooding river village·

and woven bertam palm walls. On the higher ground was a small mosque (as the Lanoh are now nominally Muslim) and a large, wall-less wooden building with a thatched roof of palm leaves. This was the *balai sewang* (in Malaysian) or *menalai ternenyoh* (in Lanoh) – a building that served the purposes of meeting place, village hall and dance hall. It formed the physical and social centre of the Lanoh village.

We pulled up in our Land Rovers next to the *balai sewang*, removed our shoes and climbed the wooden ladders to the raised floor of the hall, where the *penghulu* ('leader') of the Lanoh, Alias Bin Semedang, greeted us. We were ushered in to sit crosslegged on the floor, with Alias and some other elders. After introductions, we explained why we wanted to visit the Lanoh. Stephen carefully related his search for ancient lineages and asked Alias if it would be possible to take cheek cell samples from members of the group. I asked if I could accompany the tribe, hunting and gathering in the rainforest. Alias seemed to discuss the requests with the other elders, then turned to us. Even before the translation came back to us, we knew from his smile that it was good news. I told Alias that previous studies had already shown that other Semang tribes had ancient genes, ancient blood, and their ancestors were the first in the land. Alias was unsurprised by this; in fact, he told us that the Semang were *the* original people and everyone else had come from them.

Stephen set out his stall and prepared to collect cheek cells, opening boxes of long-handled brushes that the Semang participants would rub up and down inside each cheek. (Although the brushes looked as though they had been specifically designed for this job, they were actually produced to take samples from the opposite end of the body: cervical cells for screening tests.)

Meanwhile, I set out with a group of Lanoh girls to go fishing in the Air Bah River. We drove some distance from the village and then trekked into the rainforest. The first thing that struck me was how incredibly noisy it was: the trees seemed to be packed with highly vociferous but apparently invisible insects and birds. Down by the river, we took our shoes off and waded in, and the girls immediately started hunting for fish – with their hands and a shovel-shaped woven bamboo basket. They would turn smaller pebbles over and hold the basket downstream to catch any fish that would emerge, and fearlessly scrabble under larger rocks for fish that might be hiding there. I gradually grew braver – and managed to catch one fat tadpole. My Western sensibilities got the better of me, though, and while the girls proceeded to thread the fish they'd caught through the gills on to a forked twig – still alive – the tadpole lived to fight another day.

I sat down by a rock and watched the girls, almost completely submerged now, pushing their hands under a large boulder where, by their excited voices, they could obviously *just* feel fish with their fingertips. They laughed and splashed around in the river they'd known since childhood; they knew its twists and turns, and the boulders where fish would be hiding. And they were certainly adept at catching them: there would be fish for supper that night. But for these four young women in their late teens, it was also a group of friends having fun. It was a relic of an earlier way of living; what was work in the sense that they were serious about feeding themselves, was also something you did with friends, and

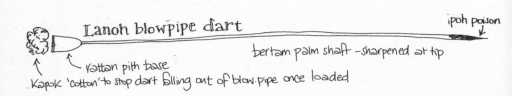

Lanoh blowpipe dart

ipoh poison

bertam palm shaft – sharpened at tip

rattan pith base

Kapok 'cotton' to stop dart falling out of blow-pipe once loaded

for yourself, friends and family. But, as more and more young Lanoh find paid work, it's a way of life that may not exist for much longer.

Later in the day I accompanied Alias and a younger man as they went on a hunting trip – armed with blowpipes. We sat at the top of a waterfall and I asked Alias about how life had changed for the Lanoh. Historically, the Lanoh would have lived in temporary, seasonal camps. The rainforest did not offer the sumptuous banquet that I first imagined it might. Hunting and gathering in this environment required specialist knowledge of which foodstuffs were edible and where to find them: it was hard work to subsist entirely on wild food, and the tribe would have ranged over a large territory. By the middle of the twentieth century, the Lanoh had begun to settle down and were semi-nomadic. Alias described how, when he was a child, the tribe would set up a camp and stay in it for one or two years, foraging in the surrounding area, before moving on. The territories were not exclusive: the broad area that the Lanoh occupied would be shared with other tribes, and they also recognised the Malay villages. At this time, the Lanoh had begun trading with the Malays, swapping jungle products such as rattan and resins for money, rice, sugar and other foodstuffs – and later, for motorbikes and televisions. They also began small-scale cultivation, of hill rice, tapioca and maize, often sowing the crops and moving away, then returning when the harvest was ready. In 1970, the Lanoh moved into Kampong Air Bah, and the nomadic way of life was replaced by a settled existence, with more cultivation of crops and less hunting and gathering. Rubber tree plantations created as part of the resettlement scheme also provided a source of income.[4, 5]

'In the past we would search for food, rattan and wood,' said Alias. 'When I went hunting in the jungle with my father, we would hunt for food, like monkeys, squirrels and birds. There were lots of things you could get in there. Now, when we go to the jungle, we would look for money. Whatever we manage to get, we sell.' The lives of the Lanoh had changed so much in just one generation. 'In the past we were free and happy. We could go wherever we wanted to and do what we wanted. We could stay in the jungle and no one would bother us. But now it is difficult.'

But Alias still knew how to survive in the rainforest: he knew which plants were edible and which poisonous, and how to stealthily creep up and dispatch a monkey or a squirrel with a blowpipe. These skills were no longer vital to survival, in a time when cash could be earned and food could be grown or bought, but in spite of this, Alias still thought it important to pass on the knowledge and skills to the next generation: it was part of the Lanoh identity.

The blowpipe was a particularly important symbol of this hunter-gatherer character and most Lanoh men owned one. Alias showed me how his blowpipe had been made, from a very straight length of sewoor bamboo with no joints. The base of the blowpipe was decorated with an incised geometric design; each Semang group used a different motif, and the design was also important to the luck of the hunt. Alias was wearing his quiver, or *lek*, of darts, tied around his waist with string. The *lek* was made of a short length of bamboo, also decorated, with a woven rattan lid. The darts were produced from bertam palm, shaved down to a point at one end and dipped in ipoh toxin. Placing a dart in the bottom of the blowpipe, Alias then pushed a small wad of kapok 'cotton' in beneath it to stop it falling back out, and the weapon was primed and ready to use. We made our way down the river, Alias cautiously and silently moving around and occasionally lifting the blowpipe to his lips when he thought he was near to a quarry. But the jungle creatures were keeping themselves well hidden that afternoon. Alias had little chance of successfully creeping up on his potential dinner when he was being followed by a considerably less stealthy anthropologist.

Abandoning the hunt, we settled for a bit of target practice. I had a go at using the blowpipe and found it to be very accurate even in inexperienced hands: the length of the blowpipe made it easy to sight a target, and even I could hit a tree trunk some 10m away with it. This was obviously very different from being able to hit a moving quarry high in a tree-top, but I felt proud of my achievement, and I was impressed with this simple but effective tool. Blowpipes were actually a fairly recent introduction: until 1910, the Lanoh had hunted using bows and arrows, but they picked up the use of blowpipes from a neighbouring tribe, perhaps as they moved deeper into the rainforest. Blowpipes have a range of good accuracy up to about 35m, compared with 100m for bow and arrow.[6] Archery is effective in more open woodland, but the blowpipe is even better suited to hunting in dense forest. The adoption of the blowpipe by the Lanoh represented a very recent example of human cultural adaptability to a changing environment.

When I returned from my trip into the forest, preparations were under way for a *sewang*: a celebratory dance. Stephen had finished his sampling and we sat down in the shade. Alias opened fresh coconuts with a machete for us and we ate a snack of roasted tapioca; for a while I couldn't work out what it reminded me of, then it came to me: roasted chestnuts.

As darkness fell, we made our way over to the *balai sewang* where people were just beginning to gather. The four musicians struck up a melodic beat on bamboo instruments, and the dancing started. Between being alternately dragged to the floor to dance, we talked about the antiquity of the Lanoh, Stephen's work on genetics in South-East Asia, and ideas about anatomically modern human origins and migrations.

Genetic analyses in South-East Asia had already revealed some surprising results. Many East Asians possess distinctive features: an epicanthic fold over the inner corner of the eye, a 'single' upper eyelid, facial flatness and shovel-shaped incisors. These features are more strongly pronounced in north-eastern Asians, so the traditional view was that people and genes had flowed from north to south in East Asia.[7] However, some anatomical studies had suggested that the flow was in the opposite direction, and this was backed up by Stephen's work on mitochondrial DNA lineages in Malaysia. The general Malay population didn't appear genetically to be part of a southern expansion of north-east Asians at all; in fact, the northern populations in China, Taiwan and Japan appeared to be descendants of South-East Asians.[8]

But the majority Malay population were not direct descendants of the original South-East Asians. Instead, the most ancient genetic lineages in Malaysia belonged to Semang tribes. Their genes recorded a near-extinction of the Semang, with a loss of genetic diversity, but the surviving lineages could still be traced back to around 60,000 years ago.[9] As I had told Alias, this meant that the Semang were 'Orang Asli' in the true sense: they were 'original people'. They represented relics of the original migration, and their ancestors were the first colonisers of this part of the world. Of the six distinct Semang groups, Stephen had already sampled four – but not the Lanoh tribe.

So another twig was about to be added to the South-East Asian part of the mtDNA tree. It may sound a little like genetic stamp collecting, gathering these populations from around the world into a bulging DNA album, but every new population sampled adds more detail to the genetic tree, more twigs. Not only that, but the dates of the branch points become even clearer, so the overall structure of the tree, from root to branch to twig, becomes more obvious, and the dating of branch points more secure. 'Looking at the Lanoh is, in the first instance, a way of testing that they are descendants of the first settlers in this area. But it's also very important to increase the number of sample points, in order to improve the dating and get a better picture from the genetics as to the date of their first arrival here,' explained Stephen.

Stephen also believed that skin colour offered clues about past migrations. The colour of our skin is perhaps the most obvious way in which we all vary from each other. And at the moment it seems to be the only variation that is completely explicable in terms of adaptation to environment. Skin colour varies according to latitude and levels of ultraviolet radiation: the closer to the equator you and your ancestors lived, the darker your skin. The further away from the equator you get, the paler indigenous people become.[10, 11, 12]

Dark skin is rich in the pigment melanin, protecting deeper layers of skin from sunburn and skin cancer. In sunny places, it seems reasonable to assume that natural selection would act to conserve genes that make skin dark, as any mutations producing pale skin would be a disadvantage. So the 'original' skin colour of modern humans was probably quite dark. But in populations moving a long way from the equator, to the cloudier, northern parts of Asia and in Europe, the selection pressure for dark skin would have diminished. And 'pale' mutations could therefore occur without being weeded out. It may be that there was even positive selection for paler skin in northern climes. We acquire vitamin D in our diets, but we also make it in our skin, in the presence of sunlight. In places with strong sunlight, dark skin is still able to manufacture sufficient amounts of vitamin D. But as people moved northwards, skin damage from sunlight became less of a pressing issue, while dark skin would have limited the skin's ability to produce enough vitamin D.

Vitamin D is important in calcium metabolism: deficiency of this vitamin leads to rickets, where bones grow soft and bendy. Long bones become crooked and the pelvis collapses in on itself, making childbirth impossible in extreme cases. So natural selection may have acted to make European skin paler: random mutations that tinkered with the production of melanin in skin cells may have offered an *advantage* to sun-deprived Europeans. And so, gradually, the population would have changed from brown to white. The same thing happened to the populations of northern Asia, and some details of the genes responsible are now known. It seems that a different set of genes mutated in Europe and in Asia, but produced a similar paleness of skin in each place: an example of convergent evolution.[12]

The Lanoh have very dark skin compared with most Malaysians. But, at just 3 degrees north of the equator, it was the Lanoh that actually have the 'right' level of skin colour to protect them from the tropical sun. Stephen believed that today's majority Malay population represented a much later incursion of people from further north in Indo-China, people with much paler skin than the indigenous Malay tribes, perhaps around the time of the LGM. Selective changes take a significant amount of time to occur.

'Twenty thousand years is not long in this context,' Stephen said; hence the too-pale-for-the-tropics skin colour of most Malaysians.

Following all this discussion about skin colour, I also wondered about the effects of sexual selection, and perception of attractiveness. In both India and Malaysia I had noticed that the people shown on billboard advertisements and posters all had *much* paler skin than the real people I had seen: those aspirational creatures were youthful, glamorous and so pale as to be almost white. Of course, back home in Britain the images of 'white' beauty we are presented with are much more tanned than the majority of Anglo-Saxons you see around the place. In Malaysia you can buy moisturiser containing whitening agents; in the UK, we tend to buy moisturisers with tanning agents. We all seem rather ungratefully unsatisfied with the skin colours that evolution has fitted to our climates for us. But Stephen thought that aspiration to a particular skin colour might be driven by more than just a desire to look 'different'. In an ethnically mixed population, skin colour was tied to perceptions of social status or wealth, related to cultural and economic history. So skin colour could act as a marker of exclusivity. Stephen also suggested that if human skin colour didn't vary, we'd have found something else on which to base exclusive behaviour.

Returning to the main theme of Stephen's genetic work, everything he and other geneticists had discovered seemed to point to an origin of modern humans in Africa. Stephen talked about his particular niche within the discipline as 'genetic phylogeography': he was building family trees based on the information contained in genes, but then relating these constructs to geography.

'Phylogeography is different from traditional population genetics because, instead of comparing mixes of genes in different populations, you follow individual genes, and find where the gene lines go,' explained Stephen. Rather poetically, he described the branching genetic lineages as ivy draped across a map of the world, with the newest branches standing out as paler green shoots from the darker green main stems. 'In phylogeography, we know when we've got new growth because there are new genetic mutations.'

The roots of the ivy were firmly planted in Africa, and a single branch, representing L3, emerged to give rise to all the secondary branches outside Africa, starting with the M and N branches: today, every non-African in the world belongs to either the M or N lineage. There are numerous, ancient, primary branches of M in India, suggesting that the M haplogroup probably arose in India.[13] But in eastern India, there was also a proliferation of younger branches of M, which Stephen thought could be related to a recovery of populations after the Toba super-eruption.

Toba again – in this part of the world it seemed you couldn't go far without coming across echoes of this ancient disaster. I didn't get to visit the site, but Stephen told me about Kota Tampan. It's a Palaeolithic site in the Lenggong Valley with stone tools dating back to the Toba super-eruption, embedded in the ash layer. These stone tools are quite crude pebble tools, worked only on one side, similar to those found with much more archaic humans. But the Malaysian archaeologists found the same sort of tools associated with the fossil remains of a modern human, 'Perak Man', dated to around 10,000 years ago, and so they argued that the older tools were probably made by anatomically modern humans as well. It's intriguing – could modern humans really have got as far as *Malaysia* by the time Toba erupted, 74,000 years ago? Stephen certainly thought so. But unless actual fossil remains of anatomically modern humans dating back to the time of Toba are found in South-East Asia, the jury will have to remain out.[14]

Our conversation came back around to the Semang – how did they fit on to that branching phylogeographic ivy? Unique mutations in the Semang mtDNA lineages marked out their branches as pale green 'new growth' on the stem. But Stephen pointed out that it was the length of these branches that was really significant. While the Semang branches were unique and specific to this part of the world, they were also extremely early branches, springing straight off the original M and N stems. This suggested that the ancestors of the Semang had been part of the rapidly moving vanguard of colonisers spreading around the coast of the Indian Ocean. In fact, that's what the branches look like all along this coastal route. The tree looks 'rake-like': the M and N stems spread along the coast really quickly: too quickly to accumulate new mutations, staying 'dark green'. It's only in the branches springing off those long M and N stems that 'pale green' shoots – new mutations in specific geographic locations – appear.

We talked about more general issues of human origins and how to define the human species. Stephen was certainly a supporter of the recent African origin. However, although he was very clear about the absence of any genetic indication of mixing between anatomically modern and archaic humans (such as *Homo erectus* and Neanderthals), he was also careful to say that this wasn't evidence of absence. In other words, modern and archaic humans could possibly have met and reproduced, but, in which case, the offspring were sterile or the lineages produced died out, and aren't seen in any people in the world today (or, at least,

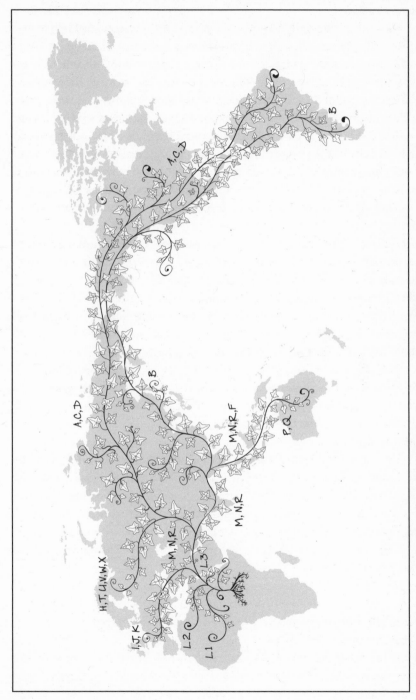

The 'phylogeographic ivy' of mitochondrial DNA lineages, laid across a map of the world

among the steadily growing number of populations whose genes have been sampled).

Although Stephen was sure that modern humans had effectively replaced earlier, archaic populations, he was quite liberal about where to draw the species line. He described the first anatomically modern humans as a small band, perhaps just 'one of the races of *Homo heidelbergensis*', with Neanderthals as another 'race'. So modern humans throughout the world today could be thought of as the human race that survived while the others died out. It was a thought-provoking idea, and I particularly enjoyed conversations with Stephen for just that reason. His knowledge about the genetic make-up of modern humans throughout the world indicated to him that a small group of a few thousand people, living in Africa around 190,000 years ago produced all the human populations alive today.

But this didn't mean that earlier and anatomically different forms like *heidelbergensis* and Neanderthals should necessarily be excluded from our species. They represented lineages that had died out, as so many other, modern human lineages have done. And this means that we could view them, their cognitive capabilities, tool-making ability, and general humanity, in a different way. They were groups of humans that are no longer with us, not other, slightly inadequate species trodden underfoot by the snootily superior *sapiens*. I think it might be the medic in Stephen, with years of applying science in a human context, to diagnose and treat his young patients, that allowed him to investigate something as abstract and mathematical as genetics at a population level, but then put a human face on it.

And so we left Kampong Air Bah and the Lenggong Valley. Stephen had made a trip to the country and people he loved, and was returning with a fridgeful of new DNA samples. And I'd learnt a huge amount in just a few days, about surviving in a rainforest, trying to keep a threatened culture alive and the ivy branches that lay across South-East Asia. My next step would take me to another bit of Malaysia, but, this time, on an island.

Headhunting an Ancient Skull: Niah Cave, Borneo

So on I flew, to the Malaysian state of Sarawak on Borneo, and headed once more into the rainforest. Niah Cave is part of a system of caverns in the Gunong Subis limestone massif, about 15km from the coast in

Sarawak. The cave lies within a national park, and we stayed at the park lodges. To get to the cave, I took a ferry across the Niah River and then walked 3.5km on boardwalks through the jungle to the cave itself.

I was excited by the prospect of visiting Niah; it was one of the places that had been on my wish list when we first started planning the series. It seemed like some archetypal archaeological site: a magnificent, vast natural cave in a mythical landscape of towering limestone escarpments, in the rainforest. And it was one of the most famous archaeological sites in South-East Asia.

Tom Harrison, then Director of Sarawak Museum, had excavated Niah between 1954 and 1967, with his wife Barbara. His first experience of Borneo had been on an ornithological expedition to the island, as an undergraduate, in 1932. He had gone there to study birds, but ended up becoming fascinated with the culture of the Dayak headhunters of Borneo, and this was the beginning of his career as an anthropologist. During the Second World War, Harrison had parachuted into the Borneo rainforest as part of a mission to recruit the native inhabitants of Borneo against the Japanese. Rather gruesomely, he successfully resurrected the practice of headhunting. Harrison stayed in Borneo until the Japanese surrendered – and beyond. After the war, he took up the post of Director of the Sarawak Museum. Tom Harrison knew of the very ancient human remains found on Java, the *Homo erectus* specimen that became known as 'Java Man', and he started digging at Niah Cave in search of 'Borneo Man'.[1]

I approached the cave through the massive, open rockshelter known as the Traders' Cave, where twentieth-century gatherers of swiftlet nests would sell their strange harvest to soup makers. There are still nest collectors working the cave today. From the Traders' Cave, I ascended wooden steps which then became an arched boardwalk, and suddenly, I was in front of the enormous West Mouth of the Great Cave of Niah. It was huge: 60m high and 180m across. I had seen photographs of it but, even so, I wasn't prepared for the sheer scale of it. I could make out wooden poles hanging from the ceiling of the cave, and I watched a nest collector climb first a knotted rope, then up a pole, to the high ceiling of the cave. He threw down swiftlet nests to his colleagues below, small, plastic-looking cups, with feathers embedded in the dried saliva. Not an appetising prospect. In recent years, nest collecting had reached such a frenetic level that the swiftlet population had dived; now the collectors are working within strict quotas, but it still seemed cruel and unnecessary, like so many luxury foods. The nest collectors, though, made good money from this bizarre Chinese delicacy.

Tom Harrison found plenty of evidence of early human use of Niah Cave, including many burials from the Neolithic, dating to between 2500 and 5000 years ago.[2] I walked around the site of Harrison's excavations in the West Mouth of the cave. The original trenches have been left open, and many Neolithic burials still lie exposed. This was quite strange: archaeologists normally remove human skeletal remains because, although bones may have survived for thousands of years, uncovering them changes their environment and they are likely to degrade. Certainly, these Neolithic burials looked a little the worse for wear. Tall trees had been cleared from the cave mouth of Niah, allowing light in for tourists to

The great cave of Niah

appreciate the sweeping grandeur of the enormous cavern. But light had also enabled green algae to grow across the limestone walls and ceiling of the cave, and all over the Neolithic burials.

But I hadn't come to see Neolithic burials: they were far too recent. Below the Neolithic cemetery, just within the West Mouth and cutting down through pond sediments, was Harrison's 25ft-long Hell Trench. This part of the excavation gained its name from the conditions in which the archaeologists were working: hot, humid and hellish. But the swelteringly hard work in this trench paid off, when, in 1958, Tom Harrison discovered the 'Deep Skull'. It was unmistakably a modern human, with a round braincase, and modern-looking browridges and occiput. From its depth, he must have anticipated that it would have been old. There was charcoal in the layer of sediment immediately above the skull, which Harrison sent off to be radiocarbon dated. When he got the results back, it transpired that he had found what was, at the time, the earliest evidence of modern humans outside Africa: the skull appeared to be 40,000 years old.[2] At the time, this date was met with disbelief: it seemed far too early a date for a modern human in Borneo.

The skull itself was normally kept in the Sarawak Museum in Kuching, but the curator, Ipoi Datan, had very kindly arranged to bring the skull back to its original findspot, and so I was to see the skull in the cave in which it was found. I carefully opened its cardboard box. The skull had been in many fragments when it was discovered, but had been glued together to make larger pieces. I carefully lifted these pieces out of the cotton wool that had cushioned them on the motorbike ride into the forest. There was a large domed piece that made up most of the calvarium, or top of the skull, a fragment of the left temporal bone, around the left ear, and another fragment of the base of the skull. The maxilla, with the upper teeth, had been left behind in Kuching, but I knew from the reports that the third molars or wisdom teeth had not erupted. This meant that this was the skull of a young person, in his or her late teens or early twenties. The base of the skull also showed signs that two of the bones had been in the process of fusing – at a joint with the grand name of the 'spheno-occipital synchondrosis' – another indication that this was the skull of a young adult.

It can be quite difficult to decide if a young skull like this is male or female. Many of the features that indicate maleness relate to the robusticity or chunkiness of a skull (as men are generally more muscly and heavily built than women), but these features are often still developing into a young man's twenties. Many eighteen-year-old men still look quite girlish, although they may baulk at this. I really notice the difference, teaching at a university, between male students in the first and third year. They grow up in all sorts of ways, but their faces really do change over those three years. The Deep Skull looked female to me, but I had to bear in mind that this was the cranium of a young adult, and so I couldn't be sure. Other researchers had reported the skull as 'probable female'.

The square shape of the eye sockets, wide nose, slightly protruding jaw and the shape of the teeth fitted very well with what the researchers expected ancient South-East Asians – the ancestors of present-day Andaman Islanders, and Aboriginal populations throughout Malaysia, the Philippines and Australia – to have looked like.[2]

Harrison had never published a full report on the site, and, in 2000, an international team of archaeologists, led by Graham Barker from Cambridge University and Ipoi Datan, who had brought the skull from the Sarawak Museum for me to examine, descended on Niah Cave.[3] Their mission was to recover information from the trenches, notebooks, photographs and excavated material from the original dig, and to carry out some new excavation as well. Harrison's 40,000-year-old date for the Deep Skull had always been controversial. Some archaeologists had suggested that his dating was flawed, others that the skull could have been much more recent, perhaps even Neolithic, and that it had somehow been pushed down into deeper sediments, making it seem much older than it really was.

So one of the key challenges for Barker and his team was to find out if the skull really was as old as Harrison had claimed. Going back to Hell Trench, they were able to confirm the exact place where the Deep Skull had been discovered. It was clear to them that the skull could not have been pushed down into the sediments in which it was found. They then applied new dating techniques on the sediments in which the skull had been found, using state-of-the-art AMS radiocarbon dating on charcoal from the sediment, and uranium series dating on the bone itself. The new dates for the Deep Skull came out at around

The vault of the Niah skull

39,000 to 45,000 years old, so Harrison's claim for the great antiquity of the skull was vindicated.

There are a couple of other sites in the region with very old fossils. The Tabon Cave on the island of Palawan in the Philippines, just north of Borneo, has yielded a skull dated to around 17,000 years, but also a tibia which could be as old as 58,000 years, although this date needs to be confirmed.[4] A premolar – possibly that of a modern human – from Punung on Java may be even older.[5] Late Palaeolithic archaeological sites in Korea have been dated to 42,000 years ago, and there are some modern human remains from cave sites that have been estimated, by dating of associated remains, to be around 40,000 years old.[6] But the Deep Skull remains, fifty years after its discovery, the earliest definite evidence of modern humans in South-East Asia, and among the oldest outside Africa.[2]

The re-excavation of Niah Cave also turned up more human bone, including a fragment of tibia and pieces of another skull. These other skull fragments were stained with ochre on the inside surface, leaving archaeologists to wonder whether it had been painted as part of a burial ritual, or even perhaps used as a paintpot.

As well as skeletal remains of the early people of Niah themselves, there was plenty of evidence of how they had lived, and the dates for human occupation of the cave went back even earlier than the Deep Skull. One of the important implications of findings from Niah Cave is that these hunter-gatherers were managing to survive in an environment that may *look* green and lush on the surface but is actually very difficult to find food in. Many plants that look quite palatable are actually poisonous, and the animals get very good at hiding in dense foliage. I had already seen the skills and knowledge needed by modern hunter-gatherers to obtain wild food in the Malaysian rainforest, and in Niah Cave there was evidence of the same sort of ingenuity and resourcefulness going back some 46,000 years.

Barker's team analysed a huge volume of animal bone left over from Harrison's excavations, giving them insights into the diet and hunting skills of the early occupants of Niah Cave. The animal bones came from layers dated to between 33,000 and 46,000 years ago. Many of the bones were burnt, probably the dumped remains of turned-out hearths – a bit of Palaeolithic housekeeping – while others had cut marks on them from butchery, so the archaeologists could be sure that humans had been involved. It seemed that the hunter-gatherers at Niah were managing to catch an enormous range of prey, from many different habitats around the cave: the bones belonged to a many different species including bearded pigs, leaf monkeys and monitor lizards.

Some of these animals wouldn't have proved too much of a challenge; molluscs could have been easily harvested from rivers and swamps, and modern hunters in the tropics have been reported taking porcupines, pangolins, monitor lizards and turtles simply by hand. I could believe this having seen the Lanoh girls plucking slippery fish out of Air Bah River.

Monkeys, though, are more difficult to hunt. The presence of monkey bones in the Niah Cave sediments showed that the humans were able to successfully hunt tree-living animals. There was no direct evidence to show that those hunter-gatherers had projectile technology, but, from what they were eating, they must have done. I thought back to the Lanoh's blowpipes: entirely made out of organic material, they wouldn't survive in the ground for very long. Anything made of bamboo or wood would be invisible to archaeologists looking for clues thousands of years later. It brought it home to me how incredibly difficult it is trying to find out what someone's lifestyle was like when all that's generally left are a few pieces of stone and bone.

The antiquity of both blowpipes and of the bow and arrow is still up for debate. There is no evidence of either being used this early, but this has to be balanced against the knowledge that most organic remains will have entirely perished. The earliest definite evidence for bow and arrow use comes from Europe, just 11,000 years ago, although some archaeologists argue that it may have been invented much earlier.[7] Bone and cartilage points were found at Niah, and it is possible that these may have been used as arrowheads.

The butchered pig bones in the cave also pointed to fairly sophisticated hunting technology – and a fondness for pork. Traditional ways of hunting pigs in modern Malaysia include using dogs and spears, blowpipes, bows and arrows, as well as ambushing and trapping. Dogs can be ruled out, as they were brought to Borneo only in the Neolithic. The other methods cannot be discarded, but without more archaeological evidence we can really only speculate about how those hunter-gatherers went about acquiring their game.

There were so many pig bones in Niah Cave that the archaeological team doing the twenty-first-century reassessment of the cave suggested that pig-hunting may have been the main reason that humans were drawn to the area. The bones were analysed to determine the age of the pigs being eaten: two-fifths of the bones to juvenile pigs. In modern Malaysia, pig populations boom and bust, following fluctuations in the fruiting of abundant tropical trees called Dipterocarps (from the Greek for two-winged fruit), which are themselves linked to weather cycles in the southern hemisphere. In years with bountiful fruit, wild pig populations can increase up to ten times in just a few months, and all that reproduction means a greater proportion of juvenile pigs around. Perhaps it was during these 'good pig years' that bands of hunter-gatherers would have set up seasonal camps in the mouth of Niah Cave, feasting on pork.[3]

The monkey bones at Niah Cave also served as clues to what the environment would have been like at the time. They indicated that the area around Niah was forested, and this was corroborated by analysis of pollen in the Hell Trench sediments, which showed cycles of alternating mountainous and lowland rainforest. At around 40,000 years ago the environment would have been humid lowland rainforest, patchier than today as the climate was drier, although there would still have been plenty of rain.[2, 8] The middle of OIS 3, between 40,000 and 47,000 years ago, was an especially warm and wet period.[6] I visited Niah Cave in the middle of the wet season; there were frequent downpours in the afternoon, and, as I sat in the cave, keeping dry and looking out at the drenched rainforest, I imagined those early hunter-gatherers doing the same. When the rain stopped, the heat quickly lifted moisture from the trees to form wreaths of mist on the steep, forested slopes opposite the cave mouth.

It also looks like the hunter-gatherers knew how to make the most of rainforest plants as well, and had learnt to detoxify yams so that they were edible. Raw yam (*Dioscorea hispida*) is poisonous enough for a few mouthfuls to kill an adult, but yam fruit and seeds can be detoxified and made safe by burying them for a couple of weeks, then boiling them, or by burying the seeds for a longer period with ash. Barker's team found pits containing ash and nut fragments, and inferred that the ancient hunter-gatherers may have been using the pit method of detoxification. There were also suggestions that the foragers had been managing the forest, using fire to clear areas: the archaeologists found particularly high levels of Acanthaceae (*Justicia*) pollen, which is among the first plants to recolonise fire-cleared areas in contemporary forest. In fact, there is evidence for wide use of burning of areas of dense wet tropical forest in South-East Asia between 30,000 and 40,000 years ago.[8]

In recent history, hunter-gatherer tribes in South-East Asia were nomadic, forced to move around seasonally in order to exploit the thinly spread resources of the rainforest. From the evidence in Niah Cave it looks as though the ancient foragers were also roaming around, but returning again and again to the cave, which provided them with a base from which they could make forays out in search of food. Most of the evidence of human activity in the cave is in the form of the animal bones left after dinner. There are very few stone tools in the cave, but, as those that have been found were made of stone originating about 50km away, it makes sense that the hunter-gatherers wouldn't have carelessly discarded tools when they left the cave. A few bone tools were also found in the archival material from Harrison's dig, from the 33,000- to 46,000-year-old layer, in among the butchered animal bones. There were six pieces of worked bone, including one that had been sharpened to a point to form an awl or hole-making device.[3] Bone tools are seen as one of those things that characterise the Upper Palaeolithic, traditionally considered part of a more sophisticated technology and culture that emerged with the Aurignacian in Europe about 40,000 years ago. However, we have already seen that this apparent 'revolution' may have had its roots much earlier, in Africa, and that the first humans leaving Africa may have already been making sophisticated tools, including mounting bone points on shafts.

Although there were no refined stone tools, or evidence of ornament or art, comparable to the European Upper Palaeolithic, Barker argued that the ingenuity, resourcefulness and forward-planning implied by the archaeology at Niah show just as 'modern' an approach to life. The foragers of Niah were certainly displaying an ability to exploit a great range of resources: they were probably trapping animals, using some sort of projectile technology for hunting, detoxifying yams and clearing areas of forest using fire. The ability to survive in the rainforest would have helped humans spread through South-East Asia, but they were not alone in the region; as they spread, they would start encroaching on the territories of earlier humans.[2]

The Hobbit: Flores, Indonesia

In the last decade the most exciting new addition to the story of human evolution has been the discovery of the remains of diminutive people who lived on the Indonesian island of Flores. It was an event that rocked the world of palaeoanthropology, as well as making news headlines. It was presented as the find of the century, but what really caught public attention was that here was evidence of a different species of human – living at the same time as modern humans. The Hobbits had been on their Indonesian island until as recently as 12,000 years ago. Although we are familiar with this idea in Europe, that there were still Neanderthals knocking around when modern humans arrived on the scene, it's something that still comes as a bit of a surprise, and even sends a shiver down the spine. We are very used to the idea that we are the only human species on the planet today (although, genetically, there is actually a good argument for chimpanzees and gorillas being included in *Homo* as well). Some even feel that we are so unlike other animals as to be a special creation. When we start finding other species that challenge our uniqueness, it unsettles that illusion. At the time of the discovery, Chris Stringer said, 'It's remarkable, astonishing, sensational, even … It challenges the whole idea of what it is that makes us human.' The idea that there could have been (and some think still *are*) other people, not quite human, sharing the planet with us, is somehow spooky. Even spookier are the myths from Flores, of Ebu Gogo: small creatures that inhabited caves, and provoked suspicion and fear among Floresian villagers.

In 1995, Mike Morwood and Doug Hobbs were on the Kimberley coast in north-west Australia, excavating an eighteenth-century site where Indonesian fishermen had boiled up sea cucumbers, ready to sell to the Chinese as a delicacy. But contact between Asia and Australia stretched back much further, way beyond historical records. The first Australians would have arrived from Indonesia. So, the archaeologists started to plot an excavation in Indonesia – to search for the early modern human colonisers of the region and the ancestors of the first Australians.[1]

The Indonesian island of Flores seemed like a good place to start: ancient stone tools had already been found there, and Mike Morwood could team up with Indonesian palaeontologists and archaeologists who were working there. And there already seemed to be something a bit intriguing, something that 'didn't quite fit' about Flores. Throughout the Pleistocene, Flores was always an island, separated by deepwater sea channels from it nearest neighbours, Bali, Lombok and Sumbawa. And it's generally thought that modern humans are the only hominins who managed to make sea crossings. But the tools that had previously been discovered on Flores seemed too ancient to have been made by modern humans. Then there was some doubt as to whether they were really stone tools at all.

Digging in 1997, the international team found definite stone tools, embedded in volcanic tuff of the Soa Basin, and got secure dates on them. The tools dated to between 800,000 and 900,000 years ago. This finding was important enough to get into *Nature*. Morwood suggested that these tools must have been made by *Homo erectus*, as that was the only hominin

known to be around in South-East Asia at that time. But that meant
Homo erectus had made a sea crossing. It was a controversial claim.

In the following years work continued in the Soa Basin, but the
team also branched out to investigate a couple of cave sites where
previous excavations had turned up some, much more recent, evidence
of modern humans, including burials and stone artefacts from the last
10,000 years. In April 2001, excavations started at Liang Bua, 'the
cool cave'. A team of local Manggarai people had been taken on as
excavators, digging with trowels and bamboo stakes, then swapping
them for sledge hammers and chisels to get through layers of hard
flowstone. The excavation went deeper than previous digs had done;
the sides of the trench were carefully shored up as they went down.
Morwood was not satisfied with stopping at layers that looked 'sterile',
or untouched by any signs of human activity: he wanted to get down
to bedrock. And he was rewarded. In deep layers, they found thousands
of stone tools, animal bones and teeth. Bert Roberts dated the remains
and found them to be between 74,000 and 12,000 years old. A strange,
small and rather curved hominin radius was the only apparently human
bone found in that first digging season.

In 2003 the digging team turned up what they thought was the
skeleton of a *Homo erectus* child: it was a very exciting find. The skull
was thick, with a sloping forehead, which fitted with *Homo erectus*,
and it was very small. The skeleton hadn't been deliberately buried; the
body had somehow ended up in a shallow pool in the cave and become
quickly covered over, so that the bones were preserved. The bones were
not fossilised and were very mushy; the team used a mixture of UHU
glue and acetone nail polish remover to consolidate the fragile bones so
that they could be lifted. But when the bones were properly cleaned up
it became evident that Liang Bua skeleton number 1 (LB1) wasn't a child
at all, but a *tiny* adult.

Peter Brown, Professor of Palaeoanthropology at the University of
New England, flew out with Mike Morwood to Jakarta to examine the
tiny skeleton. Using mustard seeds, which he poured into the skull, Peter
Brown found the braincase was astonishingly small: just 380ml. Anything
in the genus *Homo* is expected to have an adult brain size of at least 600ml,
based on previous fossils, and modern humans have brains anywhere
between 1000 and 2000ml. Big brains are a fundamental characteristic of
humans. Small brains can be caused by pathological conditions, such as
microcephaly, but even then it's unusual to end up with a brain volume of
less than 600ml. The oldest known hominin fossils outside Africa, from

Dmanisi in Georgia, dating to 1.8 million years ago, are small-bodied and small-brained (with a stature of 1.4m and brain size of 600ml), but are still nowhere near as small as LB1.

Peter Brown did not think the skeleton was pathological, neither did he think it was *Homo erectus*. In fact, he originally wanted to give it a brand new genus and species name: *Sundanthropus tegakensis*. LB1 was something very strange indeed: a tiny hominin that Peter Brown thought looked even closer to the ancient African australopithecines than to any member of the *Homo* genus. Morwood, though, in spite of the tiny brain size of LB1, thought there were enough traits in the skeleton to label her *Homo*.[1] And then there was the behaviour: tool-making is not meant to pre-date *Homo*. In 2004, the find was published in *Nature*: 'a new small-bodied hominin', named *Homo floresiensis*.[2] Morwood and his colleagues suggested that this species had derived from an ancestral population of *Homo erectus*, which had become isolated on Flores, with the whole population undergoing 'endemic dwarfing'.

The publication produced a palaeoanthropological storm and made newspaper headlines all over the world. The main debate centred on whether this really was a new species, or whether it was a pathological, modern human. Indonesian palaeoanthropologist Teuku Jakob argued that LB1 was just a microcephalic, pygmy modern human,[3] but further excavations at Liang Bua produced partial skeletons of another twelve individuals, so any idea that LB1 was a pathological one-off could be discounted. Even with a reassessment of LB1's cranial capacity as slightly larger, around 417ml, the shape of the braincase is quite unlike that of a microcephalic.[4] The debate has rumbled on, though, with a recent paper in the *Proceedings of the Royal Society* claiming that the diminutive nature of the Flores specimens could have been caused by a congenital disease, with a defective thyroid gland producing dwarfism and small brain size.[5] But the researchers making this latest claim didn't even look at the real skeletal remains; they made their diagnosis from photographs, which seems decidedly dodgy. They claimed that the pituitary fossa (the cradle of bone inside the skull in which the pituitary gland sits) looked larger than normal in the photographs: a sign of congenital hypothyroidism. But other researchers who have looked at CT scans of the skull have refuted this claim.

The pathological explanations for Flores tend to assume that the small skull is the fundamental characteristic that needs explaining. What they don't tackle is why the whole skull looks so remarkably different from modern humans, or why the limb proportions of the skeleton might be so different. An Australian team of researchers, led by Debbie Argue, compared LB1 with pygmy modern humans and other hominin species, and agreed with Morwood that LB1 was not microcephalic, and that it was different enough from both modern humans and *Homo erectus* to warrant its own, new species name. However, they also argued against *Homo floresiensis* having developed from *Homo erectus*, as it possessed features which suggested it was evolutionarily somewhere between the australopithecines and early *Homo*.[6] Careful analysis of the wrist bones and shoulder of LB1 have also shown them to be primitive and unlike those of modern humans.[7, 8]

So I was very excited indeed by the prospect of looking at the original bones for myself. On a fairly grey March day in Jakarta, I met Tony Djubiantono, Director of the National Research Centre of Archaeology. He led me down a corridor in the Centre and into a room where there was desk, a sofa and a safe. He took Tupperware boxes, containing the Flores bones, out of the safe and carried them into a larger store room where there was a table ready for me to lay out the skeleton of LB1.

I unpacked the bones in silence, watched by Tony and by a television camera. I was quite taken aback. The bones were *absolutely tiny*.

I systematically laid out the skeleton: skull first, at one end of the table, followed by fragments of vertebrae, and then arm bones, hands, pelvis, leg bones and feet, just as I would do with any archaeological skeleton in the bone lab in Bristol.

It was a very, very strange little skeleton. There was no question that LB1 was an adult: all the epiphyses (the knobbly ends of the long bones) had fused to their shafts, and there was a full set of adult teeth. But this was an incredibly small adult. It didn't look in any way pathological, and, anyway, it would have to be some weird science fiction disease to produce the mixture of traits in LB1: a dreadful affliction that caused some kind of time warp and pushed bits of the human body backwards through millions of years of evolution.

Some parts of LB1 looked quite *modern* human: the teeth in particular were set in a parabolic curve in the jaws, and looked quite like modern human teeth in shape and size. But there were things about it that were distinctly unhuman: the incredibly tiny skull, which had its widest point low down around the ears; the way the middle of the mandible was rounded off, chinless, and thickened on the inside; and the general thickness of the mandible and the wide ramus – the part of the jaw that swoops up to the jaw joint. The limb bones were chunky for their size, and the pelvic bones were a weird shape that reminded me of the flared pelves of australopithecines. It also had longish arms and shortish legs: these were not modern human proportions, but more like early *Homo*, or even australopithecines – fossils of which have only ever been found in Africa.

All my experience of early hominin skeletons up to that point was from photographs and casts. But, to me, this little skeleton looked a lot like the gracile australopithecines (*Australopithecus afarensis* and *africanus*) – who lived in *Africa* between two and a half and four *million* years ago. So what on earth was a hominin like this doing in Indonesia?

It would certainly be 'easier', and fit better with current theories, if the Hobbits *were* a pathological modern human population: an unfortunate group of people who were all congenital microcephalic dwarfs. But it seems unlikely, from both the shape of the skeletons and the date of the remains. While the LB1 skeleton is dated at around 18,000 years old, dates for other skeletal remains of *Homo floresiensis* and stone tools go back as far as about 95,000 years ago.[9] This creates problems for the scientists arguing that the Hobbits are pathological modern humans, as this is twice as old as any evidence of *Homo sapiens* in South-East Asia. So the Hobbit presents some significant challenges to the accepted thinking about patterns of dispersal and migration, and about the range of variation in the anatomy of hominins. Debbie Argue contends that the species must have been intermediate between australopithecines and *Homo* at the point it left Africa, and that it must therefore have emerged out of Africa prior to two million years ago: another – even earlier – Out of Africa. You can see why the palaeoanthropological world is reeling from the blow dealt by this Indonesian cave man. While Mike Morwood seems convinced that it is *Homo*, he also knows that the implications for hominin dispersals are profound. In his book about the discovery of the Hobbit, he even raises a question about whether *Homo* could possibly have arisen in Asia. Shocking stuff indeed.

As well as the skeletons from Flores, the stone tools throw a bit of a spanner in the works – this time for neat theories about 'who made what' in South-East Asia. Previous explanations have attributed large 'core tools' to *Homo erectus* and small-sized 'flake tool' assemblages to *Homo sapiens*. But it seems that Flores Man was doing something that brought those two 'types' of tool together as stages in a production line. Large cobbles suitable for making flake tools are cumbersome to carry around, so the hominins struck off big flakes from these cores and left them in the landscape. Then they took the flakes off with them and made smaller flakes to use as tools. This explains why the two different 'assemblages' occur in different places. But they actually appear to be part of a single process, and not two culturally different ways of making tools, used by different hominins. If that was happening on Flores, it begs questions about attributing particular tools to particular species anywhere in Asia.[9] Again, it makes me glad that I spend my time in the lab looking at bones, not stones!

So, what about the question that everyone wants the answer to: did humans and Hobbits ever meet? I interviewed two Floresian men, Gregorius Buü Wea and Anselmus La Li Wea, who told me their folk stories of small people who lived in caves in the hills: Ebu Gogo. Gregorius and Anselmus said that these human-like creatures used to be attracted to feasts, but would stay very much at the edges, not engaging with humans. But they did seem to aggravate the humans as well.

'They stole the crops that belonged to the peasants. Crops like cassava and fruit were always stolen by Ebu Gogo.'

'Were they at all threatening?' I asked.

'Not really,' came the reply. 'But they did sometimes steal children as well.'

That sounded fairly threatening to me. 'What did they want with the children?' I asked.

'They just liked children,' said the Floresians.

I asked them what Ebu Gogo looked like.

'They were hairy all over. Their faces were like those of monkeys. And they were short – less than one and a half metres tall.' This all seemed very credible. But there were some more bizarre characteristics as well.

'They had broad chests, with kangaroo-style pockets which they used to keep their stolen goods in. And the female ones had big, long breasts, right down to their knees.'

Very interesting. Well, apparently the people of Flores eventually grew tired of putting up with these strange little people who threatened their food supply and their children, and they plotted their downfall. The story went that a group of people went up to the caves where Ebu Gogo lived, and offered them rattan mats. The small creatures accepted one mat after another, and the humans set fire to the last mat before handing it up to the cave. All the mats inside the cave caught fire and the Ebu Gogo were destroyed. It was a nasty little story, but reminiscent of what happens everywhere around the world where different groups of humans take against each other.

Of course the folk stories don't prove anything. So what about the archaeological evidence? There were certainly modern humans around in the vicinity at that time. Evidence of modern humans has been found in nearby Timor, going back to more than 42,000 years ago, from excavations at the Jerimalai shelter.[10] But modern humans did not seem to reach Flores until much later, so there is no archaeological or genuine historical evidence to suggest that humans and Hobbits ever met. Although Hobbits were around on Flores until at least 12,000 years ago,[11] there is no overlap of the archaeological evidence of *Homo floresiensis* and later evidence of *Homo sapiens* on the island of Flores itself.

I asked the men from Flores if they thought any Ebu Gogo could have escaped and still be alive today.

'Maybe,' they replied.

A Stone Age Voyage: Lombok to Sumbawa, Indonesia

Even before the Out of Africa theory became widely accepted, archaeologists had suggested that early moderns would have spread across the globe most easily by following the coast. It made sense from an ecological point of view: there are rich pickings for hunter-gatherers along the seashore, and genetic evidence has provided strong support for this theory. But there was a point at which the coast ran out. Australia was always separate from South-East Asia, and separate by a long way.

Analysis of mitochondrial DNA suggests that modern humans emerged from Africa some time after 85,000 years ago, then spread 'rapidly' (in prehistoric terms) eastwards.[1, 2, 3] Studies of mtDNA in Aboriginal Australian populations suggest that the founder population arrived some time between 40,000 and 70,000 years ago.[3, 4, 5] Some archaeologists still seem to prefer a conservative date for colonisation of between 40,000 and 50,000 years ago,[6] but the consensus appears to be – based on the genetics but also on the oldest dated sites in Australia – that the colonisation of Australia happened some time between 50,000 and 60,000 years ago.[7]

At this time the environment in South-East Asia – rich with reefs, mangroves and estuaries – would have been ripe for exploitation by beachcombers with simple watercraft.[8] The sea level was about 40m lower than today: Borneo, Sumatra, Java and Bali would have been part of the mainland, forming the 'Sunda Shelf'. Early colonisers may have been able to move down through the middle of this land bridge, rather than sticking to its edges. Because running right through the centre of Sundaland would have been a long, grassy plain, a 'savannah corridor'.[9] To the south-east, New Guinea, Australia and Tasmania formed the great landmass of Sahul.

In between lay the islands of the Wallacean archipelago, corresponding to the modern-day Indonesian Nusa Tenggara islands, principally, Lombok, Sumbawa, Flores, Sumba and Timor. To South-East Asian animals these islands represented a major barrier to diffusion, and modern humans are the only large mammals who made it across.[6] So, although there was much more above water than there is today, colonisation of the islands of Wallacea and Sahul itself would have required sea crossings.

It is quite likely that the beachcombers of Sunda already possessed the means to make the sea crossing to Sahul. The evidence from Africa, 125,000 years ago, showing exploitation of coastal resources, demonstrates that humans had already developed fully modern capacities, which suggests that they would have been entirely capable of building boats. David Bulbeck, in a well-reasoned paper entitled 'Where river meets sea',[8] argued that coastal watercraft would have been enormously useful to the early beachcombers, allowing them to cross the mouths of rivers and exploit estuarine environments, to transport food and materials, while providing the sailors with protection from dangers in the water such as saltwater crocodiles. Bulbeck suggests that the richness of estuarine environments would have promoted population growth, stimulating the colonisers to push further along the rim of the Indian Ocean. He sees the wave of colonisation creeping eastwards, with regular traffic to and fro, along the entire chain of linked estuarine colonies. Behind the colonising front, the landscape would have been 'filled in' as people moved inland, along rivers and other suitable habitats – like Ravi Korisettar's chain of lakes across India.

Getting from Sundaland to Sahul would have required between eight and seventeen sea crossings, and at least one trip of more than 70km.[6] Two routes have been proposed for the colonisation of Sahul: a northern route through Sulawesi and a southern route along the Nusa Tenggara islands. The northern route holds advantages (though of course the colonisers were not equipped with this sort of foresight): each island is visible from the next, all the way to New Guinea, the islands are well watered by the monsoons, and the prevailing winds blow in a helpful, easterly direction.[10] The southern route is more parsimonious, in that it requires the least number of sea crossings. But even considering the 'best route' with the benefit of hindsight like this creates a bizarre impression of those hunter-gatherers sitting on a palm-fringed Indonesian beach and planning their migration to the south. Perhaps it's even too simplistic to think of the early colonisation of Sahul as a single event, via a single route. There were a scattering of Wallacean islands close to Sahul, and

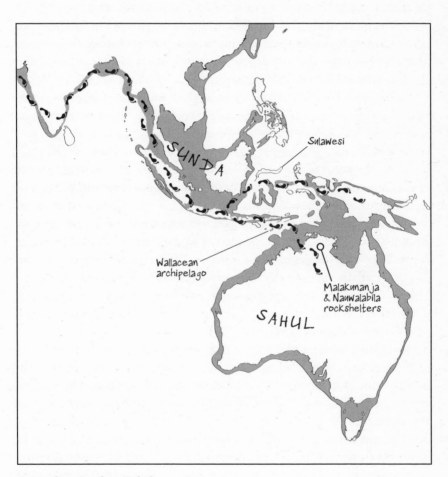

Routes from Sunda to Sahul

colonisation would probably have happened in a much more haphazard way, with multiple arrivals from the nearest islands.

So what about archaeological evidence for early colonisers of the Sunda Shelf and Sahul? Finding sites that might be close to the proposed front of colonisation has proven difficult: there are quite a few Palaeolithic sites on islands close to Australia, but most are dated to less than 30,000 years old,[6] when what we're really looking for – based on the genetics – is something between 60,000 and 75,000 years ago.[3] As in India, rising sea levels means that many sites that were right on the coast back then may now lie up to 130m beneath the waves, and kilometres out to sea. In fact, there is not a single archaeological site that shows occupation of the northern rim of the Indian Ocean or the Wallacean islands between 75,000 and 60,000 years ago. The oldest evidence of modern humans in South-East Asia comes from Niah Cave, around 42,000 years ago. But while this site may give us fascinating glimpses into the lives of the early hunter-gatherers in Sundaland, it is far too recent to represent the forefront of the wave of colonisation. Sue O'Connor argues that the rockshelter site of Jerimalai on East Timor is strong evidence for a southern route of colonisation of Sahul, but it also dates to about 42,000 years old, at least 10,000 years after the first evidence of humans in Australia. Jerimalai is still important, though: at the moment it is the earliest evidence for modern humans out on the Wallacean islands. Although there are no human skeletal remains, and the stone tools from the site are quite basic, shell beads and fish hooks, fish and turtle bones and marine shells are tell-tale signs of modern humans. Bones from large pelagic species like tuna suggest that those early Timorese may have even been using boats for fishing.[7] Flores is a bit problematic, though; it has been extensively investigated, but there's no evidence of modern humans there until 10,000 years ago, and it seems like a necessary stepping stone on that southern route.

We can only guess what the vessels that carried the first Australians might have been like. It's entirely conjectural, as there is no actual archaeological evidence of watercraft from this long ago – anywhere in the world. Even at Jerimalai, the use of boats is only inferred from the big fish on the menu at the rockshelter.

In the absence of direct archaeological evidence, I settled for an experiment. Robert Bednarik is an experimental archaeologist who enjoys the challenge of putting conjecture to the test. He has built a series of rafts, using only materials and techniques that would have been available to Palaeolithic raft-builders. It seemed reasonable to assume that these

Families in the village of Nhoma, Namibia

!Kun and //ao hunting in the northern Kalahari

Foolhardy perhaps, but I was determined to experience the African wilderness for myself; so here I am, surrounded by thorn branches, preparing to spend a night out in the bush in Namibia

Examining a cast of the Omo skull – in the original find-spot in Ethiopia

The lunar landscape of the Kibish formation, Ethiopia

Young man at Kolcho village, Ethiopia

Buna painting my face at Kolcho village

View of the Omo River

Muda – the young man from Kolcho who took me to meet his friend Chowli – and me

Where it all began: the 'Lucy' Gazebo and Restaurant, Addis Ababa, Ethiopia

The wooden steps down to the caves at
Pinnacle Point, South Africa

Cave 13B at Pinnacle Point

A camel in the verdant Wadi Darbat, Oman

Jeff Rose on the trail of the ancient toolmakers at Djebel al-Qara, Oman

*Archaeologists excavating in trench 22 at Jwalapuram, India, finding stone tools even deeper
– and therefore earlier – than the 74,000-year-old layer of ash from the Toba eruption*

The Penghulu (leader) of the Lanoh, Alias Bin Semedang, in the Lenggong Valley, Malaysia

Traditional house on stilts in Air Bah, Lenggong Valley, Malaysia

Stephen Oppenheimer in the Lenggong Valley

Boardwalk under limestone cliffs on the approach to Niah Cave in Borneo, Malaysia

Ipoi Datan unpacking the Niah skull – still the earliest, definite evidence of modern humans in South-East Asia

View from the mouth of Niah Cave, with the rainforest steaming after the afternoon's rain

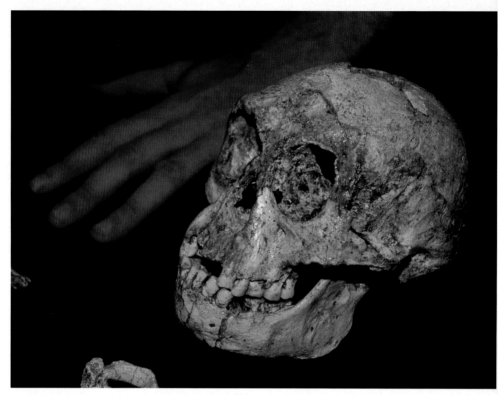

The tiny skull of the Hobbit (LB1) found in Flores, Indonesia

An experiment with a 'Stone Age' raft looks like it's heading for trouble, given the stormclouds looming over our destination: the island of Sumbawa in Indonesia

Uncovering the Willandra footprints in Australia, conservator Colin McGregor opens a hole in the membrane and carefully removes the protective sand

Flying over the flooded Northern Territory, Australia

Gershom Garlngarr's painting of animals and spirits tumbling into a cleft in the ground in Gunbalanya, Australia

Garry Djorlom, one of the artists from Injalak Arts & Crafts Centre, standing on its namesake, Injalak Hill, Australia

The painting of the aboriginal creation mother, or 'Yingana', which is hidden away on a rockface at the top of Injalak Hill

View of Gunbalanya from the top of Injalak Hill

early watercraft would have been rafts, the simplest possible solution to carrying people on water. In terms of materials, Robert has used something that would have been abundant in Palaeolithic South-East Asia, just as it is today: bamboo. And in the last decade he had carried out a series of experimental sea crossings between the islands of Wallacea, culminating in a successful thirteen-day raft voyage from Timor to Melville Island, near Darwin.[11, 12]

I met Robert on a beach on the east coast of Lombok. He had drafted a small army of local fishermen to build a bamboo raft to his specifications, and it had taken them three days to complete. It was an impressive-looking creation: two layers of thick bamboo trunks running the length of the raft, braced with cross-pieces, all lashed together with rattan. On the sides, double lengths of bamboo formed raised seats for paddlers. It looked pretty seaworthy, but I knew that Robert had also had a few unsuccesful attempts at sea crossings.

'So far we have built seven rafts, had five actual attempts at crossings, and three were successful,' Robert told me.

'But that's quite a few failures, too,' I noted.

'Well,' Robert laughed, 'we shouldn't be talking about the failures but the first one was a total failure, a catastrophic failure – because we had no idea about the design that was required. At the time there was no scientific knowledge whatsoever about the design of seagoing rafts.'

'So how about this one? Might it sink?'

'No, a raft is unsinkable – that's one of the great things about rafts. Boats can sink, rafts cannot.'

But this raft was a bit different from the ones Robert had built before. 'This is the first time that green bamboo has ever been used in one of these experiments. On all previous occasions the bamboo was fully dried, for about six months,' explained Robert. There was one other new variable in the experiment. 'All previous attempts have been made with men only, and this is the first time that a woman will be on board – so that's something new as well,' Robert joked.

The experimental green bamboo raft, with me aboard, was ready to be launched – but not before it, and all the raft-builders and paddlers (including me) had been blessed by the local mystic, or *dukan*. First, she made coloured patterns, using seeds and what looked like ground coconut, on a miniature raft, and balanced a small stone egg in the centre. Then she took the raft down to the water's edge and pushed it into the breakers. Coming back to the full-size raft, she prepared a yellow paste and then anointed each of us in turn: on the forehead, cheeks and top of the breastbone. She walked around the raft, anointing it as well. Then she twisted together several lengths of yellow cotton to make a long string, which she cut into shorter pieces with a knife, and tied a length around each person's left wrist. Everyone was quiet and respectful during this ceremony, which lasted over an hour. It was fascinating to see what looked like an ancient animistic cult still going strong on what was, officially, a Muslim island.

With the blessing over, we prepared to launch the raft. First, the awning that had covered the construction site, high up on the beach, was cut down to avoid the back of the raft hitting it as we pushed it down the steep slope to the sea. Thick lengths of bamboo were brought and placed in front of the raft to act as rollers, and we started heaving the thing down the beach. There were about thirty men, and me, pulling the raft. We would drag it a few feet then pause, rearrange the rollers and have another go. Eventually, we were in the breakers, but the raft was still firmly on the beach, with the waves washing through it. It was incredibly heavy. We pushed it further and further, and finally, like the moment a plane heaves itself off the runway and into the air, the raft let go of the beach and was afloat. About fourteen of us jumped on and started paddling furiously, still in the breaking waves, to get it out and safely beyond white water.

Quite soon, the manic activity of the launch was over: we were out, on calm sea. It had been a successful launch, and the raft, to my relief, had not begun to fall apart. Robert was disgusted by the very suggestion that his raft might falter at this first hurdle. He had designed the raft for eight people, so we were reduced down to a lean, mean team. There were five Indonesian fishermen: Muhammed ('M') Suud, Idrus, Malaburhana, Narno and Ama Ros, and an Indonesian translator, Muliano Susanto ('Tokyo'); Robert (as captain) and I completed the crew. We settled down to paddle over to Sumbawa – which looked as if it was hundreds of miles away, all misty on the horizon.

'The weather conditions seem to be quite ideal,' Robert reassured me. 'There's barely a breeze, and the sea is very flat, as you can see. Our main obstacles are really the currents. The currents in any sea strait anywhere in the world are unpredictable. If we go too far south we are going to be in trouble.'

We had cast off at 7.25 a.m. 'How long is it likely to take?' I asked, even though we were only ten minutes into the journey. 'That's a good question,' said Robert. 'We'd all like to know that! If you want an estimate – with these conditions … it should take between six and nine hours.'

Every half hour or so, a pair of paddlers would swap sides. This was a hot tip from Robert, having made a few raft voyages in his time, and having learnt that always paddling on one side was tiring. The raised seats were essential: I was sitting in an upright position with my legs out on the 'deck' of the raft, with the waves gently lapping over the sides and cooling my feet. The deck was covered in woven palm mats, and we stashed our supplies, including medical kit and life jackets, near the middle of the raft, where they would get least wet. I was pleased to find the wooden paddle very ergonomic: with one hand grasping the handle at the top, and the other low down near the blade, I could really drive the paddle through the water. It was satisfying and even more so when Robert consulted his GPS and told us we were doing a good three knots – however, in the next breath, he pointed out that this was probably mostly due to the current.

After a few changes of position, I found myself sitting directly in front of Robert; I had a brief rest from paddling and swivelled round to ask him about archaeology, raft-building and human origins. Quite quickly, it became clear that Robert didn't think that it was just modern humans who had the wherewithal to build ocean-going craft. He thought that seafaring may have begun more than a million years ago – with *Homo erectus*.[11, 12] This is an extraordinary claim. Going to sea is generally seen as something that is part of modern human behaviour – although, of course, the Hobbit on the island of Flores challenges this preconception.

But although other species of human may have reached some of the Wallacean islands, getting all the way to Sahul would have required sea crossings of a different magnitude. Along the proposed southern route, each island would have been visible along the chain until the last crossing, and although migrating birds and smoke from bushfires may have provided a suggestion of the land that lay ahead, Sahul itself would not have been visible from Timor. While I felt that the sea crossings implied by the presence of earlier humans on Flores could have been accidental, perhaps on natural mangrove rafts ripped from the mainland in storms, the colonisation of Sahul seemed to indicate that something a bit more sophisticated was going on. In fact, some archaeologists have even said that Flores is the exception that proves the rule: it must have been a fluke, otherwise *Homo erectus* and *Homo floresiensis* would have quickly colonised the Wallacean archipelago, and Sahul.[13]

As I questioned Robert further, it emerged that he had a deeper philosophical reason for believing that *Homo erectus* had modern capabilities. He believed that *Homo erectus* in Asia had evolved into *Homo sapiens* locally.

'By about half a million years ago you have one species grading into a different species,' said Robert.

'So you're saying that *Homo erectus* got out from Africa, spread throughout Asia, then, independently, in lots of different places, just turned into what we now recognise as modern humans?' I queried. 'And don't the fossils *and* the genetics tell us that there's a family tree – rooted in Africa some time between 100,000 and 200,000 years ago?'

'Ah – the god of genetics,' replied Robert. 'I don't believe it. The root of the tree is two million years ago.'

This is the multiregionalist view: that modern humans evolved, right across Africa, Europe and Asia, from earlier human populations already living in these places. In order to accommodate the unity of the human species today, the multiregionalist argument invokes gene flow between evolving populations, keeping the whole thing together. This is a strange idea for many reasons, not least because this concept of an almost global species changing so much and keeping together as a species is *weird*. Speciation usually happens in a small population, reduced from a larger population by catastrophe, or perhaps separated from it geographically. Or speciation can occur on the very edges of a large population. But the idea that a widespread population could, as one, change in a unified direction, seems highly improbable. A biologically more likely scenario – and indeed, one that fits the evidence – is a single origin followed by a dispersal of the species.

'Ten years ago, there was no support for multiregionalism whatsoever,' said Robert. 'But support has grown. The only major remnant of strong support for African Eve is in England. You're making a programme based on an outmoded theory.'

This sounded rather extreme to me, and I found this a bizarre point of view, as what actually seemed to be happening in palaeoanthropology was a growing consensus, based on both palaeontological and genetic research, in support of a recent African origin for modern humans. Robert appeared to be presenting the exact opposite of the sea change in (most) expert opinion that had actually taken place. So there we were: an Out of African and a Multiregionalist, stuck on a bamboo raft together for what would be ten and a half hours.

Despite our different views, Robert was an excellent captain. He kept an eye on everyone on the raft, and, when we needed to make decisions, it was democratic. He knew that the Indonesian sailors understood these waters and he took counsel from them before making a decision. Out on the very calm, open sea, with the tropical sun beating down on us, the most pressing dangers were not waves or sharks, but heat stroke and exhaustion. We had ample supplies of water and encouraged each other to keep drinking, shouting '*air minum!*' (one of the few Indonesian phrases I had learnt: 'drinking water').

I felt quite isolated on the raft, even though I had the assurance of knowing that two small motor boats and one larger support boat were always around. They kept out of our way most of the time, circling us every now and then to film progress and check our water supply, but, generally speaking, we were on a long leash and felt independent. It was still good to know, though, that if something did go wrong we had help nearby – not like those Palaeolithic nautical adventurers.

After about five hours on the raft, Robert to start talking to me about how he had got started in archaeology. I was interested because I knew he wasn't affiliated to a university. He had been a successful businessman for many years, then after early retirement he had decided to invest his time and money indulging his interest in archaeology – specifically in rock art and experimental nautical archaeology. He boasted about the huge numbers of papers (over a thousand) he had published in his short career. It all sounded very impressive, and it wasn't until much later, when I was in Australia talking to rock art expert Sally May, from Flinders University, that I discovered Robert had published a good many of these articles himself, in his own journal.

Although Robert exhibited a slight tendency to self-aggrandisement, and seemed to have a general disdain for academic archaeology, I actually found him quite likeable. Underneath the often brash exterior there was a more personable man, someone who was fascinated by the past and who believed that the human spirit – ingenuity, creativity and adventure – had very ancient roots.

We paddled on and on. For hours, Lombok's beaches appeared very close and Sumbawa still looked blue and very far away. It became quite disheartening to glance back, so I concentrated on what was ahead – rolling blue sea – and had a chat to Tokyo about sharks. I asked him if there were any in these seas, and he said that there were, and asked if I liked shark. This was a slight misunderstanding. I wasn't interested in finding sharks to eat; I was more worried about them eating me.

But we didn't see any sharks, and the sea was very calm, and the sun was shining. As we reached the middle of the channel between the islands, the water became a dark, dark blue, almost purple. The physical work of paddling provided a constant rhythm, and I felt myself drifting into meditative calm. Perhaps this journey was going to be much easier than I'd anticipated. Messing about in Palaeolithic boats seemed simple.

But then, as we got a little closer to Sumbawa, things started to change. First, the surface of the water became a little troubled. Soon after, the wind perceptibly picked up, and little white horses started to appear on the choppy water. And clouds had gathered over the mountains of Sumbawa, which looked closer now, but decidedly unwelcoming. The clouds spread, the sun went in and the sea got rougher.

On the raft we all put life jackets on and paddled with renewed energy. Sumbawa was so close now that we could make out sandy beaches. It looked as if there were plenty of places to land the raft. Although we were trying to direct the raft, we were really at the mercy of the currents, which became unpredictable as we neared land. Robert kept reassessing our position in respect of which bit of beach we were headed for, and we paddled accordingly. As we got closer, we could even see some houses – a little fishing village perhaps – along the beach we were approaching. The end was in sight. After about nine hours of paddling, fatigue was setting in, but there was an air of relief pervading the crew. The fishermen sang songs to help the paddles through the water, and, when they stopped, I taught them old Scout songs that my dad had taught me as a kid. 'We're riding along on the crest of a wave' went down well, but not as well as 'Gingangooly', which they all thought was hilarious. Especially when I told them, via Tokyo, that it didn't even mean anything in *English*.

As we continued on our course, we started to enter a bay with a long headland just to the south of us, and I spotted some very large breakers peeling off it towards us. I pointed at the waves and made wide-eyed faces and the fishermen laughed. They weren't worried a bit – we were far enough away from them – then it gradually became apparent that we weren't getting any closer to the shore. We were paddling very hard, but staying opposite the same part of the headland. Then, suddenly, we moving closer and closer to the rocky headland and *backwards* – right towards those breakers. We were caught in a very strong rip current, and there was absolutely nothing we could do. I radioed the support vessels, which were keeping a close eye on us now, while staying a safe distance from the breakers themselves. We stopped paddling and let the current take us – not into the breakers, but right back out to sea again. It was as though Sumbawa had lured us in and then spat us back out.

The current was now taking us south, down the coast of Sumbawa, so the idea of landing on the first beach we had spotted was abandoned. As we rounded the headland, we could see a row of small beaches separated by rocky spurs. Any one of them looked like a good landing site, if it hadn't been for the almost continuous reef break flanking them. We needed to find a gap in the break that matched up with a beach – but there wasn't one for a good two more kilometres down the coast. We were all very tired by now.

But we paddled on. We had come so far, it was impossible to imagine doing anything but finishing the trip, and by landing the raft, not abandoning it. Then we got stuck again: we had been paddling past the same headland (or trying to) for half an hour. Dusk was fast approaching, and it was clear that we would have to elicit some help from the twenty-first century. One of the support boats came up and we threw them a line, and were towed a few hundred yards, out of the rip.

And there it was: a gap in the breakers. Paddling furiously, we negotiated our way through unbroken waves, with great barrels of white water crashing down either side of us, but just running out before they got to us. It felt as if the raft was heaving itself over the waves, and you could see the whole thing twisting and bending as it rode the swell. But – it rode it. As we neared the shore, we were into smaller breakers, with waves rushing up between the bamboo and soaking us. We all stayed on the raft and continued paddling, until we were well into the shallows. Then everyone leapt off, and tried to manoeuvre the suddenly very unwieldy and dangerously heavy craft up and on to the beach. Safely on the sandy beach, I actually jumped up and down and hugged Robert Bednarik. It had taken us ten hours and twenty-five minutes. We had made it: our Palaeolithic voyage had been successful.

It had been an amazing experience. Most of the journey had been extremely easy – much more so than I'd expected. But in those last two hours it was as though the sea had decided to remind us who was boss. We had certainly proved the principle: that it's possible to cross open sea on a bamboo raft made with stone tools, without a sail or a motor. It did make me wonder about the intention behind those early voyages, though. Just as we had been taken by currents and swept down the Sumbawa coasts, perhaps rogue currents had carried early fishermen, plying their craft along the coast, to new lands. The prevailing winds and currents between Indonesia and Australia would certainly have facilitated such an unintentional trip. Equally, those natural forces would have favoured a deliberate crossing, just as they do today: illegal immigrants can make their way from Timor to the Kimberley coast in three days, in motorless craft with small sails. Simulation studies suggest that the colonisation of Sahul would have been more purposeful than the old 'pregnant woman aboard a log' scenario, and that founding populations probably had some contact with parent populations.[14] Several researchers have argued that the most likely scenario was a process of disorganised colonisation, with numerous small bands of people from various Wallacean islands arriving in dribs at drats, at many points along the northern Sahul coast at different times.[8, 10]

What would they have found when they got there? They would have landed on a great coastal plain, some 200km wide. The vegetation may have been quite familiar to humans from the Wallacean islands. There was an expansion of tropical forest in Australia during OIS 3, peaking around 50,000 years ago,[15] but the animals would have been distinctly unfamiliar: strange-looking marsupials which had evolved along their own routes, in that isolated continent. Most of those animals would have been the marsupials still existing in Australia today, but others were truly monstrous: they included a giant constrictor, *Wonambi naracortensis*, an enormous carnivorous lizard, *Megalania prisca*, a huge emu-like bird, *Genyornis newtoni*, a rhinoceros-sized marsupial, *Diprotodon optatum*, and the 3m-tall giant kangaroo, *Procoptodon goliah*. Those giant 'megafauna' have all become extinct, and their demise seemed to coincide all too neatly with the arrival of humans.[10]

It was time to make that journey to Australia myself.

Footprints and Fossils: Willandra Lakes, Australia

I flew over miles and miles of red land, criss-crossed by straight roads. It was immense, empty and dry. I flew into Mildura, a strangely cosmopolitan town in the middle of a semi-desert, late in the afternoon. That evening, I met up with archaeologist and Mungo National Park officer, Michael Westaway, along with a team of archaeologists who had arrived from various corners of Australia to revisit the lakes.

The following day, we piled our gear into Land Rovers and drove 113km north-east, and were very quickly on dusty, unsurfaced roads in a desert-like environment. The three Land Rovers kicked up a great cloud of dust behind them.

After a few hours we turned off the road and drove down a hill on to a flat plateau. Most of the drive had been over a low-lying landscape, with gentle rises and dips, so the hill was quite strange, and once we had driven down it I could see that it was actually part of a long, curved ridge stretching off into the distance, like a white wall in an otherwise featureless landscape. We pulled up at a cluster of low buildings: we had arrived at Mungo National Park Visitor Centre.

Michael Westaway pointed out the ridge that I had spotted as we approached, and told me that it was called the 'Walls of China'. It may have been named by Chinese labourers who had worked on the sheep farm in the nineteenth century. The odd feature curved right around us, in a wide sweeping arc. There is no water in the Willandra Lakes system today: it all dried up thousands of years ago. What I was looking at on the horizon, that curving ridge, was a prehistoric lake shore. The flat plain was the dry lake bed. And then, among the scrubby bushes, I spotted my first kangaroo.

The Willandra Lakes system includes the dried-up relicts of nineteen interconnected lakes, including Lake Mungo, where the Visitor Centre was situated. I was rather confused about the history of the lakes: they had been wet for much of the Ice Age, and were dry now, which seemed the wrong way round. I was used to the broad concept of things being generally, globally, drier during glacial periods, with lower sea levels when so much water was locked up as ice, and conversely wetter, with higher sea levels during warmer periods, when the ice melted. But local geography is also important: during the Ice Age, the lakes were fed by run-off from the glaciers of the Eastern Australian Highlands. It seems that, when the world warmed up, the glaciers melted away, and, by about 18,000 years ago, with no water left to feed the lakes, they dried up.[1]

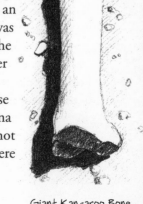

I stayed in the lodge at the Visitor Centre, situated between a small museum and an old shearing shed. My room was basic but comfortable, containing the essentials for a good night's sleep in the middle of Australia: a bed and an air conditioner. There was another building containing a kitchen and a dining room, where someone had painted an enormous kangaroo on the wall, about 3m tall. This was an artist's impression of *Procoptodon goliah*. It wasn't the most accomplished painting I had ever seen, but, rather unnervingly, its eyes did follow you around the room.

So, had the earliest Australians killed off these ancient beasts? It seems that humans and megafauna may have co-existed in Australia for 10,000 years, if not more. Some researchers think that the megafauna were

Giant Kangaroo Bone

hunted to extinction – although there are no megafauna kill-sites to support this argument. Others have suggested that climate change was the real killer, with the cold dryness of the LGM finishing off the huge Pleistocene animals of Australia. But recent dating of fossil megafauna suggests that many species of megafauna became extinct across Australia between 40,000 and 50,000 years ago – at least 20,000 years before the LGM. So these dates do suggest that humans might have had something to do with it, either killing off the giant beasts directly or else by disrupting the ecosystem in some way.[2] Some archaeologists believe that there is another indication of human impact on the environment at just this time, with a decline in forest starting around 45,000 years ago, linked to burning of large areas,[3, 4] but it's impossible to know if it was people who started the fires.

The next day we set off again and ventured on to even rougher tracks, winding through the bush until we reached the site where the archaeologists had converged. They had come on a conservation mission: to check the preservation of the Willandra footprints. And these were very precious footprints: they were around 20,000 years old. Several Aboriginal elders were there at the site as well, overseeing the operation. Some of the younger archaeologists were themselves Aboriginal people, trained to recognise archaeological features in the landscape and to look after their own heritage. The footprints themselves were found, in 2003, by a young Aboriginal woman, on an archaeological training day. Steve Webb, an anthropologist from Bond University in Robina, Queensland, had been taking a group of trainees on a field-walking exercise, teaching them how to recognise ancient artefacts and features. They had actually ended up in the wrong area, but Webb had decided it was as good a place as any just to do some training. Then one young woman, twenty-six-year-old Mary Pappin Junior, spotted a footprint. Webb realised that this couldn't have been a recent footprint: it had been buried under sediments, and revealed by the wind blowing away the sand over a much harder surface. And there were more: in all, eighty-nine footprints had already been revealed by natural erosion.

When the site was excavated, more footprints were found: 563 of them, over an area of 850m².⁵ The prints were preserved in a layer of very hard, silty clay. When they were made, the clay must have been moist, perhaps soon after rain. Then it dried hard before being covered by windblown sand. Twenty thousand years later, the wind was undoing its work and revealing these usually ephemeral traces. But the same wind was also threatening to destroy the exposed prints, literally sand-blasting them away.

'Once the site was excavated and exposed to elements, it almost immediately started to deteriorate,' Michael told me. 'First it was sand-blasted by the wind, and then it suffered a nasty freeze-thaw event. We watched the footprints change before our eyes – over just a few months.'

The archaeologists had to come up with a plan to preserve the precious trackways; they decided to cover the prints back up, with 65 tonnes of sand, pinning a fabric membrane over the surface of the sand to stop it blowing away. But before they did this, they made a record, by digitally scanning the whole site.

The archaeologists had come back to assess how well the conservation plan was working, and I was extraordinarily lucky to be visiting the site when this health check of the footprints was being carried out. In any other week of 2008 I would have seen only the membrane covering the site. But there I was, watching the archaeologists cutting neat holes in that protective fabric and carefully digging and sweeping out the sand from a few chosen footprints.

'We're reopening just a few of the prints to see if the weight of that sand is distorting the footprints,' said Michael. The re-exposed footprints certainly looked good to me: the toes were clearly defined, and you could even see where the muddy clay had squelched up between them.

I had seen some ancient footprints before, on the coast of Formby in north-east England, but these Australian prints were much, much older. Footprints can be very useful from an archaeological, anthropological point of view. The very ancient footprints from Laetoli in Tanzania have been useful in reconstructing the way australopithecines walked – and indeed, prove that they *did* walk on two legs.

Footprints of early modern humans tell us where people were, and a little about past societies and social behaviour. But I actually think most of the 'scientific' information to be gleaned from them is fairly obvious stuff: people in the past (including children) walked or ran, in straight or curving lines. You can work out stature, too. Not exactly mind-blowing. Although, rather strangely, there is one rather intriguing trackway at Willandra that appears to be that of a one-legged man – moving at some speed!

But, to my mind, the footprints do something more than just provide us with data: they link us back to long-forgotten people in a very personal way. They are records of a brief moment in someone's life. There is something quite profound about looking at something which is usually such a fleeting sign of another human's presence, and knowing, without a shadow of a doubt, that a person walked just *there*, where you are standing, all those thousands of years ago.

So how, you may be asking, did those archaeologists manage to date those footprints? The answer: OSL. By taking samples from the sediment just under and just above the footprints, and using optically stimulated luminescence to measure the amount of time that quartz grains had been buried for, an age range could be determined. The footprints had been made some time between 19,000 and 23,000 years ago.[1]

So what had those prehistoric Aboriginal people been doing there? It looks like there was a group of people, of different ages, moving around the margins of the lake, and precisely what they were up to as they made those footprints we can only guess. But we do know that, at the time, the lakes would have been a good place for hunter-gatherers to hang out, with plenty of fish, shellfish, waterbirds and game to hunt.[1]

But it was a rare combination of factors that meant that their footprints had lasted until the present.

'We don't find footprints all over Australia, and the preservation of the footprints here is still a bit of a mystery – but there would seem to be a couple of favourable factors here,' explained Michael. 'One factor is the clay itself: it contains a mineral called magnesite, which is fairly rare, and seems to help mould the footprints perfectly. And the clay must have been wet when the people walked across it. Then, soon after, windblown sand covered the footprints over – and locked them in for some twenty thousand years.'

For Michael and the other archaeologists who had come up with the conservation plan, it was looking very reassuring. On site, they were able to compare the re-exposed footprints with photographs from the original excavations. It looked like very little deterioration had occurred: the sand and membrane were definitely doing their job. They had also brought along a laser scanner to take detailed scans of the footprints, to compare with scans taken when the prints were first exposed. Once each of the six or seven exposed footprints had been scrutinised, photographed and scanned, they were carefully covered up again with sand, and the membrane drawn back over and secured with cable ties.

There is other, more ancient evidence of humans living around those lakes. In fact, Willandra Lakes were famous long before the footprints were discovered. In 1968, the remains of a cremated skeleton were found, eroding out of the dunes that fringe Lake Mungo. Geologist Jim Bowler and his colleagues had been collecting shells and stone tools from the Walls of China, and had found a collection of burnt bone fragments that he first thought were the remains of some early Australian's dinner. But, on further inspection, it looked like the bones could in fact be human.

The archaeologists had been expecting to work on surface finds and now they had an excavation on their hands. They collected up the loose pieces of bone, then removed a block containing bones which had been naturally cemented together with calcium carbonate or calcrete. The specimens were carried off site in a suitcase.

In the lab, anthropologist Alan Thorne carefully removed the calcrete from the bones using a dental drill; they were definitely human. The fragments represented the remains of two individuals, named, rather predictably, 'Mungo I' and 'Mungo II'. Number two consisted of just a few fragments. Mungo I was also fragmentary, but there were enough pieces – about 25 per cent of a skeleton – to be able to tell that these were the cremated remains of a lightly built young woman. The skull was fragmentary, but what was there looked similar in many ways to modern Aboriginal Australians.

Careful analysis of the bones provided some insight into the funerary ritual. The body had been burnt while still complete, but the bones of the spine and the back of the head had largely escaped the flames of the pyre. After the cremation, the bones had been smashed up, a funerary practice which was still happening in Australia and Tasmania until quite recently. The bone fragments had then been scooped into a shallow pit at the lakeside. Initial radiocarbon dating of shells associated with the cremation suggested it could be around 30,000 years old: the oldest human remains to be found in Australia at that time.[6]

Alongside the cremation, the archaeologists found stone tools, animal bones and hearths, which provided important clues about the lifestyle of 'Mungo Lady' and her people. They ate a varied diet, catching golden perch in the lake and eating shellfish collected from its margins. They also ate emu eggs, small birds and bush meat. The archaeologists examining the clues left on the lake shore decided that the evidence pointed to a few visits to that particular site, by a small group of perhaps one or two dozen people. They suggested that it was a camp that had been used for a few seasons before either being abandoned or getting covered up by sand and mud.

The diet and lifestyle of these ancient Aboriginal people seemed quite similar to that of Aboriginal Australians around the Murray and Darling rivers in the nineteenth century: camping by rivers and lakes in spring and summer, eating fish and shellfish, and moving into the bush in winter, where they would catch and eat a range of marsupials.[6] It seems strange in some ways that this ancient lifestyle persisted in Australia, whereas, in different parts of the world, people settled down and started farming. Whatever pressures there were in other places, to give up that nomadic lifestyle and settle down, it seems they were either absent or were somehow overcome in Australia – without necessitating a major change in subsistence strategy.

Six years after the discovery of Mungo Lady, Jim Bowler found another burial in the dunes. A few days later, Jim Bowler and Alan Thorne excavated the burial. This time, the body had been buried, with the hands clasped over the pelvis, and covered in ochre. Its position in the dune again suggested that the burial was about 30,000 years old. This skeleton, Mungo III, was described as being lightly built, but nonetheless male, and was quickly dubbed 'Mungo Man'.

In 1999, Alan Thorne had another go at dating Mungo Man, this time using ESR and uranium series techniques on the skeleton itself, as well as OSL on sand grains from sediment below the burial. All the results were much older than the previously published dates for Mungo Man. It looked as if the burial could be over 60,000 years old.[7] This was an incredible date: it would make Mungo Man the oldest known modern human outside Africa.

Jim Bowler was not convinced by this very early date, and published a paper saying so, entitled: 'Redating Australia's oldest remains: a sceptic's view'.[8] He called into question the methods that Thorne had used. He wrote that Alan Thorne had made erroneous assumptions about the composition and erosion of the sediments that Mungo Man had been buried in, which made his ESR and uranium series dates untenable. And he didn't mince his words when he came to the OSL dates, which Alan Thorne had obtained from samples of sediment not directly next to the skeleton, but from about 400m away.

A few years later, Jim Bowler and a team of scientists published some new dates for Mungo Man and Mungo Lady. Samples of sediment from the burial sites were sent to four different labs for OSL dating, and all came back with dates of around 40,000 years ago.[9] These were older than the original radiocarbon dates, but much more believable than the dates published by Thorne's team. Mungo Man and Lady were still the most ancient Australians known.

Even when I got to Mungo, I wasn't sure if I was actually going to be able to see the human remains. Ancient skeletons have become symbols of identity for modern Aboriginal people, who feel – and understandably so – that their own heritage and beliefs have been largely disregarded, even trodden underfoot, in the recent past. Colonialism has a lot to answer for. As some attempt to redress the balance, the official policy in Australia was now that prehistoric human remains, however old, should be returned to communities of indigenous people,[10] but the implementation of this policy could be quite difficult – but perhaps not for obvious reasons.

It was very interesting talking to Michael Westaway, who had worked for some years on the repatriation of human remains. His experience was that Aboriginal communities often recognised the value of archaeological and anthropological investigation in revealing information about their heritage, and, indeed, saw it as a shared project. Aboriginal groups were keen for ancient human remains to be kept by museums for future research, as long as they were treated with respect.

'I was stuck in a difficult position,' he said. 'I had a quota of human remains to be returned to communities, but the Aboriginal groups I was working with often wanted us to keep the remains.' It was an interesting conflict: between being seen to be doing the right thing and doing what was actually right, in dialogue with local communities.

At Mungo, the situation was even more complicated, with political tension between the three separate tribal groups, the Barkindji, Ngiyampaa and Mutthi Mutthi, meaning that it had been very hard to achieve a consensus about how and where the human remains should be kept. Mungo Lady had been officially returned into the custody of local Aboriginal groups in 1992. In practice, this meant that the remains were locked away in a strange, painted safe in one of the store rooms of Mungo museum. In contrast, Alan Thorne had been entrusted to curate the remains of Mungo Man in Canberra. These were both temporary arrangements, and Michael Westaway told me about how he was helping the local Aboriginal groups in planning a permanent 'keeping place' at Mungo, where the bones could reside, with the three traditional tribal groups maintaining control over who had access to the remains.[11] It seemed like a perfect solution to a difficult problem, and I was aware of similar schemes – for much more recent bones – in the UK, where a consecrated ossuary could be used to hold boxes of skeletons, which would still be available for study. Michael was also keen on the idea of a 'keeping place' at Willandra as he felt that it would further strengthen the engagement of the Aboriginal community in any future research.

The air was thick with politics at Mungo. Although the elders of the Barkindji and Ngiyampaa tribes gave their permission, in principle, for me to see the remains of Mungo Lady, it seemed that the key to the safe had been given to a certain Badger Bates, who lived hundreds of kilometres away from Mungo. But the elders had agreed to me seeing Mungo Man – and so had Alan Thorne.

So – with a joyful heart – I left dry, dusty and fly-ridden Mungo and headed for Canberra. I was looking forward to meeting both Mungo Man and Alan Thorne. Both were controversial figures in Australian prehistory. Mungo Man's antiquity was hotly debated as, in fact, was his sex. Although he had been initially reported as male, Peter Brown had subsequently pored over his bones and decided that there wasn't enough of him preserved to be sure about his sex.[12]

As well as his very old dates for Mungo Man, Alan Thorne was also well known for his 'two wave' hypothesis of Australian origins. He believed in regional continuity, specifically that *Homo erectus* in nearby Java had evolved into a robust type of modern human, which had made its way to Australia. Then a second wave of more gracile modern humans had moved in, and hybridised with the first population. In his scheme, lightly built Mungo Man was part of that 'second wave', whereas some more robust specimens from Australia had retained characteristics of the earlier, more robust and locally evolved people.

I was sceptical about this theory, but fascinated to meet Alan Thorne himself – and to see the evidence. In a small laboratory in the Australian National University in Canberra, Alan Thorne unpacked Mungo Man from various boxes, under the watchful eyes of Aboriginal elders. The skeleton was very slightly built, and, just as Peter Brown had commented, the parts of the pelvis that would have been so useful in determining 'his' sex were damaged and missing. The arm and leg bones were slender, the joints small and the skull was quite female-looking, although the mandible was quite strong-jawed and masculine. I made a mental note: 'indeterminate sex'. But whether man or woman, this was certainly the skeleton of an anatomically modern human, and a very ancient one.

Then Alan took brought out the WLH 50 skull ('WLH' stands for 'Willandra Lakes Human': another important fossil find from those strange, dusty, dried-up lakes). This was a specimen that had been discovered sitting on the surface of the shoreline of Lake Garnpung, north of Lake Mungo, in 1980.[10] As it was found 'out of context', sitting on the ground surface, it is very difficult to pin a date on WLH 50. It has been dated to about 14,000 years ago by uranium series dating, but about double that by ESR, so the jury is out.[13] For Thorne, this was an important specimen in his argument that *Homo erectus* in South-East Asia had evolved into *Homo sapiens*.

I looked at the skullcap, sitting there on the table. It had large browridges and a sloping forehead. The whole shape was long and low; it looked fairly *erectus*-like. I was quite taken aback. But then I picked it up. The bone was very thick and heavy. There are some illnesses that produce thickened bones, like Paget's disease; this is a strange affliction, of unknown origin, that causes increased bone turnover in the skeleton. Normally, the flat bones of the vault of the skull have a layer of trabecular, or 'spongy', bone sandwiched between an inner and outer 'table' or plate of dense bone. In Paget's disease, the skull bones become thicker overall, while the tables thin down and the entire bone takes on a strange spongy appearance. But I could see from the broken edges of the skull bones that the marrow space sandwiched between the dense inner and outer plates of the skull looked normal: not like Paget's. I still thought it might be pathological. Alan was quick to crush this suggestion.

'We've analysed it and had a hard look at it, and we can't find any indication of pathology,' he told me. 'It's a very thick – but normal – skull. And there are other sites in Australia that have that thickness of skull bones, like the site that I excavated at Kow Swamp.'

Milford Wolpoff, Professor of Anthropology at the University of Michigan, and perhaps the leading proponent of multiregional evolution, analysed WLH 50 and compared it with other ancient skulls, finding it to be closer to archaic crania from Ngandong, Java, than to the anatomically modern skulls from Skhul and Qafzeh.[13] He argued that this was evidence of regional continuity in South-East Asia and Australia. (The Ngandong crania are a bit of a mystery in themselves; some believe them to be late-surviving *Homo erectus*, others think they are archaic *Homo sapiens*.[14] Recent uranium series dating of these fossils places them at between 40,000 and 70,000 years old.)[15] Chris Stringer compared WLH 50 with the Ngandong crania, and with anatomically modern crania from Skhul and Qafzeh, and more recent Aboriginal crania from South Australia. He found that WLH 50 clustered more closely with the modern specimens than with the Javanese fossils, even though it was certainly very long and narrow. He was slightly cautious about his conclusions, though, because he also thought WLH 50 may be pathological.[16]

Other investigators have also suggested that WLH 50 might be diseased in some way. Biological anthropologist Steve Webb, of Bond University, wrote: 'The unusual development of the vault structures in this individual has few, if any, equals among other hominins or more recent populations from around the world.'[17] So I wasn't completely mad after all. Webb suggested that the expansion of the marrow space in the skull bones may represent a response to anaemia, and that perhaps this was even a genetic form of anaemia. I was reminded of Stephen Oppenheimer and his research into thalassaemia and malaria resistance in South-East Asia; perhaps this adaptation to one of the most aggressive tropical diseases had already evolved in the ancestors of the first Australians.

The main argument against Alan Thorne's 'two wave' hypothesis is that the robust and gracile skulls he sees as representing two different populations could actually be male and female, from the same group of people. Peter Brown's analysis of WLH 50 concluded that, while the extreme thickness of the skull was bizarre and very possibly pathological, many of its other features were 'not unusual for an Aboriginal male'.[18] And, while Alan Thorne stuck to his guns, saying that gracile 'Mungo Man' *was* a man, others were less sure.[19] Perhaps the differences between WLH 50 and 'Mungo Man' could be explained if the former was male and the latter female.

Some researchers had suggested yet another idea: that the extreme robusticity of skulls like WLH 50 and Kow Swamp, which date to 14,000 and 20,000 years ago respectively, was neither due to inherited archaic features from Javanese *Homo erectus* nor to accentuated 'maleness', but, instead, represented some sort of adaptation to the extreme aridity of the LGM. And then the people of Australia became more gracile with the improvement of the climate in the Holocene.[20]

But Alan was very sure that the skulls he was showing me not only supported a 'two wave' theory of the peopling of Australia, but were also evidence for regional continuity in South-East Asia and Australia. He was sceptical about genetics and its contribution to this field of enquiry: he thought too much was missing from the modern gene pool to be able to usefully construct the past. He was much happier arguing from the fossil evidence, and he believed that what other people called *Homo erectus, heidelbergenis, neanderthalensis* and *sapiens* were actually all part of one single, but very variable, species.

'I think there was one Out of Africa,' he said. 'I don't believe in *Homo erectus*. It's just an old-fashioned name for the earlier part of modern humans. Today, we're a polytypic species, with so many variations. And there's enormous variation in the fossil record as well as among living people.'

It was a view of human evolution that seemed strange to me. I left Alan Thorne's lab convinced by the deep ancestry of the Aboriginal people of Australia but also unwilling to concede that they might be so different from me as to have developed from Javan *Homo erectus*. I wasn't convinced by the fossil evidence for this, although there was certainly something rather odd going on with those robust Australian skulls. I couldn't quite come to terms with Alan's very inclusive approach to human origins, fitting everything from *Homo erectus* onwards into one species. And I thought the genes did have something useful to say.

In fact, my next port of call was to explore what genes could reveal about the earliest Australians. I made my way to Sydney, where I tracked down Sheila van Holst Pellekan, in her lab at the University of New South Wales. Sheila had done a huge amount of genetic work with Aboriginal populations, but her achievements in genetics would have been impossible without her ability to engage individuals who were naturally very suspicious of the motives and intentions behind the research. For Sheila, the Aboriginal people who gave her their DNA were more than donors or subjects in the investigation: they were recipients of the knowledge it revealed. It was true community engagement with science. And, like Raj in Cape Town, she was very clear about the main message contained within the genes: that we were all part of one, fairly young, family, with a recent African origin.

'I think the overwhelming thing, looking at mitochondrial DNA from different people from all over the world, from Australia, Europe and Africa, is that nearly all of it is the same,' said Sheila, as she took me through the results of some of her analyses.

'It's only little changes every now and then which give us clues as to how the family tree fits together.'

Sheila showed me some sequences of mtDNA, from a range of people with European, African and Australian ancestry. She scrolled through hundreds of As, Cs, Ts and Gs, looking for the rare differences between populations. But the vast majority of the sequence in each person was identical. 'We're all so similar inside,' she said.

The work carried out by Sheila and other geneticists, looking at Aboriginal mtDNA, has shown that all the Australian lineages represent branches of the M and N macrohaplogroups, confirming their recent African origin pedigree.[21] And, going back in time, those lineages all coalesce 40,000 years ago, suggesting that Australia must have been colonised by at least that time.[22] Comparison with other populations has shown that Aboriginal Australians are very closely related to the indigenous people of Melanesia and New Guinea. Indeed, from the mtDNA tree it appears that Australia and New Guinea were colonised in a single wave, some time around 50,000 years ago.[21]

There may not be any fossil remains of humans in Australia dating that far back, but there are some very ancient traces of modern human behaviour. In the Northern Territory, archaeologists have found traces of modern humans that seem to go back as far as 60,000 years ago. This early evidence for modern humans in Australia comes from rockshelters where stone tools and traces of pigment use have been found. There are three rockshelters that have become famous (and infamous) as the earliest archaeological sites in Australia, all in the Northern Territory: Malakunanja, Nauwalabila and Jinmium.

In the 1970s, archaeologists excavated two rockshelters in the Alligator rivers region of Arnhem Land, Malakunanja and Nauwalabila, as part of the World Heritage assessment. The initial radiocarbon dates from these sites were fairly old, at around 20,000 years. In the 1980s, the rockshelters were re-examined by Rhys Jones and luminescence-dating specialist Bert Roberts, and new dates were published for both sites in the nineties. Roberts and Jones claimed that the rockshelters contained evidence of humans from as long ago as 50,000 to 60,000 years.

Malakunanja, a rockshelter in a cliff face near Magela Creek, a tributary of the East Alligator River, features some very recent rock paintings, including depictions of a man wearing Western clothing, rifles and a wheel, but the rockshelter seems to have been used for thousands of years. The ancient artefacts found buried in the sediments included stone tools and evidence of pigment use: grindstones and ochres. Thermoluminescence dating of the sediment in which the oldest artefact was found put it around 61,000 years ago.[23] Nauwalabila I is a rockshelter in Deaf Adder Gorge, about 70km south of Malakunanja, formed by a massive, fallen sandstone slab. The archaeologists dug down through 3m of sandy deposits, and found man-made stone flakes all the way down to the clean red sand at the bottom. As well as the stone flakes, they also found a piece of ochre which had been ground down, producing flat facets on its surface – again, evidence of pigment use. OSL dates suggested the earliest occupation levels in the rockshelter dated to between 53,000 and 60,000 years old.[24]

These very early dates for colonisation of north Australia have been questioned by other archaeologists, including Jim Allen and Jo Kamminga (who discovered Malankunanja and also dug at Nauwalabila; I was to meet Jo later in China).

Critics of the early dates from Malakunanja II and Nauwalabila have raised concerns about the accuracy of themoluminescence and OSL dating, and, in addition, have suggested that the artefacts could have moved downwards in the sandy sediments, ending up in earlier layers that they didn't really belong to. In 1998, James O'Connell and Jim Allen argued that dates of around 35,000 to 40,000 years for the first colonisation of Australia would make more sense in the context of the existing archaeology between Africa and Australia.[25] Six years later, the same authors had revised their date of colonisation back to 42,000 to 45,000 years ago, based on archaeology within Australia, although this is still very much a 'short chronology'.[26] It's a conservative position, and certainly ignores the possibility of the loss of earlier archaeological sites beneath the sea. In contrast, David Bulbeck seemed content with a date of 60,000 years ago for the first colonisation of Australia: it sat well with his suggestion of a rapid dispersal of humans around the rim of the Indian Ocean to Sunda and Sahul.[27] However, the hard evidence is still lacking for those who doubt the dates from Malakunanja and Nauwalabila.

Rhys Jones and his colleagues published replies to the criticisms of their dates for Malakunanja and Nauwalabila, arguing that the stone tools were found orientated flat, not angled, so were unlikely to have edged their way downwards in the sediment. The dates from the two rockshelters corroborate each other, and, at higher levels within the rockshelters, the luminescence dates tallied with radiocarbon dates, suggesting that the former were indeed reliable.[24, 28] I had bumped into Bert Roberts in India, where he was collecting samples at Jwalapuram. He had seemed utterly convinced that the first Australians could have reached the northern coast of the continent by soon after 60,000 years ago, and he was sure of his methods, and of the conclusions. But still, there are many archaeologists who do not believe the evidence to be either robust or conclusive.

In 1996, an article was published in the journal *Antiquity* that caused a huge splash in the Australian media, and sent a strong wave of scepticism spreading through the archaeological community.[29] Another rockshelter in the Northern Territory, Jinmium, had been excavated, and the thermoluminescence dates for layers bearing quartz flakes were extraordinary: 116,000 to 176,000 years old. However, just two years later, the excitement was over when the site was redated and found to be less than 10,000 years old.[30]

Mulvaney and Kamminga[10] used Jinmium as a cautionary tale, to warn both professional archaeologists and the media not to grasp so readily for the 'Holy Grail of Australian archaeology': the earliest date of colonisation of Australia. So, are Malakunanja and Nauwalabila the real thing? I'd like to think so, but I'll keep an open mind as more evidence comes to light. It's clear that the critics are not yet convinced. At the moment, those two rockshelters hold out the tantalising possibility that Australia may have been colonised 60,000 years ago, about 20,000 years before the first evidence of modern humans in Europe. But more archaeology and more dates are needed to clarify the situation.

Art in the Landscape: Gunbalanya (Oenpelli), Northern Territory, Australia

The traces of ground ochre at Malakunanja and Nauwalabila suggest that rock art could have a very deep history in Australia. Pieces of ground-down ochre are enigmatic and frustrating in many ways: like finding a pencil stub but not being able to find what was written or drawn with this implement. There are some more recent but still ancient rock paintings in north-west Australia that have been dated to more than 20,000 years old – but nothing as old as the 'crayons' from Malakunanja and Nauwalabila. The sophisticated, figurative style of the rock art in north-west Australia has links with paintings in Borneo dating back to more than 10,000 years ago. Mike Morwood argues that the particular complexity of the art and language families of this region provides an additional reason to suspect that Australia was colonised from the north.[1] Links between styles of ancient rock art may be a little flimsy, but there is no doubt that rock art has deep roots in Australia – and it is still important today.

I was very keen to meet some practising Aboriginal artists and find out more about what their art meant to them. In many places, Aboriginal art has become popularised, sanitised and commercialised, but I was going to a place where (I hoped) the old traditions were being kept alive. When I flew to the Northern Territory it was the middle of the wet season. Sites like Malakunanja were very difficult to get to, but so were towns and villages. For months during the wet season, the settlement of Gunbalanya (also called Oenpelli) was effectively an island, surrounded by water as creeks flooded and rivers broke their banks and coalesced. The only way in was by plane – so I caught a small Gunbalanya Air Cessna. We took off from Darwin, and flew about 300km east over a wet, green landscape where drowned trees stood up like white matchsticks in the flooded creeks, and engorged rivers short-circuited their meanders to flow northwards to the sea. We landed on a tiny airstrip at Gunbalanya, and I headed straight for Injalak Arts & Crafts Centre.

Gunbalanya is a community of about a thousand people, the vast majority of whom are Aboriginal. Most are Kunwinjku people, a discrete tribal group in western Arnhem Land, with their own language. They call themselves 'Bininj', and refer to white people as 'balander' (which may be a derivative of 'hollander', after the Dutch settlers). About a quarter of the locals in Gunbalanya were artists: painters, sculptors and basketmakers.

The Arts Centre, a single-storey building on the edge of the town, comprised a gallery, office and studio spaces. In the gallery there were beautiful paintings on bark and cartridge paper, painted didjeridus, burial logs, and piles of baskets. In the office, I met the director, Anthony Murphy, who managed the finances, arranged exhibitions and generally kept the thing going as a commercial enterprise. Through a door at the back of the office I was into the studio spaces. They were grubby and basic, but adequate. Aboriginal artists sat on concrete breeze blocks, painstakingly painting Dreamtime stories and animals on large sheets of cartridge paper.

Graham Badari was painting a frieze of fruit bats on a blue-green background. Over the days that I was there, the black shapes of bats appeared, and then their bodies came alive with fine white edging and hatching.

Northern Territory
Australia

'Have you seen them?' he asked. 'They sleep in the day then start flying out as the sun goes down.' I saw the bats, exactly as he said. At dusk, they left their perches in the tall trees and came alive, swooping down the streets of Gunbalanya.

Gershom Garlngarr was painting an epic tale of death and destruction, on a red ochre ground. A black cleft in the earth had opened up and was swallowing creatures: turtles, echidnas and crocodiles; even Mimi spirits (who inhabit the rocks) and yawk yawks (mermaid-like water spirits of the billabongs and creeks) were being sucked to their fate. Anthony said that a similar painting had been used as a cultural objection – in court – against the opening of another uranium mine in the Northern Territory. This was powerful art.

It reminded me of observations and conversations contained in Bruce Chatwin's book about his travels in search of the Aboriginal 'songlines', ancient tracks that criss-cross Australia, a 'labyrinth of invisible pathways which meander all over Australia', and which are passed on in song. The pathways are also known as the 'Footprints of the Ancestors', and Chatwin describes how these tracks are intimately linked with Aboriginal creation stories, which 'tell of the legendary totemic beings who had wandered over the continent in the Dreamtime, singing out the name of everything that crossed their path … and so singing the world into existence.' The songlines also link up sacred places across the landscape. Chatwin met Arkady, who was mapping these sites, and who said, 'To wound the earth … is to wound yourself, and if others wound the earth, they are wounding you. The land should be left untouched: as it was in the Dreamtime when the ancestors sang the world into existence.'[2]

I walked through into a larger room, which had originally been used for screen-printing but was now acting as a dormitory. The trestle tables for laying out screens were being used as beds, and a group of artists were taking time out, lounging around and – bizarrely – watching reruns of *The Bill* on television. Outside the back of the Centre I found Wilfred Nawirridj, sitting on the floor and grinding a chunk of ochre on the concrete floor, ready to colour the background of a new painting. The ochres were gathered from around Gunbalanya, and determined the artists' traditional palette: largely red, brown and yellow. The ritual of the painting started with the collection of the materials: the sites from which the ochres were gathered may be spiritually important to the artist, and various colours may also have ritual significance in the paintings. White clay and black charcoal were used for fine line work, for edges and crosshatching. Some artists were also excited by new colours available to them in tubes of gouache: purples, blues and greens, just as I had seen Graham using in his bat painting. Traditionally, a paste made from tree orchid roots was used as a bonding agent, and Wilfred showed me where an orchid had been deliberately seeded on to a tree just behind the Arts Centre. But Wilfred was mixing his ground ochre with a modern alternative: PVA glue. He thinned the mixture with water then applied it to cartridge paper with heavy strokes of a thick brush. The Injalak artists have been using heavy cartridge paper – more familiar to watercolourists – since the 1990s. They had traditionally used sheets of bark, but these were only harvestable during the wet season, and were also less stable over time.

Gabriel Maralngurra, one of the founders of the Arts Centre, was also outside, sitting at a table and starting to paint in white on a red ochre ground.

'Is there a story in this painting?' I asked, as he carefully laid down white hatching against the red ochre ground.

'Yeah. This one is about the flood and then the fresh water,' he replied. 'It's from the beginning. These are the escarpments,' he said, pointing out the hatched areas.

'There's the waterfalls, fresh water, and that's the sea. That's the rainbow serpent – making creeks. That's when the creation started.'

Ngalyod, the rainbow serpent, was one of the powerful creator spirits of the Dreamtime. In his painting, Gabriel was illustrating the great flood brought by Ngalyod, when fresh water started to flow. Meandering black lines in his painting expanded into billabongs. He was painstakingly filling in the flood with hatched white lines, or *rarrk*, using a fine brush with just four long fibres – made of sedge.

'How long does it take you to do a painting?' I asked Gabriel.

'Probably take me about a week and a half,' he replied.

'How long have you been painting for?'

'Oh … since I was twelve.' I asked him if anyone had taught him to paint.

'I used to sit with my grandfather,' he said. 'Sit next to him, see him paint, tell stories. That's how I learnt from him.'

'Are there any children learning to paint at the moment?' I asked.

'Yeah, there are. Schoolkids. Some of them even made a clay-mation about the Turtle and Echidna story.'

(The tale of the Turtle and Echidna was a bizarre just-so story about how those animals originated. They had once both been women. One woman was meant to be looking after the other's children but somehow ended up eating them instead. The children's mother was understandably annoyed, and threw a stone at the greedy woman, which had stuck to her back and she became a long-necked turtle. The turtle, in turn, threw spears at the outraged mother, who became an echidna.)

All the artists I had met at the Arts Centre were men. 'Are there any women artists?' I asked Gabriel. 'No, not really. The women make baskets – secret women's work!' he smiled. The baskets themselves were works of art. Woven from pandanus leaves, they were coloured with natural pigments which echoed the ochres used traditionally in the paintings, but with the reds, browns and oranges combined with green and yellow plant dyes. There was a great range of different basketwork in the Arts Centre, from large, flat trays and mats to string bags and small, fluted dilly bags. These little bucket-shaped bags were both practical and symbolic; they could be used to carry bush food or sacred objects. Just as Gabriel had learnt to paint from his grandfather, young girls would sit with their mothers and grandmothers, watching them weaving and listening to their Dreamtime stories. They would also learn the practicalities of how to collect pandanus, bush string, pigments and bush food.[3]

Gunbalanya is special not only for its thriving community of artists, but as an area containing a very rich concentration of rock art 'galleries', which can be visited only with permission from the traditional, Aboriginal landowners. I was lucky enough to be getting a personal guided tour by Wilfred and his cousin, Garry Djorlom. Gunbalanya lies on a coastal plain, just to the west of the 'Stone Country' of the Arnhem Land escarpment. The Arts Centre is named after Injalak Hill, an outlier of the escarpment, separated from the settlement by a lily-fringed billabong during the wet season.

I went up on Injalak Hill with Wilfred and Garry. Getting there was a feat in itself; we had Land Rovers which we could drive into the creek up to the tops of their bonnets, but there would be parts that were too deep. We had taken a flat-bottomed boat with us as well, to navigate the last section of the billabong, which would have been fine if the water had stayed deep, but sometimes we were trying to float in less than a foot of water. It would get too shallow to use the motor, then the bottom of the boat would scrape along on the gravel, like nails down a blackboard, and we'd be forced to get out and wade alongside the boat, pushing it. This was rather nerve-racking as I'd been told that there were crocodiles around. We kept a beady eye open and I was very pleased not to see any.

Eventually, we were back on dry land. Wilfred and Garry led me through tall grass (we were now on the lookout for snakes) and to the foot of Injalak Hill. We made our way up a narrow, winding path, between boulders which became larger as we climbed higher up the hillside. About halfway up, we paused on a flat outcrop and surveyed the landscape. We were looking down on the billabong, the lush green landscape, and Gunbalanya beyond. Garry told me about hunting in the bush. Although apparently living a settled existence, many of the Aboriginal families of Gunbalanya still went off into the countryside from time to time to find bush tucker. Garry said that people would do this when they'd had enough of eating bad, Western food from the supermarket.

Whole families would head off into the bush to live more naturally, as their ancestors had done. The land looked lush and bountiful: it had always been a good place for hunter-gatherers, with abundant food all year round, and rocky escarpments providing shelter from tropical monsoons.

We climbed further up the hill, and the ground levelled off. We were now in a clearing with a large, overhanging rock on one side. Garry and Wilfred gestured upwards: I looked at the rocky ceiling above me and an amazing sight met my eyes. The overhang was *covered* in paintings. It had been painted over and over: a great palimpsest of images stretching back

Injalak Hill

who knows how many hundreds or thousands of years. The last paintings were done in the 1960s, and as far as Garry and Wilfred were concerned the earliest paintings went back into the Dreamtime, the time of their ancestors, when the world was young.

All across this rock face were paintings of animals. Garry pointed out the different species to me: echidna, crocodile, turtle, kangaroo, snake, and many, many types of freshwater fish, including whiskered catfish, fat barramundi and slender long tom. After all, this was Fish-Dreaming (Injalak) Hill. Beyond a kangaroo's head, a human-like arm reached out. The colours – red and yellow ochres, white and black – and the patterns including the hatching, or *rarrk* – were familiar from the paintings I had seen down at the Arts Centre, as was the 'X-ray' style of the barramundi, showing its spine.

On first inspection, the paintings seemed to be about what animals were available in this rich environment: a sort of *à la carte* menu for hunter-gatherers. But although this was one level of meaning, the various animals also stood for other things, including features of the landscape. Aboriginal mythology was very firmly tied to geography. Creation stories described how land and waters had arisen, but also how particular elements of the countryside had come into being. Some ancestral beings had even changed into parts of the landscape, to become sacred sites.

Garry described how the children of Gunbalanya would still be brought up here to learn about their environment and their heritage from elders. The rock paintings were so much more than decoration: they were a method of communicating ideas. And, as I had already learnt, this tradition was much more ancient than writing.

'Always pass it on, pass it on. That's the real importance of painting,' said Garry.

Higher up on the hill, I followed Garry and Wilfred down narrow corridors formed by cracks between enormous boulders. A niche in one rocky wall formed a canvas for a very special piece of rock art. No animals here – this image was purely mythological. 'This is a picture of the creation mother: Yingana,' Wilfred told me. He loved nothing better than telling a story. He had a wonderful voice: slow and measured, and, quite strangely, English-sounding. I later found out he had been brought up by missionaries (Gunbalanya was originally a mission town). He sounded like an English country vicar telling parables from the pulpit.

'She came from the coast, travelling,' he said. 'She is carrying lots and lots of dilly bags.' The woman was pictured wearing a headband, from which were hanging fourteen or so stripy dilly bags, and Wilfred said that each one contained a baby.

'Once she came, she put down one baby, and gave him a language and a skin-name, a clan. Then she came to another place. The second one, she gave him a different clan, skin-name, language. Then she kept travelling. The third one, she gave him a different language, different moiety. She kept travelling, putting down these babies – the ancestors of us.'

The story went on. Yingana was seeding Australia with its original people.

'I've been trying to find out how people first got to Australia,' I said. 'I've been looking at fossils and genes. We think that perhaps the ancestors came from the north, from across the sea.'

'From the north – yes. Creation mother may have come ashore from Macassar way, from somewhere in Indonesia. That's my guess.'

This was fascinating. But it was probably wasn't the amazing coincidence it at first seemed. Wilfred was an educated man. He knew that Macassar (Indonesia) lay to the north, but he was also prepared to incorporate this knowledge into his understanding of the traditional stories of the origin of his people.

But other Aboriginal stories also seem to have their starting point in the north. Bruce Chatwin tried to trace 'songlines', the sacred songs that describe the journeys of ancestors across the landscape of Australia, and he wrote:

> The main Songlines in Australia appear to enter the country from the north or the north-west – from across the Timor Sea or the Torres Sea – and from there weave their way southwards across the continent. One has the impression that they represent the routes of the first Australians – and that they have come from somewhere else.

We emerged from between the clefts in the massive boulders on to an eagle's-nest viewpoint overlooking Gunbalanya and the hills around: Turtle Dreaming Hill and Magpie Goose Dreaming Hill. I sat near the edge, with Wilfred and Garry, looking out at the land their ancestors had inhabited. I had rarely felt such a strong sense of place. And that flowed from Wilfred and Garry. The hills, creeks and billabongs belonged to them in a loose but powerful way, because their ancestors had lived here, and because they had walked all over this landscape and knew it so well. Aboriginal people in places like Gunbalanya still go on walkabout, travelling to visit friends and relations – as so many of us do – but also to remind themselves of ancient stories and their landscape. Wilfred and Garry's sense of belonging in a landscape seemed not to be tied to one particular place, but to the idea of a journey across the countryside. It seemed like an ancient – but not at all primitive – idea, perhaps something that many of us, settled in a very sedentary existence in villages, towns and cities, feel we have lost.

So I had found evidence of the first Australians. I had followed the faint traces of the first colonisers all the way to Sundaland – and Sahul. And I had met people in Australia who had taught me something about what humans are, passing on knowledge from generation to generation through art, and still feeling the need to roam in the landscape.

3. Reindeer to Rice:
The Peopling of North and East Asia

Shanghai tea room

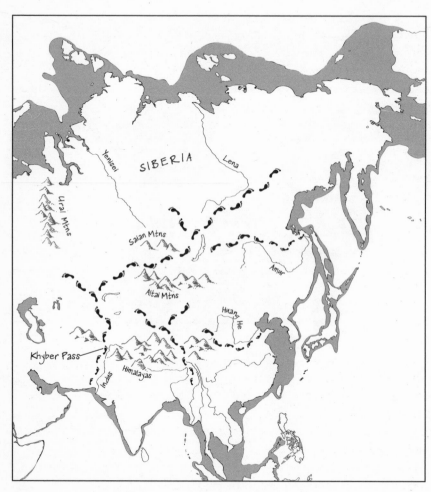

After the initial colonisation of the southern and eastern coasts of Asia (not shown on this map), Asian genes (mitochondrial and Y chromosome) reveal colonisers spreading up the rivers, making their way north of the Himalayas and reaching Siberia.

Trekking Inland: Routes into Central Asia

The southern route out of Africa scattered people along the coast from India to South-East Asia and Australia. But what about the rest of Asia? There was a vast expanse of land to the north and east. Today that land is mainly occupied by two vast countries: Russia and China. But in the Palaeolithic it was one great wilderness, teeming with wildlife: prime hunter-gatherer territory.

I was on the search for the first Russians and Chinese. How did they penetrate the continent and just how far north would Palaeolithic technology let them venture? I was also interested in faces, and the origins of characteristic East Asian features, and this enquiry would take me back to that debate about replacement versus regional continuity, a recent African origin of modern humans or multiregional evolution.

Analysis of mitochondrial DNA in northern Asia has shown that all the lineages can be traced back to M and N in southern Asia (see page 101). This means we can be sure that people moved out of Africa, along that southern, beachcombing route, and then flowed northwards to populate northern Asia.[1] Populations spreading northwards from the South Asian coast could have followed rivers inland – but a massive barrier stood between South and Central Asia, stretching all the way from Afghanistan in the west to China in the east: the Himalayas.

In *Out of Eden*, Oppenheimer[2] described potential routes around and through this mountainous barrier, allowing access to Central Asia, following rivers inland. The Indus may have led colonisers up to the Khyber Pass, skirting the Himalayas in the west. Colonisers could have tracked northwards along the rivers of South-East Asia. If the beachcombers had continued up and along what is now the Chinese coast, they could have then moved westwards, along what would become the Silk Road, just north of the Himalayas. People could have also spread in from the west, from the Russian Altai.

Once populations were established north of the Himalayas, the vast expanse of Central Asia and Siberia lay open to colonisation. Siberia itself is huge, covering about 10 million km² and stretching from the Altai and Saian mountains in the south, to the shore of the Arctic Ocean in the north; from the Ural Mountains in the west, to the Pacific in the east. The environment here was about as different from the tropics, where early modern humans had flourished, as it could possibly have been. There would have been many new challenges facing the early pioneers: a lack of plants to eat, extreme cold and sometimes even a scarcity of wood for building shelters or burning for warmth. Colonising Siberia would have required new adaptations, new ways of hunting: a whole new suite of survival skills.

Mitochondrial DNA from populations across modern Siberia contains a record of a complex process of colonisation, with a mixture of lineages from European and Asian branches, presumably relating to incursions from both the west and the south, from modern-day Mongolia. Genetic diversity is greatest among Altaians, suggesting that modern humans have been in that region the longest. Unlike the long, thin, rake-like distribution of mitochondrial DNA branches in South-East Asia, which maps so well on to coastal dispersal from west to east, the branching bush of mtDNA in Siberia gives us less insight into the actual routes into the icy north.[3]

Considering that Central Asia is so huge, the archaeological evidence for early modern human colonisation north of the Himalayas is scarce. The earliest archaeological evidence comes from a site far north of the Himalayas: Kara-Bom in the Russian Altai. Kara-Bom is an open-air site, situated at the foot of a steep cliff, close to tributaries of the Ursul River. The site was excavated during the 1980s and early 1990s. Digging down through five metres of sediments at the foot of a slope, the archaeologists found evidence for three phases of occupation. The deepest layer contained Middle Palaeolithic, Mousterian tools, including Levallois cores and flakes as well as finished tools including points, side-scrapers and knives: it looked like a fairly standard Neanderthal toolkit.[4] In the next layer up at Kara-Bom, the archaeologists found the clear signature of modern humans: early Upper Palaeolithic tools, including prismatic cores left over from blade manufacture, and plenty of blades, retouched on one or both sides, as well as end-scrapers, side-scrapers and burins. Above that again were late Upper Palaeolithic tools including microblades.

In the 1980s, archaeologists used conventional radiocarbon dating to age the Upper Palaeolithic layer at Kara-Bom, and reported a date of about 32,000 years old. In the 1990s the archaeologists used the new, more reliable method of AMS radiocarbon dating, on charcoal from the microblade layer, and obtained a date of about 42,000 years old.[5, 4]

The animal bones also excavated from Kara-Bom provide us with a good idea of how rich the area was in terms of animal life: there were bones of horse, woolly rhino, bison, yak, antelope, sheep, cave hyena,

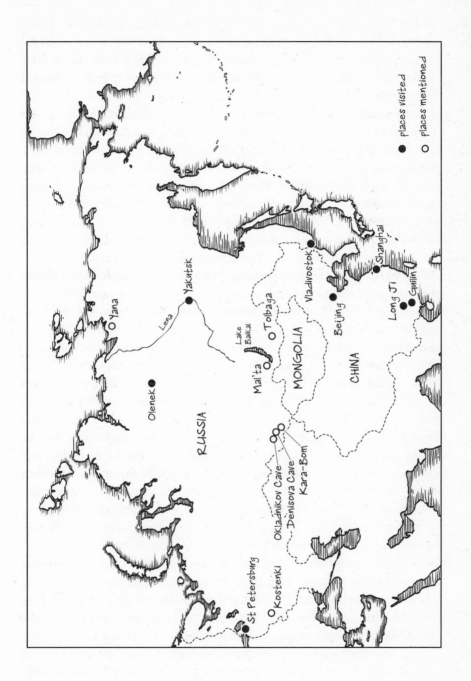

Places visited
Places mentioned

Shanghai
Guilin
Long Ji
Beijing
Vladivostok
Tolbaga
Yakutsk
Yana
Lena
Lake Baikal
Mal'ta
MONGOLIA
CHINA
Olenek
RUSSIA
Okladnikov Cave
Denisova Cave
Kara-Bom
St Petersburg
Kostenki

grey wolf, marmot and hare; it sounded as if it would have been a good place for the Palaeolithic hunters.

In fact, there is a scattering of Upper Palaeolithic sites in southern Siberia, stretching as far east as Lake Baikal, as well as a couple further south, in Mongolia.[6] So far, Kara-Bom has produced the oldest date for modern humans in Central Asia, but several other sites in southern Siberia are nearly as ancient, dating to between 30,000 and 40,000 years ago. Although some are cave sites, most are open-air camps, like Kara-Bom. These Upper Palaeolithic sites have characteristic stone tools, but also artefacts made from bone, ivory and antler, as well as deer-tooth pendants. In terms of evidence for art, there's really not much from the Siberian Upper Palaeolithic. Just a few sites have produced objects that could be construed as artistic artefacts: a stone disc coloured with red ochre, a possible carving of a bear's head on a woolly rhinoceros vertebra and an ivory sphere.

Judging by tools in deeper layers, quite a few of the sites seem to have been used by archaic humans, perhaps Neanderthals, long before moderns got there. Ted Goebel places the colonisation of southern Siberia by archaic humans in the Middle Pleistocene, somewhere between 200,000 and 100,000 years ago, with modern humans moving in to occupy the area between 45,000 and 35,000 years ago. Could Neanderthals have still been living in Siberia when modern humans arrived? Dates and ancient DNA results from Okladnikov Cave in the Siberian Altai suggest that this may have been a possibility. Mousterian tools were found in the cave, but of course those could have been made by modern humans or Neanderthals. Some researchers suggested that teeth found in Okladnikov Cave, dating to around 40,000 years ago, were Neanderthal, but others disagreed. There were also some fragments of bone, too smashed up for us to be sure what species they belonged to. But geneticists recently managed to extract ancient DNA from those bits of bone – and they did indeed turn out to be Neanderthal. This is extraordinary because it suggests that the Neanderthals penetrated a lot further into Asia than previously thought, and that there may well have been overlap with modern humans, as in Europe.[7] (We don't know if those Neanderthals and modern humans ever met each other, though, and this is a question that will be pursued in the next part, The Wild West: The Colonisation of Europe.)

Although it is clear that archaic humans, including Neanderthals, lived in the mountainous regions of southern Siberia, they did not seem to get any further north, into the subarctic and Arctic regions. And the first modern humans to live in the area were also limited to southern Siberia.

It appears that these first colonisers were, as Goebel puts it, 'tethered to places on the landscape', where there was suitable stone for making tools and plenty of animals to hunt. There's no evidence of long-distance trade networks: tools were made from local stone. In this respect, it's a similar picture to Europe in the early Upper Palaeolithic. But there is a distinct difference in the types of tools being made. The stone tools from the early Upper Palaeolithic of Siberia, and from China, too, are a bit of a strange 'mixed bag' when compared with the European toolkits. There are 'modern-looking', light-duty tools like small end-scrapers, borers, points and burins, but there are also very old-fashioned looking tools like side-scrapers, Mousterian-like points and sometimes even hand-axe type tools. Some archaeologists have argued that this shows a lack of technological development in isolated large populations, but others have said it can be explained functionally: the toolkits reflect different patterns of hunter-gatherer subsistence in particular environments. This functional and ecological interpretation also gets us away from the idea that humans were always on a quest to make 'better' tools. Instead, they were making what they needed to survive in particular place.[8]

On the Trail of Ice Age Siberians: St Petersburg, Russia

I travelled to St Petersburg to find out more about Ice Age Siberia. It was mid-spring and the River Neva ('Nyeva') had almost completely broken free from winter ice. Just a few plates of ice clung to its edges and floated like diminutive icebergs under its bridges.

In a restaurant in St Petersburg I met up with Russian archaeologist Vladimir Pitulko to find out about an exciting site in the far north-east of Siberia. Until just a few years ago, it was thought that modern humans hadn't got as far as the Arctic Circle until after the LGM, around 18,000 years ago. But Pitulko had been excavating a site which suggested that the early modern human colonisers had adapted to the extreme conditions, and spread further north than any archaic humans had managed, right into subarctic and arctic Siberia, well before the LGM. The site was called Yana: it was an Upper Palaeolithic site frozen into the permafrost.

In 1993, geologist Mikhail Dashtzeren found a foreshaft, or spear end, made from woolly rhino horn, in the Yana Valley. (Foreshafts allow speedy replacement of broken spear points, thought to be a great advantage when hunting large game.) The artefact was the first sign of a Palaeolithic site eroding out of the permafrost – a site to become known in full as Yana RHS (Rhinoceros Horn Site). Excavations began some years later, in 2002. The archaeologists found stone tools made from flinty slate, including side-scrapers and end-scrapers. These are all based on flakes: there is no sign of blade manufacture from Yana. There were vast quantities of animal bones, mainly reindeer, but there were also mammoth, horse, bison, hare and bird bones. Nearly all the bones showed signs of scraping. Two further foreshafts were found, made from mammoth ivory, and a bone punch, or awl. The finds were radiocarbon dated to around 30,000 years ago.[1]

The date of Yana places it just on the transition between a 'warmer' period, when larch and birch forests would have covered northern Siberia, and a cold phase, when the landscape became treeless tundra. The average temperature would have been colder than today.[1,2]

The date and location of Yana are significant: modern humans were in the Arctic well before the LGM. And the stone tools and ivory foreshafts preshadow some of the earliest implements found in the Bering land bridge region – and the Americas.[1]

Pitulko was returning to dig at Yana that summer, and I made arrangements to fly out and meet him there. But it wasn't to happen. In the end, the vagaries of Russian airline schedules thwarted my plans, and my chance of seeing the evidence for the Ice Age inhabitants of the far north was to be cruelly denied.

Between 26,000 and 19,000 years ago, the world was cooling towards the Last Glacial Maximum (or 'LGM'). In northern Europe this meant the advance of ice sheets. In Siberia, the drying climate created a vast, almost unimaginably arid plain: the 'mammoth steppe'.[3] The steppe was *so* dry that many plants and animals were driven to local extinction in the north. Flora and fauna that survive today in the Arctic tundra, including humans, gradually retreated thousands of miles to the south, or to the far north-east, to the land exposed between Asia and North America: Beringia.[3] Although the Upper Palaeolithic people of Siberia had become experts in surviving in extreme conditions, colonising the Arctic, they were now put under immense pressure. The number of archaeological sites dwindles as the Ice Age was reaching its climax around 19,000 years ago.[4]

Most of Siberia would have been locked in permafrost, but areas of southern Siberia, around the Transbaikal and Yenisei rivers, would have been slightly milder. Human and animal populations may have survived in these refugia while most of Siberia was an Arctic waste. The Siberian environment around the LGM is hard for us to imagine: it was very different from anything in existence now. There were combinations of plants and animals living on the mammoth steppe that we just don't see today. Imagine this vast, treeless plain. All the vegetation was low-level, mostly grasses and sedges, no trees. And in it there were some animals which we recognise as cold-loving creatures such as reindeer and Arctic foxes, but alongside them were others that we'd associate with much warmer environments today, like cheetahs, hyenas and leopards. Ice Age Siberia was a place of extremes, with colder winters but warmer summers than today.[3]

Around this time there was also a change in culture in Siberia, typified by a site called Mal'ta, about 80km north of Irkutsk.[5] The site was discovered in 1928, by local peasants, who found some bones along the road to Moscow. When archaeologists arrived to dig the site, they uncovered the remains of a Palaeolithic camp with substantial, semi-subterranean houses. They retrieved more than 44,000 stone tools, and over five hundred artefacts made from bone, ivory and antler. Among these were spectacular pieces of art, including some thirty human figurines and fifty carvings of birds, in mammoth ivory.[6] New radiocarbon dates from Mal'ta place the site at around 21,000 years ago – practically right on the LGM.[5]

The artefacts from Ice Age Mal'ta are now kept in the Hermitage in St Petersburg and this was another reason for my visit to that beautiful city in European Russia. I entered the magnificent building on the banks of the Neva through a back door, then navigated my way through galleries full of much more recent, spectacular artwork. At the end of a long corridor I approached a large wooden door and waited to be admitted. I was met by a curator, an extremely smart and petite Russian woman named Svetlana Demeshchenko.

Behind the first door was another door. I followed Svetlana through it, and then up a spiral staircase until we reached the upper level and the door into the archaeological offices and stores. This was a different world from the grand state rooms of the Hermitage, with their polished floors, neoclassical sculptures and gilded pillars. There were corridors flanking by tall wooden cupboards, faded posters of archaeological exhibitions, and I caught glimpses of offices, in one of which were archaeologists poring over papers, surrounded by piles of dusty books and a jungle of houseplants.

Svetlana led me into a room lined with shallow wooden drawers and cupboards, and she started unpacking artefacts. She took a velvet-covered board and laid out a beautiful bone necklace that had been found with a child's skeleton. There was an oblong of mammoth ivory inscribed with a stippled spiral pattern on one side, and with three snake-like lines on the other. Svetlana said that it had been proposed to be some kind of shamanic map, with the central hole in it representing a connection between worlds above and below the physical realm of existence. It was certainly a mysterious object, but it is surely impossible to know whether the design was anything more than decoration.[6, 7]

Nestling in tissue paper in boxes were miniature figurines, like small ivory dolls. Some were thin and rangy, whereas a couple were a bit more buxom, and reminiscent of the European 'Venus' figures. Some were naked, with carved breasts, but others appeared to be clothed. One was clearly wearing a hat, and her body was scratched, textured, suggesting fur clothes perhaps. Bone awls and needles have been found at several sites from this period, indications that the technology to make clothes was in place. In fact, there are eyed needles dating to about 30,000–35,000 years old from Kostenki, in Ukraine, and Tolbaga, to the south of Lake Baikal. A needle from Denisova Cave in the Altai has been reported as being even older, perhaps as much as 40,000 years old, although this may be an over-estimate.[8] But even without any of this material evidence (forgive the pun), the technology to make clothes could have been inferred: it would have been impossible for the Ice Age Siberians to have survived without substantial fur clothing.

The feet of several figurines were pierced through, as if they had been intended to be worn as pendants – though they would hang upside-down. And then there were two ivory, bird-like figures, only about 6cm long, with outstretched necks and stubby wings. Were they geese? Or swans? I imagined those ancient Siberians, huddled in tents against the growing cold of the Ice Age, carving these objects by the firelight. Was it just something to do on those long, cold evenings, or did these female figures and birds hold some meaning for those people? Could they have been mythological or shamanic emblems? Some anthropologists, arguing from analogy with recent ethnographic studies, have suggested that the pierced figurines and birds represented spirit helpers that were designed to be attached to a shaman's costume.[6, 7] Their meaning has long since been lost, but they are very beautiful objects.

Mal'ta and the other sites from Siberia between 30,000 and 20,000 years ago show us that the hunter-gatherers in this region were surviving, and, given the artistic outpouring

carved ivory
figurine
from Mal'ta

in Mal'ta, even flourishing, as glacial conditions set in. And the similarity between tools and art produced in Siberia and Europe at this time suggests that there were communication networks linking communities across a vast area. People seem to have been moving around the landscape, using large base camps, with smaller hunting camps like satellites around them. The base camps were often a long way from any outcrops of rock ideally suited to making tools – presumably their location was determined more by the proximity of animals to hunt. But this meant that stone tools either had to be made from less than ideal, coarse rocks, or that finer stone had to be carried long distances. Many of the stone tools at Mal'ta were small blades, or 'bladelets', made from correspondingly miniature cores. Perhaps it was a need to economise, to make the most of scarce supplies of decent stone, that drove the toolmakers to make smaller and smaller blades?[4]

The animal bones from Mal'ta show that these Ice Age Siberians were hunting a wide range of animals: woolly rhinoceros, mammoth, bison, reindeer, horse and red deer, as well as hares, Arctic foxes and wolverine, and geese, gulls, grouse and ptarmigan. There were a great many reindeer antlers at Mal'ta, but these might have been scavenged for use in house-building, as reindeer naturally lose their antlers each year. At numerous places on the Russian plain, around and immediately following the LGM, and including the famous site of Mezhirich, mammoth tusks were used in the construction of huts.

Leaving the Hermitage, I crossed over the River Neva and made my way to the back door of the Zoological Institute of the Russian Academy of Sciences. Another museum, another back door, and another corridor with store rooms full of bones. The Institute is famous for its mammoths. In the exhibition hall upstairs there were stuffed mammoths, towering over dioramas of more reasonably sized mammals. There were even mummified mammoths, which had survived, frozen in permafrost, to the present day. Downstairs in the store rooms there were stacks of enormous bones, huge skulls, and piles and piles of tusks. I stood next to a femur, and the mammoth thigh bone reached up to my chest. The animal it came from would have dwarfed me. Mammoths would have been formidable prey for those ancient hunters on the steppe.

Mammoth remains, from the Pleistocene, have been found all over Central Asia, from the Arctic Ocean in the north to Mongolia in the south. Collections of mammoth bones often turn up in riverbanks, from animals or skeletons that have been swept up by rivers and redeposited. But mammoth bones and tusks have also been found at archaeological sites, alongside evidence of human activity and occupation.[9]

The interaction between humans and mammoths on the plains of Siberia, in Europe and in North America, has been widely debated. Certainly in Siberia it seems that humans were using mammoths, presumably eating their meat, and using their bones and tusks to build houses, carve tools and make art.[5] In western Siberia, there are a few sites with evidence of 'mammoth processing', where practically complete skeletons have been found along with stone tools and traces of fire. But it is impossible to tell if these mammoths had been hunted or had been collected as frozen carcasses. Other sites suggest that the ancient Siberians were choosing to camp near collections of old bones and tusks, which they could then gather to use. Such collections in the banks of lakes and rivers represented mammoths which may have died hundreds or thousands of years earlier, maybe by falling through the ice.[9]

The last mammoths appear to have inhabited the far north of Siberia up until around 10,000 years ago. So what finished them off? For some researchers, the answer is obvious: humans. But this really does depend on humans having actively hunted mammoths in Siberia. From research that has tried to assess fluctuations in mammoth and human populations, it seems that people made very little impact on mammoth populations, at least until well after the LGM. During the Pleistocene, human population sizes were small, and were concentrated south of the main ranges of the mammoths. Mammoth populations actually appear to have expanded after the LGM, during the cold snap called the Younger Dryas, but at around 13,000 years ago their numbers started to dwindle, and by 11,500 years ago they had all disappeared. Now, this does coincide with the continued expansion of human populations; perhaps, for an already shrinking and stressed mammoth population, a modest amount of hunting by humans might have been enough to tip the balance towards extinction.[10]

There were major changes in the climate and environment at this time: the world was warming up and the mammoth steppe was disappearing, and for some researchers this is enough to explain the extinction of woolly mammoths and other Pleistocene megafauna.[11, 12] As far as archaeological evidence from Siberian middle Upper Palaeolithic sites is concerned, there is often no way to tell if mammoth remains have come from hunted or scavenged animals.[5] In fact, there is no definite evidence of mammoth-hunting in Siberia, but there is plenty of evidence of humans collecting bones and tusks from already ancient mammoth 'death sites'.

The idea of those Ice Age hunters as big game specialists, or exclusive 'mammoth hunters', does not stand up to scrutiny. The hunters appear to have been generalists; large mammals like woolly rhino, mammoth and bison are actually rare in archaeological sites. Medium-sized animals such as reindeer, red deer and horse are much more common. And the hunters were also bringing back small game like fox and wolverine, as well as birds like geese, gulls, grouse and ptarmigan.[5, 12] Wolf bones are also common – but this may be domestication rather than predation: perhaps first evidence of man's best friend.[12]

So, for the time being the question of whether it was climate change, overkill by humans or a mixture of the two that eventually did in the mammoths is far from settled.[13, 14]

After the worst of the Ice Age, while megafauna like the mammoths had disappeared for good, other plants, animals and humans started to spread northwards again as the climate warmed. But using a word like 'warm' to describe Siberia is somewhat misleading, as I was about to find out: I was going to stay with the reindeer hunters of the north.

Meeting with the Reindeer Herders of the North: Olenek, Siberia

I was heading for the coldest inhabited place on earth: northern Siberia. Winter temperatures there can fall to below −70 degrees. First, I flew to Yakutsk, where, as I stepped off the plane and inhaled the icy air, I felt my bronchi constricting in protest. It was about −20 degrees. At the airport, I met my guide, anthropologist Anatoly Alekseyev. From the airport we drove through snowy streets, past ramshackle houses that were visibly subsiding into the permafrost, and past a huge statue of Lenin in the square, his right hand outstretched. Although I'd heard Yakutsk described as a modern Klondike run by diamond merchants, there was little outward evidence of this wealth. There were lots of new bars and casinos, and very smart women in chic fur hats and high-heeled boots, but the whole place still felt very run down. Arriving at the hotel, and stepping from the icy cold into an overheated foyer, I was clearly entering a different culture from European Russia. There were woven panels of horse hair depicting long-bearded old men and reindeer. People's faces – including Anatoly's – were also very different here from those in St Petersburg or Moscow. Gone were the large noses and long faces. The majority of people were now much more oriental-looking, with broad cheeks, narrow eyes and small noses.

The following day we caught another, smaller prop plane and flew further north, from Yakutsk to the village of Olenek. We were sharing the aircraft with a crowd of very smart-looking Siberians: men in suits and women in exquisite long fur coats and hats. The flight took us over a snowy landscape sparsely covered in larch forest. We followed the meandering River Lena, icebound and covered in snow, northwards, then peeled away to the west.

The plane landed at Olenek, and from my seat just behind the wing I could see the wheel make contact with the runway, kicking up an impressive plume of snow. And as we slowed to a halt we could see a welcome party assembled on the runway: a group of women dressed once again in long fur coats, one in a crimson coat lined with white fur, like a female Santa Claus, and a circle of dancers dressed in traditional fur outfits that looked almost Native American. One woman held out a round loaf with a small pot of salt embedded in the middle of it; I tore off a piece of bread, dipped it in the salt, and ate it. Children ran up with necklaces made of reindeer antler and hung them around our necks. I had arrived in Olenek on the eve of their annual reindeer festival, as had many others from the region, including diamond-mine owners and politicians – 'Big Fircones', as the Russian phrase puts it.

Anatoly and I somehow got swept along with the diamond óligarchs, and ushered into the Regional Administrators' office, where we were treated to a discourse on the progressive changes being made in the village. Although I had never been to Russia before this trip, the whole meeting felt rich with Soviet overtones. When the meeting drew to a close, I was given a key-ring emblazoned with a badge of the Reindeer Festival. Then Anatoly and I took a rugged little Toyota van across the frozen River Olenek to our lodging on the opposite bank.

We were staying in Marina Stepanova's single-storey wooden house, with a woodshed and an outside toilet near the fence, flanking the driveway. There were steps up the porch and a heavy, felt-edged front door (which had to be shut quickly as you passed through so as not to let the heat out of the house). Inside, a small hallway, with hooks for coats, led straight into a room with a tiny kitchen on the right and a dining area on the left. There were two bedrooms at the back of the house. The house was toasty warm: a central wood-burning stove was kept alight during the day, to melt ice for water as well as providing heat, and at night hot water flowed into radiators in each room from a remote boilerhouse. Marina was staying with family, though popping in to cook us meals – very generously letting us take over her house for a while.

As Anatoly and I settled in, Piers Vitebsky appeared: a great bear of a man, he would help me understand Evenki culture as I experienced it over the coming days. Piers was an anthropologist who had specialised in shamanism, spending time with tribes in India and northern Siberia. He headed up Anthropology and Russian Northern Studies at the Scott Polar Research Institute at Cambridge University, and it was through Piers that Anatoly had developed his interest in anthropology and become a historian of his own people.

Marina Stepanova's outside facility

Marina's dinners were generous and unchanging. There were hunks of white bread, biscuits and sweets, arranged on a cake stand in the centre of the table. We had cups of tea made with hot water from electric urns, and cranberry juice. There were small bowls of carrot and cabbage salad, and, when we sat down, huge steaming dishes of potatoes and reindeer meat would appear. Essentially, every time we left the house for any length of time, when we returned this spread would be awaiting us. I was reading Piers' book on his experiences living with the Eveny reindeer herders (Anatoly's people), where the camp cook would always have dinner ready in the tent for the herders returning from working in the cold. Marina was maintaining the custom in the village.

Later that day we crossed back over the frozen river to visit the museum in Olenek, where Piers gave me a guided tour of local Neolithic pot and stone tools, scale models of Evenki 'sky burials' with coffins on stilts, items of clothing stitched together from reindeer skin, and shamanic paraphernalia – including a shaman's coat adorned with iron animals and complete with reindeer-hide 'tail'. The shaman's assistant would apparently hold this two-metre-long tail to anchor the shaman and pull him back to earth after he had been flying in a trance. I had hoped to see a shaman, but the Soviet regime had made them particularly rare and shy. That night, though, we went along to a concert in the village hall, where, along with folk rock and pop, we were treated to full-on shamanic kitsch with a folk singer dressed in a quasi-traditional costume, beating a skin drum, accompanied by reindeer dancers. This was the opening night of the Reindeer Festival.

The following day, I went to the festival itself, held on the frozen river. Two 'chums', tepee-like tents, had been erected within an arena framed with coloured flags. People were out in their best furs: little girls dressed cutely, head to toe in white fur, women in long fur coats, men in fur jackets. Teams of reindeer waited patiently inside the enclosure, tethered to their sleighs ready for the races and some of the animals were wearing beautiful bead-embroidered headdresses and trappings. I don't think I'd ever seen a real, live reindeer before arriving in Siberia. Now there were reindeer *everywhere*, looking like mythical creatures in the bright snow and sunshine. Throughout the day there were various reindeer races: children riding reindeer, women racing single-reindeer sleighs, men racing double-reindeer sleighs on 3 to 8km tracks up and down the frozen river. People had come to the festival from all over Yakutia and the neighbouring district of Zhigansky, a geographically enormous area, something like the size of Britain and France combined. Some of the reindeer had literally flown in – by helicopter. It was still, as these festivals would have been when all the reindeer herders were truly nomadic, an opportunity for a scattered population to come together, and, in particular, for young men and women to meet up.

I went up to the river cliffs above the festival to get a panoramic view of the sleighs as they approached the finishing line, the arena and the chums. Then we headed back to Marina's where dinner was waiting for us. But we weren't stopping long. The time had come for the next leg of my journey, to an even more remote place. All around Olenek, reindeer herders lived in mobile camps – and I was going to one of them, some 70km from the village.

I packed up my gear in a kitbag and wrapped myself up in layers and layers of merino wool thermals, fleeces, a jacket and, finally, a reindeer fur coat. My feet had got cold in the afternoon, walking around in my Baffin boots. Marina looked at them with disdain and produced a pair of reindeer boots, so I stuffed my feet into two pairs of thick woollen socks and then into the furry-inside-and-out boots. On my head: a black woollen hat, two buffs around my neck, pulled up over my nose, and sealed with a pair of ski goggles so that no skin showed. I tugged a hood trimmed with wolf fur firmly down and around my well-wrapped face, and pulled on two pairs of gloves: an inner silk pair and an outer, fleece-lined, wind-proof pair. Then I walked stiffly out of the house and down the steps of the porch to where some of the men we'd seen racing earlier in the day were loading our gear on to sledges behind snowmobiles.

We had planned to leave for this journey at six o'clock. Piers had warned us about the difference between 'intentional time' and 'real time' in northern Siberia. It was gone nine o'clock when we left and the orange sun was resting on the horizon. Piers, Anatoly and I were each to travel with a reindeer herder, on separate snowmobiles pulling sledges. I jumped on to a sledge, sitting with my back against a pile of bags. Then we set off – bumpily – along the snowy road, heading west into the setting sun. My goggles started to ice up almost immediately; all I could see were occasional glimpses of trees in the upper margins of the goggles, but most of my field of vision was a uniform yellow-grey. We hurtled along, with the sledge riding up over the bumps and crashing down on the other side.

It wasn't long before we started to spread out, travelling far apart, as the air darkened and cooled around us. It was like a strange exercise in selective sensory deprivation: all I could see now was dark grey, and my ears filled with the chainsaw-like drone of the snowmobile. As tempting as it was to sleep, I had to be an active passenger: as we veered, rolled, pitched and jolted along, I concentrated on hanging on to a thin rope binding down the tarp under me. Failure to hang on would mean coming off at the next bend or bump. And as I was effectively blind, I could not anticipate when the sledge would try to buck me.

The cold gradually seeped in through all the layers of thermals and fur. My feet were distinctly chilly and my toes started to feel numb, so I had to focus on keeping all my toes moving as well as hanging on, and every now and then I thrashed my legs up and down on the tarpaulin-covered sleigh, to encourage warm blood to flow down to my extremities.

After a few hours, the snowmobile stopped and I raised my goggles again. Darkness had properly fallen by now, and we were in a long, snowy valley with larch forest on each side. I drank some hot water from my thermos. After a few minutes, Piers and his reindeer herder arrived and stopped to check that I was all right. The air was thick with alcohol fumes: Piers' driver seemed to be keeping himself going on the long, cold journey with the help of a couple of bottles of vodka. I got off the sledge and sat on the back of the snowmobile instead, and prepared myself for the final leg, even though we were still apparently at least two hours away from the camp. I had taken my gloves off for about two minutes to find things in my stiff, frozen rucksack, and my fingers were already hurting. I wriggled them inside my mitts, where I had also stuffed a small bag of hand-warmer granules, and my fingers started to sting and warm up.

Back in that dark grey space behind the iced-up goggles, I felt myself drifting into a strange internal world, as though one part of my brain was looking after the physical challenge of holding on while another part let me revisit earlier parts of my journey. In my head I was examining enormous, dusty bones in the vaults of the mammoth museum in St Petersburg, and drifting around rooms of French Impressionist, Flemish and Italian Renaissance paintings in the Hermitage. I thought about family and friends back home, and my garden where daffodils and primroses were coming into flower. I felt a very, very long way away.

We seemed to be going up and down over hills and very bumpy ground. Other snowmobiles – carrying more herders returning from the festival – would catch up and then overtake us; their headlights would turn my world a lighter grey as they approached, then it would go dark again when they passed. A couple more hours and we stopped. I hoped we had reached the camp, but when I pushed up my goggles we were still in the snowy woods. My driver had stopped for a cigarette. My hands and feet were very, very cold by now. I swapped my gloves for enormous boxing-glove-like down mittens. It would be harder to hold on, but my fingers would be less cold. Before we set off again, I looked up at the sky and saw skeins of bright light flowing and dancing across the backdrop of the stars. It was stunningly beautiful.

The last half-hour of that journey was the longest I have ever experienced, and in that time I started to despair. It was too cold, too far, and there was no way out of this situation. I couldn't put my hand up and say, 'All right, I've had enough now – take me home.' I was very tired and very cold and my entire being was simply focused on getting through, minute by minute. My fingers and toes were numb, and I had started to shiver. I had reached the limit of my protection against the elements. There was nothing else I could put on. A polar expert had once advised me never to go further than one kilometre away from 'base' in the Arctic without some means of warming up: at the very least a sleeping bag, ideally a tent and stove. I had none of these, and I was effectively alone in the darkness; I couldn't talk to my driver, and I was separated from Piers and Anatoly. I just had to hold on to the hope that we would arrive at the camp very soon.

We came to a sudden halt in darkness and silence. I pushed my goggles up: we had reached the camp. I stumbled into a tent where a recently lit stove was emitting a very dim light and a little heat. I had just endured the most extreme experience of my life. I understood the fragility of existence in such a harsh environment. I had never been so grateful for shelter and warmth.

The journey had taken six hours. For another two hours it was difficult to find out where I was meant to be sleeping – and then another Marina arrived: Marina Nikolaeva. This formidable woman ran the reindeer herders around Olenek. People suddenly started scuttling around to sort things out, then Marina found a tent for me, made sure a stove was lit, and I gratefully went off to bed, huddling into my down sleeping bag. The stove went out in the night, and I awoke a few times with ice on my eyelashes, and around the edges of my sleeping bag where the moisture in my breath had condensed and frozen. Each time, I pulled the cords of my sleeping bag tighter and snuggled further down.

I woke up in an orange tent, glowing with the sunshine outside but icy cold. Marina's husband came in to light the stove, and I pulled on layers in preparation for emerging into the outdoors. When I did, the nightmarish dark wood of the night had gone and I was in a clearing in a sparse larch forest, in bright snow and brilliant sunshine. It was still less than –20 degrees, but the sun felt warm on my bare face. A thousand-strong herd of reindeer moved around in the forest close to the camp. Several reindeer for riding – *uchakhs* – were tied up in the camp itself. Men were sawing wood for fires and tinkering with sledges and snowmobiles; children were riding reindeer round the camp.

That afternoon I rode on a sledge pulled by two reindeer, with the camp leader, or brigadier, Vasily Stepanov. He had loaded his rifle and wedged it under a reindeer skin on the back of the sledge, and I perched quite nervously on top of it. We moved off down the hill from the camp, at the front of a small caravan of reindeer sledges, and went looking for traces of wild reindeer. Before long, Vasily halted the caravan and pointed out the mashed-up snow where wild reindeer had passed. As wild and domestic reindeer are essentially the same animals, I presumed he knew that these were wild tracks simply because we were far from the camp and his own herd. There were other tracks in the snow as well: Arctic hare and larger prints that may have been wolverine, as well as trails of birds' feet, only just sunk into the top of the snow. (Stepping off the sledge, which formed a track about a foot deep in the snow, I plunged down to three or four feet.) A small flock of five white ptarmigan had flown off up the hill as we had stopped.

There were elements of the Evenki lifestyle that were still close to those of their hunter-gathering ancestors: they were still nomadic, and still hunting. But there were also modifications, some very recent. The Evenki were not just hunters of reindeer; they were reindeer herders as well. It is thought that domestication of reindeer happened relatively recently, perhaps in the last 3000 years.[1] So why continue hunting reindeer if you already have domestic animals? The answer to this question seemed to lie in the fact that the impetus driving the domestication of reindeer was not to secure a food source, but rather to obtain a more efficient way of hunting food: the ancestors of the reindeer herders domesticated reindeer to ride them, in order to hunt wild reindeer. Piers thought this was probably the only example of a species being domesticated in order to hunt wild members of the same species.

Siberia became part of Russia in the seventeenth century. The Russians had moved in along the great rivers, building wooden forts as they went, suppressing the indigenous tribes and taking the whole of Siberia within a generation. Traditionally, domestic reindeer had been used for transport, milk, and as decoys to attract wild reindeer into ambushes,[2] but now the herds were also used as livestock, although hunting of wild reindeer persisted.

The herding and hunting tradition went through the mill of Sovietisation in the twentieth century, when northern Siberia was seen as a huge meat farm. Indigenous reindeer herders were compelled to become large-scale meat producers: herds were expanded and clusters of nomadic ranges were incorporated into enormous 'farms' producing reindeer meat for industrial mining towns in the far north. Starting in the 1930s, villages, like Olenek, were set up to process the meat and to house the women and children. The reindeer herders had operated in family groups, but these were broken up according to Soviet principles: the men, as Soviet workers, would run the camps, as working collectives, or 'brigades' (of which there were three around Olenek), while women and children would live in the village. Some families had always tried to stay together, though that still meant that the women and children would be out in the camps during the school holidays and back in the village for the winter. I was fortunate to visit the Evenki camp in spring holiday time, when the camp felt very family orientated. The main family in our camp was the Stepanov family, of whom Vasily was the patriarch and brigadier.

In the post-Soviet era the need for reindeer meat had diminished as the industrial towns closed down and demand dissipated, but now Siberia was being mined for another commodity: diamonds. Diamond money meant new schools and community works in Olenek, and a renewed demand for reindeer meat on a commercial scale. The Stepanovs' thousand-strong herd of reindeer was there to provide meat for sale and reindeer for riding, but reindeer meat for the camp itself was still mostly obtained from hunting wild reindeer. In fact, there had been a local decline in domestic reindeer numbers so that hunting of wild reindeer was actually on the increase. The balance between pastoralism and hunting

Siberian reindeer 08

had shifted back a little, but wild reindeer were a force to be reckoned with: the population was booming, with a huge wild herd building up on the Taimyr Peninsula to the east. Wild herds would occasionally sweep in and 'abduct' domestic reindeer. Spending time around the reindeer in the Stepanov herd, this was entirely understandable; the deer for riding were quite tame, but most of the herd seemed very much like wild animals, as though they could quite easily revert to that state in fullness, at any time.

Piers had spent the last twenty years studying Anatoly's people, the Eveny, a sister group to the Evenki, both indigenous tribes of Siberia and the Altai. All of these tribes in northern Asia shared a common pattern of subsistence, being nomadic or semi-nomadic hunters and herders. Some of the tribes, like the neighbouring Yakuts, were traditionally semi-nomadic horse and cattle breeders and spoke a Turkic language. The Yakuts, with their southern Siberian cattle-breeding style of subsistence and Turkic language, are probably the remnants of a recent northwards migration of a southern Siberian steppe population, impelled to move north as the Mongol empire expanded in the thirteenth to fifteenth centuries AD.[3, 4]

The Eveny and Evenki people traditionally spoke Tungusic languages (although most we spoke to were now using the more widespread Yakut language), and were herders and hunters of reindeer. Genetic analyses have shown that indigenous Asian and American populations are a separate branch from European and African populations, and, furthermore, that the Asian population has two main branches: one grouping Siberian populations (including the Evenki) with indigenous Americans, and the other relating to people in Central and South-East Asia. These two branches seem to have diverged between 21,000 and 24,000 years ago, although this seems too late to reflect the earliest colonisation of north and Central Asia.[5]

Having spent so much time living with reindeer herders in Siberia, Piers was able to tell me that I shouldn't ask Vasily about his hopes for the hunt. And in the end it seemed that the wild reindeer were too far away, and we were too late to be able to catch up with them that day, so we all returned to camp.

When a hunt was unsuccessful, meat could be obtained from the domestic herd. I found this quite difficult. There was an incredible sense of life and death in this place. The reindeer were such beautiful creatures that it seemed awful to consider killing them. But of course I knew that was exactly what the Evenki depended on for survival in this climate. This dreadful battle between surviving and dying, respect and killing, was deeply embedded in their culture. The Evenki revered reindeer, especially wild ones. Later that evening, Vasily described to me some of the rituals associated with the hunt.

'Before going on a hunt, I always feed the fire and feed the spirit of the locality, before setting out. When a hunt has been successful, I give some of the meat to the spirits and after we finish the deer, all the bones, and the head and the antlers go up on a platform.'

The bones would be placed up high, facing east to ensure reincarnation, like a smaller version of the sky burial for dead humans.

'The animals should be reborn again and again, so that our children and grandchildren have enough to eat.'

The ethics and morals attached to hunting extended to other members of the group, as well as to the animals being hunted. There was an element of karma, of 'what goes around, comes around' as well. Marina said, 'You must be respectful to nature, or you won't be successful next time.

You should also always share your catch with people close to you, or people who don't have much. Whoever does this is guaranteed success [in the hunt].'

I was well aware that my survival on the sleigh ride through the night, and the integrity of my fingers and toes, had depended on my reindeer coat, hood, mitts and – perhaps most importantly – boots, and the generosity of the Evenki villagers who had lent me these clothes. Here I was: a vegetarian who would never buy fur, standing in the snow in Siberia dressed almost entirely, from head to foot, in the skin of a dead animal. But I was borrowing these clothes, and the deer had not been killed for me. Neither was the one that was about to meet its end. I walked over to the opposite end of the camp while a reindeer was lassoed, gently pressed to the ground, then stabbed swiftly through the heart.

I waited, then went over just as the Evenki men were starting to expertly skin the deer. Its eyes were still bright. But very quickly, it was changed from dead animal into a pile of meat: gutted and jointed with surgical precision. Anatoly took a metal mug and dipped it into the cavity of the reindeer's abdomen, bringing up a cupful of fresh, warm reindeer blood to his lips. He offered it to me; I refused as politely as I could.

The skin was laid out, fur down, on the snow. I noticed strange, bean-sized objects stuck to the inside of the hide around the shoulders; Piers later told me that these were the grubs of warbleflies, which lay their eggs under reindeers' skin. Later that afternoon, I watched three Evenki women, Valya, Tanya and Zoya, making reindeer boots from a prepared skin. They chose the fur from the legs of the reindeer to make the upper part of the boot, and sewed together patches from the reindeers' feet to create the soles. The sewing thread was also from the reindeer – long fibres pulled from a dried ligament.

The finished boots were both pieces of art and functionally formidable. It seemed amazing that these relatively simple fabrications could outperform my Baffin boots – but they did. Reindeer fur provides fantastic insulation: there is an outer coat of long guard hairs, and dense under-wool next to the skin. The guard hairs contain cells with such large air spaces within them that they appear to be hollow under the microscope – and this is what makes reindeer fur so effective.[6]

The meat from the slaughtered reindeer was swiftly converted into dinner. Marina was very sanguine about the slaughter.

'Whoever likes the blood can drink it,' she said. 'The liver is also good hot, and you can eat the eyes, which are delicious and good for you. We always eat the raw brains – they're very tasty and healthy, too.'

Meals with the Evenki consisted of a lot of reindeer meat, mostly boiled, with the water forming a sort of fatty, reindeer broth. Sometimes there were also bowls of chopped reindeer fat and, once, a bowl of pieces of frozen reindeer milk. I was lucky in that there was also some bread, small triangles of processed soft cheese in foil wrappers, and bowls of bon bons (which I suspect were laid on for guests). I had brought some packet meals with me, and the Evenki looked on in mild disgust as I poured boiling water into these concoctions. One of the children was brave enough to try some – and the other children ran away from him, screaming.

The meat-rich diet of the Evenki seems somewhat bizarre and unhealthy from a Western perspective, but there's evidence that it's just what the Evenki need in their extreme environment. Our bodies produce heat all the time, as a by-product of metabolism, and the Evenki have been found to have a very high metabolic rate, probably due to high levels of thyroid hormones. A study of thyroid hormone levels in the Evenki suggested that there was a correlation with total energy and protein intake. A high proportion of the Evenki's energy intake came in the form of protein and fat, not surprising given their reindeer-rich diet. It seems that eating a lot – in particular, a lot of *meat* – may spur the thyroid gland into producing more hormones. The result: raised metabolic rate and heat production. It's as though the body is so well supplied with fuel that it can afford to 'waste' some as heat – except that, in northern Siberia, that 'wastefulness' is itself important to survival.[7, 8]

A diet like the Evenki's should set the heart disease alarm bells ringing, but, in spite of their meat-rich diet, the Evenki appear to have paradoxically low levels of 'bad cholesterol' in their blood. There are probably a number of reasons for this, including a genetic predisposition to low cholesterol, a high metabolic rate and a physically active lifestyle, all of which should help to keep 'bad cholesterol' down. Studies of other northern indigenous people have also shown strangely low rates of heart disease; a high consumption of fish containing omega-3 fatty acids may also play a role. However, very sadly, there have been recent reports of rising rates of heart disease in Siberian and Alaskan natives, as they move away from traditional lifestyles. The modern lifestyle diseases of heart disease and diabetes are spreading into the far north.[9]

———

That night, Marina and her two children shared the tent with me. There was plenty of room; we slept close to the edges (but not too close as it would drop to −40 degrees outside during the night), on a raised floor of slim larch logs. The stove occupied an off-centre position. Marina's husband was sleeping somewhere else but he would dutifully come in periodically to check on the stove and replenish the pile of larch logs. We fell asleep in such warmth, with our faces almost roasting in the heat from the stove, that I stripped off most of my layers inside my down sleeping bag, but left them in the bag in case I needed them later in the night. I did.

We awoke in a cold tent. The temperature inside the tents quickly dropped to match the ambient temperature outside as soon as the life-giving stove burned out. But before long Marina's husband had coaxed the stove back to life and I could contemplate emerging from my sleeping bag and getting layered up, ready to step outside. It felt warmer again, but it was hard to tell if this was a real change in external conditions or if I was acclimatising to my new environment. Certainly, I was now out and about in −20 degrees, thinking how pleasantly mild it was, very different from my initial reaction to the same temperature when I had stepped off the plane in Yakutsk. But the Evenki were walking around in far fewer layers than their recently acquired and rather less self-sufficient companion. Many of the snowmobile drivers on the journey to the camp had kept their heads covered with reindeer hoods but their faces had been bare the entire way.

Cold adaptation in humans is a tricky subject. It's very difficult to be sure if what you're looking at in terms of anatomy is an adaptation to, rather than a consequence of, an environment. Short stature and limbs certainly make sense in a cold climate as this reduces the surface area to volume ratio, making it easier to keep body warmth in. But short stature may also be the result of cold stress as the body is growing, in other words, a by-product of cold rather than an adaptation to it. Short limbs, though, may be a true anatomical-physiological adaptation to low environmental temperatures: something which is *inherited* rather than acquired as a child grows.

In the 1960s the anthropologist Carlton Coon and others proposed that facial characteristics such as narrowed eyes, epicanthic folds, small noses and broad, flat faces – i.e. East Asian, or what were then described as 'Mongoloid' features – were specific cold adaptations, protecting the eyes and creating fewer projecting 'corners' to get cold. But at the other end of the landmass, *large* noses are put forward as cold adaptations in Neanderthals and modern Europeans, designed to warm icy air before it's drawn into the lungs. And if East Asian features are cold adaptations, why haven't northern Europeans ended up looking the same? The theory starts to look decidedly shaky.

It seems unlikely to me that the environment could have been such a powerful sculptor of our bodies and faces when a fundamental characteristic of modern humans is the use of culture to buffer ourselves from such pressures. Being able to sew fur together to create protection from the elements would have been essential for the initial colonisation of northern Siberia, as it clearly still was for day-to-day survival in this extreme environment. Looking at the poor little girl who had ridden in on a sledge the same night as me, with an enormous chilblain blister on her right cheek, it was also clear that the Evenki were not immune from the cold. And, beyond intuition and anecdote, various researchers have presented anatomical and physiological evidence to show that East Asian faces cannot be the result of cold adaptation. Steegman published a series of papers along these lines in the sixties and seventies, including a report of a physiological study where he had compared the surface temperature of the face in Japanese and European people, at zero degrees, and found absolutely no difference in the thermal responses;[10] indeed, he wrote, 'If anything, the thin and hawk-like visage of the European is better protected from cold than that of the Asiatic.' Evolutionary biologist Brian Shea[11] looked at the facial anatomy of Eskimos; he suggested that the internal architecture of the nose and sinuses might show some evidence of cold adaptation, but concluded that there was nothing to support the general idea of Asian faces being 'cold-engineered'.

Having eliminated cold adaptation, we are still left with the question of why (or indeed where and when) typically East Asian features arose. I will return to these questions later in this chapter, with the fossil and genetic evidence that I explored in China.

Staying with cold adaptation for a minute, though, there does seem to be some interesting recent research suggesting that there *may* be some adaptive changes in northern populations – not in faces, but deep inside cells. There are very few examples of definite Darwinian or genetic adaptations among modern humans. Sickle cell anaemia and Stephen Oppenheimer's thalassaemia in South-East Asian populations are rare examples, and the links in the chain are understood: the gene(s) responsible, the effect on phenotype (the observable characteristics: in these examples, the effect on blood), and the way in which a mutation confers its selective advantage (protection against malaria infection in these cases).

The boosting of thyroid hormones and metabolic rate discussed earlier is a short-term, physiological mechanism that allows the body to effectively turn excess food into heat, not an example of a Darwinian adaptation to cold. The proposed genetic adaptation to the cold is related to the efficiency of mitochondria. In these minute 'power stations', of which there are thousands per cell, dietary calories are transformed into a package of energy that can be used by the cell (adenosine triphosphate, or ATP). Mitochondrial DNA contains the genes coding for just thirteen proteins, all of which are employed in energy production. Doug Wallace of the University of California, aficionado of all things mitochondrial, has studied how genetic mutations in mtDNA could alter the efficiency of mitochondria. A less efficient system produces less ATP per calorie, and loses energy as heat. So here comes the adaptation: Wallace argues that, in the tropics, mitochondria tend to be very efficient and generate little heat, whereas, in the Arctic, mutations make the mitochondria less efficient, and they produce heat.[12]

So while I was relying on chemical reactions inside bags of hand-warmer granules to keep my fingers going, it seems that the Evenki may have been benefiting from their own internal heat generation. And the short-term physiological response stimulated by the Evenki's meaty diet would have further amplified that effect: thyroid hormones mainly work on mitochondria. Native Siberians have higher metabolic rates than non-natives on a similar diet. Even if I'd eschewed my vegetarianism and eaten reindeer meat, I still wouldn't have been able to compete in the production of metabolic warmth.

It does seem that modern humans are the only hominins to have colonised the Arctic and subarctic regions of the far north. This feat of survival may have depended on a whole range of adaptations, biological, behavioural and cultural, which together made it possible for humans to flourish in Siberia, and heat-generating mitochondria may be among those adaptations. But one other adaptation seems even more important: it was reindeer hunting, providing a meat-rich diet and fur for cold-weather clothing, that paved the way for colonisation of the north.

Back to the Evenki, and the camp was getting busy as the brigade prepared to move house. It was remarkable how quickly a chum could be taken apart and packed up ready to be moved off on reindeer-drawn sledges. Once the stove had been dismantled and taken away, it was time to start on the tent itself. I helped to untie the leather strips holding the reindeer-hide covering on to the larch skeleton of the chum. Tricky little knots involved removing my gloves to work with bare fingers for a few minutes each time before the cold got to my fingertips and I had to plunge them back into my mitts to recover. Nevertheless, the hides came off quickly and we then folded them up and tied them on to sledges. Then the tepee-like framework of larch poles was taken apart. All the poles were simply resting together, apart from the final three, which were tied at the top to create a tripod, which gave the chum its strength. These were toppled and separated, and then all that was left was the floor of larch branches. It was interesting to reflect that there would be no archaeological trace of this temporary dwelling.

The twenty-foot larch poles were tied up and attached to the back of sledges, and we were ready to set off: a caravan of reindeer sledges moving through a snowy landscape. The female leaders of the herd had been lassoed and came with us on the caravan, and as I looked back along the snowy valley my eyes met an extraordinary sight. The *entire herd* was following us, surging like a great wave through the woods. You could hardly separate individuals: it was like a flood of fur, hooves and antlers, flowing down the valley behind us. And this, of course, was the whole point of the nomadic lifestyle of the Evenki: their reindeer needed to migrate, to move on to fresh pastures, however strange it may have seemed to think of the lichen buried beneath the snow as 'pasture'. But as soon as we stopped the herd was busy digging down through the snow with hooves to get at their food. Piers said it was difficult to work out who was leading whom. The reindeer herd periodically felt the urge to migrate; their human companions needed to anticipate this and direct their movement accordingly

At a new site we dug out a circle of snow with wooden spades, and the air was full of the smell of freshly cut larch as small trees were brought over and stripped of their branches to create a floor for the chum. The three foundation poles went up first, and then more poles laid up against them, about twenty in all. Then we unfolded the hides and tied them in place. Reindeer skins were spread on the larch branches inside. Just ten minutes later, we had something that looked like home.

But having nomadised with the Evenki, it was time to continue my own migration. The temperature was dropping by the day and Vasily advised us to travel out that afternoon, so Piers, Anatoly and I packed up and readied ourselves for the snowmobile caravan again. This time, in daylight, it was much warmer than the trip out (this was relative by now: it was still less than −20 degrees), and I even risked going without goggles. Riding, facing backwards, on the sledge, I waved to the Evenki children as we left the camp, and had one last view of the reindeer herd among the

Setting up the chum

trees, and then we were off into the woods and snow. The landscape was beautiful: we rode for a long time in a valley alongside pink cliffs, and I saw rolling snow-clad hills stretching off into the distance. Eventually, the windows of Olenek came into view, shining out in the distance like a cluster of beacons, reflecting the orange light of the setting sun.

Before flying out, I spent one more night in the village, at Marina Stepanova's once again. And just before I left the following morning, the other Marina, head of the reindeer herders, who had travelled back with us, sniffed both my cheeks and gave me the reindeer boots as a gift.

The Riddle of Peking Man: Beijing, China

Having explored the genetic and archaeological evidence for the peopling of North Asia, and having experienced the deathly chill of the taiga for myself, it was time for a change of tack, to track down the first modern human East Asians. But in the Orient, I was going to be walking into one of the biggest controversies in palaeoanthropology, because the prevailing theory in China is that the modern Chinese are descended from *Homo erectus* in China. Chinese palaeoanthropologists claim that the available evidence supports regional continuity: an unbroken line of descent from the archaic humans that made it to East Asia over a million years ago, and they even claim that the features so characteristic of modern Chinese faces are already there in ancient Chinese *Homo erectus* fossils. This stands in direct opposition to the more widely accepted theory of a recent African origin for all anatomically modern humans, across the globe. Social scientist Barry Sautman has argued that the Chinese state has used palaeoanthropology in this way to support racial nationalism and foster a sense of 'Chineseness'.[1]

I left Russia, flying from Yakutsk to the severe, grey, end-of-the-world airport at Vladivostock, and on to Beijing. Once again I was meeting two great figures from the world of palaeoanthropology: one in the present and one from the deep past. I was to become acquainted with Professor Xingzhi Wu and Peking Man.

The argument over Peking Man and the ancestry of the Chinese people has a long history, and progress in the debate has been stymied by a lack of open scientific communication between East and West, the language barrier further cemented by political tensions. Anthropologists from the West have had limited access to both specimens and ideas from China.[2] But things are changing. There are now British and Canadian researchers working away on Chinese dinosaurs in the Institute of Vertebrate Palaeontology and Palaeoanthropology (IVPP) in Beijing, and Chinese researchers are publishing more and more in international journals.

I met Professor Wu at Zhoukoudian, the place where the Peking Man fossils had been found. It felt as if I'd come on some sort of pilgrimage to one of the ancient places. Zhoukoudian is about 50km south-west of Beijing, in a landscape dominated by limestone massifs riddled with fissures and caves. In 1921, a Swedish geologist called John Gunnar Andersson was visiting Zhoukoudian, and a local resident took him to a cave that was reputed to be full of 'dragon bones'. He realised that the bones of the mythical creatures were in fact fossils, and, throughout the 1920s and 1930s, extensive excavations were undertaken at the site by an international team of scientists. Ancient human teeth, and then parts of skulls, started to emerge from the ground, and the hominin was christened *Sinanthropus pekinensis*. The fossils have since been reclassified as *Homo erectus*, and the finds from Zhoukoudian represent the largest sample of specimens of this species from a single locality.

Professor Wu led me through a tunnel-like cave into a deep, steep-sided pit on the side of Dragon-Bone Hill. I was amazed to find out that this pit, which was around 40m in depth, had been entirely dug out as an archaeological excavation. It must have been a mammoth undertaking. What was originally a cave in the side of the hill had filled up with sediments, and the roof had fallen in. As the archaeologists had dug down and down, they had found layers full of limestone blocks, fossils and stone tools. The first skull of Peking Man was found in layer 11, in sediments 20m below the surface.

The skulls of what was then called *Sinanthropus pekinensis* (literally 'Peking China-Man', or, as it was to become known, Peking Man) were studied by the German anatomist Franz Weidenreich, who proposed that they represented ancestors of modern Chinese. To him, the ancient skulls possessed features that linked them with modern populations. Weidenreich took the fragments that had been recovered in the dig at Zhoukoudian, and, with his assistant Lucille Swan, made a reconstruction of an entire skull, which was to become the icon of the newly discovered species.[3] The idea that Chinese people had a lineage stretching back a million years, in China, led Chinese archaeologist Lin Yan to proclaim that the Chinese are 'the earth's most ancient original inhabitants'.[1]

Wu and I walked up, out of the pit of Locality 1, and up on to the north-east slope of Dragon-Bone Hill, where we could look down into the excavated Upper Cave, Shandingdong. In the 1930s, excavations in this area had yielded three well-preserved skulls of modern humans, as well as perforated animal teeth, pebbles and shells. The Upper Cave skulls were also examined and described by Weidenreich.

Then, in late June 1937, Chinese and Japanese troops clashed in the town of Wanping, about thirty miles from Beijing, in what was to become known as the 'Marco Polo Bridge Incident', and effectively the beginning of the Second World War in the East. By the end of July, Beijing had fallen to the Japanese. Excavations at Zhoukoudian ceased, and the precious Peking Man fossils were packed up into crates to be sent for safe-keeping to America. But they never made it. The tragic loss of the Peking Man fossils is one of the great mysteries of palaeoanthropology. There are all sorts of stories about where the fossils may now reside (if indeed they still exist). Suggestions include the fossils having been removed to a museum in Taiwan, sent to the Crimea on a Russian ship, or having been kept in a hospital in Beijing.[4] So, when I said I was to meet Peking Man, I was actually going to see casts of the original fossils.

Professor Wu and I returned to Beijing, to the IVPP, where the casts of Peking Man and the Upper Cave skulls are kept. We entered a room where one wall was lined with lockers from floor to ceiling. A table covered in a deep red cloth stood in the centre of the room. But then we both had to leave the room and stand in the corridor while a security officer removed the specimens from the lockers.

'I am not allowed to see the drawers out of which the skulls come. I will stand back and wait here.' The location of the specimens was kept secret even from Professor Wu, such was the nature of this magnificently elaborate security system.

'He keeps the key so that nobody knows the number,' said Professor Wu.

'So you don't know the number?'

'No. I don't want to know. If I know, and if it is lost, then I have the responsibility. But now I don't know anything. I have no responsibility,' Wu smiled.

While we were waiting out in the corridor, I asked Professor Wu how he had become interested in palaeoanthropology. It turned out that he had qualified as a medical doctor, but that, at the time, China had needed more medical teachers, so he was instructed to become a lecturer in anatomy. Anatomy and palaeoanthropology have always been closely allied professions, and so when excavations restarted at Zhoukoudian in the 1950s, Wu naturally became involved.

He asked me about my background, and I was really pleased to be able to tell him that I too had trained as a medical doctor, then became a lecturer in anatomy, and developed an interest in palaeoanthropology. I felt like an apprentice in the presence of a grand master. At the age of eighty, Professor Wu was still coming to work every day (on his bike) at the IVPP.

Once the specimens were out of the lockers, we could re-enter the room. Six or seven of them now sat, neatly lined up, on the table. There were some modern human skulls, and casts of *Homo erectus*. I recognised the cast of the original Weidenreich/Swan reconstruction but I had also read about a newer reconstruction put together by Ian Tattersall and Gary Sawyer from the American Museum of Natural History, New York, and Professor Wu had arranged for this to be taken out as well. There were also two thick fragments from the top of a skull.

'These are original fossils of Peking Man,' he said. I was taken aback. I knew the story of the missing crates of fossils in the Second World War, and I had not expected to see any *actual* fossils, just casts.

'I honestly thought all the specimens had been lost,' I said.

'After the war, in the 1950s, we carried out some new excavations, and we got some new specimens,' explained Wu.

So I was holding in my hands fragments of a skull of an earlier human who had lived in China some one million years ago. It was quite a strange moment: there was something about the physicality of holding something I knew to be truly ancient. The vast depths of time through which these fossilised bones had passed, and then to end up in the present as some sort of magical talisman that could give us the power to know the past, made me feel almost giddy. A great range of dating techniques has been used to pin down the dates of the layers at Zhoukoudian. Most suggest that the hominin fossils come from layers dating to between 400,000 and 250,000 years ago. But the latest uranium series dates suggest that the fossils may be more than 400,000 years old, and perhaps as much as 800,000 years.[5, 6] I laid the fragments back down on the red cloth. This was Chinese *erectus*, the real thing, with its massive brow and sloping forehead.

Having got over the shock of seeing Peking Man, I turned my attention to the reconstructed casts – and the face of Chinese *Homo erectus*. The Weidenreich/Swan and Tattersall/Sawyer reconstructions were quite different looking. Professor Ian Tattersall had wanted to make a new reconstruction using more fragments from the face than had been used in the Weidenreich model. This had been possible, because, although the majority of original fossil fragments had been lost, casts had survived – and there were fragments of twelve separate skulls to work from. The original reconstruction had been based on just three pieces of skull: the skullcap, part of the right side of the mandible and a fragment of the left maxilla (the bone that bears the upper teeth and forms much of the cheek), and photographs and casts of this reconstruction didn't make it obvious which parts were real and which were 'restored'. Tattersall and Sawyer had wanted to keep artistic licence to a minimum, by using as many fragments of facial bones as possible in their reconstruction.[3]

Professor Wu held up the Weidenreich/ Swan reconstruction and started to point out the features that he

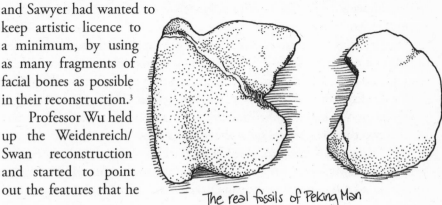

The real fossils of Peking Man

thought to be evidence for regional continuity. He indicated the flatness of the bridge of the nose, and the shape of the edges of the cheekbones. Then he pointed to these features on his own face and mine: it was true that he had a much less pronounced bridge of the nose than I had, and much wider, flatter cheekbones. But I couldn't really see any similarity between Professor Wu's face and that of Peking Man, even though it seemed that the Weidenreich/Swan reconstruction had somehow accentuated these Chinese-looking features. On the Tattersall/Sawyer skull, the nasal bones were more prominent, the face taller and narrower across the cheekbones, and the jaw more protruding. In general, this reconstruction looked less 'Chinese' and more like other *erectus* specimens around the world.[3]

The shape of the front teeth was another feature that Wu considered as evidence of regional continuity. Weidenreich had proposed that shovel-shaped incisors were a regional trait shared by Chinese *Homo erectus* and many modern Chinese. But this tooth shape is also seen in African *erectus* and in Neanderthals, so, while it may be an archaic trait, it is not specifically East Asian. And the shovel shape seen in the teeth of some living Chinese people seems to be developmentally different from the shovelling of the archaic teeth, even though it looks superficially similar – in which case it can't be used to argue for a connection at all.[7]

Then we looked at the modern human specimens, including casts of two of the Upper Cave skulls. Their dates haven't really been pinned down, and range between 10,000 and 30,000 years old, based on radiocarbon dating of animal bones in the same layer. Some archaeologists argue that the Upper Cave skulls might be burials that have been moved around by animals, and so seem older than they should be – preferring the lower estimate of around 10,000 years old.

'There are many common features among them,' said Wu, indicating the Upper Cave skulls and the Peking Man reconstruction. 'I think it is most probable that the Upper Cave man are the descendants of *Homo erectus* man.'

To me, the Upper Cave skulls looked undeniably modern, but they didn't really look East Asian. Neither had the characteristic flattened nasal bones or tilted-out cheekbones. In fact they both looked quite similar to European modern humans: Cro-Magnon man. And the Peking Man skull looked completely different.

Professor Wu had failed to convince me, by showing me the skulls, that there was any evidence here for regional continuity in China. But it's not that easy. It was clear to me that interpretation of the characteristics of these skulls could be very subjective. Professor Wu, who had spent a lifetime studying these skulls, was utterly convinced by the features, which seemed so obvious to him.

Chris Stringer, a prodigious analyser of skulls, has tackled the problem of quantifying differences and similarities between archaic and modern skulls. Rather than describing and counting up 'archaic' and 'modern' features in skulls, he prefers to take measurements and objectively compare differences in skull shape in that way. The regional continuity model as put forward by Weidenreich proposes not only that *Homo erectus* in China is a direct ancestor of modern humans, but also that later archaic fossils represent an intermediate or 'Neanderthal' stage between *erectus* and *sapiens*. Specimens that seem to fit into Weidenreich's 'in-betweener' stage of human evolution in China include fossil skulls from Maba and Dali.[2] The Maba cranium was discovered in 1958 in Guandong Province in southern China, while the Dali cranium was found in 1978, in Shaanxi Province. Uranium series dating of a cow tooth from the site gave a date of around 200,000 years ago for Dali, but it's not clear how close the tooth was to the skull, so this date should be treated with some caution.[5] Maba has been reported to be around 150,000 years old.

Stringer included the Maba and Dali skulls in a study where he took skull measurements from a range of archaic and modern skulls from Africa, Europe and East Asia, and then mathematically compared the 'shape distance' between skulls to show how closely related they might be. He also looked at the three anatomically modern skulls from the Upper Cave at Zhoukoudian. The results showed that the archaic African skulls were the 'best shape ancestors' for modern human skulls, including those from East Asia, and that the Dali and Maba skulls were very different from modern skulls, and so were not convincing 'in-betweeners'.[8] It seems they might represent East Asian populations of *Homo heidelbergensis*, or even Neanderthals, that were later replaced when modern humans arrived.[9]

It is clear that analysing skull shape is a very complicated business. Anatomical features don't just vary from one population (or species) to another: they also vary *within* populations. Just to make things even more complicated, the variation within populations is often greater than that between them.[2, 7] If you're looking for differences between groups, you have to pick your features carefully. Another major problem is that we don't yet understand how different anatomical features of the skull are connected or related to each other. Such connections could really skew the data one way or the other, whether we are just 'counting up' features or indeed taking measurements and trying to do something a bit more objective, as in Stringer's analyses.

These connections are a bit difficult to get your head around (again, forgive the pun), but just imagine that, for instance, eating a really crunchy diet from a young age could produce widespread effects on your skull. For the sake of argument, let's say it would make the corners of your mandible more pronounced, give you a stronger browridge to resist the powerful forces of the chewing muscles, and maybe even affect the whole shape of your skull. (This isn't entirely hypothetical: there's good evidence to suggest that a switch to softer diets in the last thousand years is linked to smaller face size.[10]) Now, if I compared your skull with that of someone who ate much softer food, there would be several features that looked different, all of them linked to diet. If I compared your skull to an ancient fossil, from an early human who also ate hard food, then you might look more like them than your soup-drinking counterpart would. I could count up the features and find at least three where your skull was similar to the fossil, but this wouldn't mean that you were closely related, just that you ate a similarly tough diet.[7]

As well as acquired characteristics which may be linked by some aspect of function, there are likely to be genetic connections between various skull features. There's no way that each tiny feature of your skull is neatly controlled by a separate gene; one gene might affect a whole suite of features in different parts of your head. So again, just counting differences between skulls isn't going to give you a real idea of how closely, genetically related two populations or species are.

The genetics of morphology is hugely complicated, and an area with which researchers are really only just starting to get to grips. Genes don't operate independently; they work as a team, with proteins sticking their oar in as well. Geneticists can look at whole genomes now, but it is like having a book in a foreign language (in this case, spelt 'AGTCTGTTAATCCGG' etc ...), where we are only just beginning to understand what a few of the words mean. Some of them are about chemistry inside cells, but others dictate anatomy. Somehow, those gene-words create a conversation telling one fertilised cell to multiply and change and multiply and change until an adult human being is produced. Picking apart the complex tapestry of development, and finding out which genes are responsible for each motif, is a hugely exciting area of research in the twenty-first century.

As many features of skull shape may be tied together by function or by genes, it means that huge, inclusive lists of features linking modern and archaic skulls (or not) are misconceived. It explains how it's possible for such long-list-making studies to have been used on both sides of the argument, both for and against the regional continuity. This doesn't mean we should give up on using morphology to help work out our evolutionary past, but we do have to be careful about how we go about it. And at the moment, while we're still finding out how morphological features are related through function and genetics, all we can do is try not to use sets of features that seem to have a tendency to occur together.[2]

Of course, while the relationship between genes, function and morphology is still being elucidated, the genes of living people offer us another powerful tool for reconstructing human lineages and migrations. I asked Professor Wu what he thought about genetic studies that suggested all modern humans had a recent African origin. He was quite sceptical about the potential for building evolutionary trees based on genes, and especially about the power of genetic studies to predict dates of divergence of lineages. 'Different genetic studies even disagree on the age of the last common ancestor,' he said. 'Putting a date on it using the molecular clock assumes a constant mutation rate, which we can't be sure about.'

It's true that different genetic studies have produced different predictions of age of a last common ancestor – but most agree that the date lies between 100,000 and 200,000 years ago.[11]

It was clear that Professor Wu had much more faith in what he considered to be the hard – fossil – evidence, and he was absolutely sure that the fossils pointed to regional continuity. For him, as for Alan Thorne, *Homo erectus* and *Homo sapiens* were not even separate species, but subspecies. Wu preferred the labels *Homo sapiens erectus* and *Homo sapiens sapiens*. He argued that one form had gradually changed into the other over time, without speciation, and that the unity of the species across the world was maintained through gene flow between populations. Wu himself had proposed this theory, of 'continuity with hybridisation'.

But there does seem to be a significant temporal gap between archaic and modern human fossils in China. The most recent archaic fossils, from Xujiayo, date to around 100,000–125,000 years ago. The oldest well-dated modern human remains in China, including a mandible and limb bones, were discovered in Tianyuan Cave, about 6km away from the main site at Zhoukoudian. AMS radiocarbon dating placed these fossils at 39,000–42,000 years old.[12] The next oldest modern human remains in the Far East are some leg bones from Yamashita-cho, Okinawa, dated to around 32,000 radiocarbon thousand years old (about 37,000 calendar years), and the Upper Cave skulls, at around 10,000 to (at a push) 30,000 years old. And, on balance, it seems that most investigators believe that the gap between archaic and modern Chinese fossils is not only temporal, but also morphological and genetic. Having seen the fossils and casts of Peking Man, I was not at all convinced that I had seen the ancestors of the Chinese.

But there was something else in China that Wu believed supported his theory of regional continuity: stone tools. And here, I had to admit, he had a point. In Europe, the arrival of modern humans on the scene was marked by a clearly new 'archaeological signature', a sudden change in stone tool technology, with the appearance of the Upper Palaeolithic. In the East, modern humans seem to have been around for a long time before a distinct toolkit appears.

An Archaeological Puzzle: Zhujiatun, China

From around 1,000,000 to 30,000 years ago, the stone tools of East Asia are predominantly fairly crude, Oldowan-style pebble and flake tools. The archaeology of this part of the world is particularly strange, when viewed from a European perspective. The Acheulean, with its classic hand axes, doesn't really happen, and neither does the Middle Palaeolithic.[1, 2] People just seem to carry on using really basic pebble tools that even Neanderthals would have considered crude. In 1955, this moved American archaeologist Hallam Movius deprecatingly to call the East 'a marginal area of cultural retardation'.[3]

It's not until 30,000 years ago, in the late Upper Pleistocene, that the Upper Palaeolithic appears in China, with more sophisticated tools like end-scrapers, burins, blades and microblades, as well as tools made from bone and antler. But that transition happened some 20,000 to 30,000 years after genetic estimates for the arrival of modern humans in East Asia, and a good 10,000 years after the first fossil evidence of modern humans in China.[4] Before then, the tools the modern humans were making were no different from the tools made by earlier archaic humans in the East. It was this persistence of pebble tools – the so-called 'chopper-chopping' technology – that Wu interpreted as archaeological evidence for regional continuity. If all we had to look at were the stone tools, then regional continuity does seem like a reasonable explanation.

But if the balance of evidence favours a recent African origin for modern humans, and if we take those dates from the genetic and fossil record, suggesting a colonisation of South-East and East Asia by (presumably ingenious and adaptable) modern humans between 40,000 and 60,000 years ago, then why were they putting so little thought into their tools? Were they really culturally retarded, or had Movius missed the real technology of Palaeolithic East Asia?

Palaeolithic archaeology is a difficult area just because so little is left behind. As I had seen in Siberia, it was possible to erect a comfortable home, live in it, then move on without leaving any clue for future archaeologists to uncover. Archaeological clues in prehistory are rare and precious. As for tools, anything made out of something we would now consider as biodegradable – wood, other plant materials, animal skin – is by its very nature going to disappear from the archaeological record.

Some archaeologists believe that the riddle of stone tools in East Asia could be explained by something that is both biodegradable and ubiquitous in the region. I met up with Australian archaeologist Jo Kamminga and we made our way to the small village of Zhujiatun in southern China, to find out more about this theory.

Jo had armed himself with a selection of typical pebble and flake tools. I had a look at one of them.

'This is a fairly crude tool,' I observed.

'Well, this is what we find in China, and South-East Asia, and in Australia as well. It's not beautifully shaped, but it's incredibly sharp,' said Jo.

'But at the end of the Palaeolithic in Europe, people are making really sophisticated stone tools – so what's going on here?'

'Well, we're in a different world here. Firstly, you don't have the big cobbles of flint that are found in the chalk and limestone areas of Europe. The chert here comes in smaller nodules, and the material is not so nice to work. You can't make those fine points. But the most important thing is that we have a different climate and vegetation here. You have a different range of raw materials available: not only stone, but also bamboo. It's a very resilient material and it can be used to make the sorts of things you might be using stone for in other places.'

So perhaps those most basic of pebble tools are just enough to make more sophisticated tools out of plant material, and, perhaps, out of that most prolific subfamily of grasses in East Asia: bamboo. Bamboo is used so widely in the East today that it

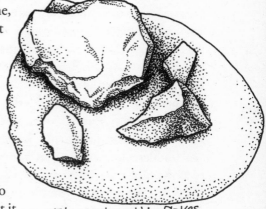

Jo's crude pebble flakes

is easy to imagine it would have been seized upon by the first colonisers. But there would be no trace of it left in the ground, only that of the stone tools which had been used to shape it. Although Movius had concentrated on pebbles from which flakes had been struck, leaving a sharp edge on a heavy tool, perhaps he had missed the point. Certainly, the flaked pebble could be used as a crude 'chopper', but the 'waste flakes' also had useful cutting edges.

'Why would you go to all that trouble of making a sophisticated stone tool,' asked Jo, rhetorically, 'when you can just take a piece of bamboo, use that as a knife – and throw it away when you've finished? Because it's everywhere.'

He was right. Zhujiatun was surrounded by bamboo forest. From a distance, the hillsides looked feathery: the wind blew through the bamboo leaves as through a field of corn. Jo and I were going to try a bit of experimental archaeology. We used a large, very crudely sharpened cobble to bash at the base of a bamboo trunk. The bamboo was thick, about 15cm in diameter, and I prepared for some hard physical work to fell it. But after just a few minutes of bashing, the bamboo fell. With a little twisting, it ripped away from its base, and we had the raw material we needed to try making some 'Palaeolithic' bamboo tools.

We took our length of bamboo down to the village, where this material was still being used to make all manner of things. There were piles of long, thick bamboo trunks lying around, ready for construction. We were invited into a house in which there were piles of bamboo baskets stacked up, and an old man was in the process of weaving one out of long, bendy strips of bamboo epithelium, or 'bark'. In the yard, small ducklings had nestled together under a bamboo cage.

Jo and I set to work on our bamboo tools. Using very basic, unretouched stone flakes, we pared down slivers of bamboo and quickly made sharp-edged 'knives'. I was surprised at how quick and easy the process was, though sceptical about just how sharp and strong my bamboo knife would prove to be. The family had a chicken carcass ready for supper, and, using my brand new bamboo knife, I made short work of butchering it, separating drumsticks, wings and breasts from the carcass (I may be a vegetarian but I'm also an anatomist). The use of bamboo knives has been documented right across the Pacific. Ethnographic studies have also shown that bamboo may even be used in preference to stone tools: the stone-adze-makers of Irian Jaya in New Guinea, for example, like to use a simple piece of split bamboo for butchering.[5] And although my green bamboo knife had done the job well, Jo said it would have been even sharper had it been dried bamboo.

It was clear that bamboo could be used to make excellent cutting tools as well as being suitable for houses, baskets – and even, as I had already seen, rafts. It was a wonderfully versatile material. (You could eat it, too. Bamboo became one of my favourite dishes on my visit to China, not as the more familiar, delicate white bamboo shoots, but as inch-long pieces of thick shoots, rather like asparagus but crunchier – and delicious.) However, the fact that bamboo *could* be used to make efficient tools isn't proof that it was used. Neither is the fact that bamboo is still being used in some places to make tools.

chickens in Zhujiatun

So was there any archaeological evidence of bamboo tools?

'Well, no, not of the bamboo itself,' admitted Jo. 'But there is evidence of bamboo being *worked* – as well as other materials like rattan and palm wood – because there's a high silica content in these materials. Using a stone tool on bamboo leaves a very distinctive polish on that tool.'

This polish could be seen with the naked eye; Jo had brought along with him some stone flakes that he had used to cut bamboo, and I could see the polishing quite clearly, next to the cutting edge.

'How long does it take for this type of polish to develop?' I asked him.

'It starts to form immediately,' he replied. 'You can see the polish appearing within the first few minutes of cutting bamboo or rattan.'

Employing microscopic use-wear analysis, it was actually possible to find out *exactly* what had been cut: different materials leave different tell-tale marks on stone tools.[6] Under a light or electron microscope, minute abrasions and polishing on tool edges provide clues that allow the material that was cut to be identified. This means that archaeologists should be able to tell if stone tools have been used to cut into bamboo, or rattan, or other materials. It may also be possible to find evidence of bamboo tool use as well. A study of experimentally produced cut marks on bone showed that it was possible to tell the difference between cuts made with bamboo knives or stone flakes, using scanning electron microscopy, which produces a very detailed, 3D image.[3]

Jo had found examples of polish from cutting rattan on archaeological stone tools from a rockshelter in Timor. But those tools had been only a few thousand years old.

'What I'd really like to do is look at the very early, Chinese stone tools,' said Jo.

As more Asian Palaeolithic sites are discovered, it is clear that the toolkits are more variable than was apparent to Movius in the 1940s, when he drew a simple distinction between Acheulean hand axes in the West and chopper-chopping tools in the East. His scheme also lifted the tools out of their environmental context: it didn't take into account what the tools may have been used for, or the raw materials that were available to the toolmakers. Archaeologists now caution against ranking various toolkits in terms of how sophisticated the tools may appear, and also argue that Palaeolithic toolkits are generally more diverse than previously thought, reflecting intelligent adaptations to a variety of environments.[5]

However, even with new discoveries and the emerging impression of a wider variety of stone tools in East Asia, there is still a clear distinction between the tools of East and West. Bamboo technology provides a tempting explanation for that difference. Bamboo was everywhere, whereas it may have been difficult to find sources of good quality, workable stone in the rainforest. Modern rainforest-based hunter-gatherers are mostly vegetarian and occasionally eat small animals – which would have been easily butchered with bamboo knives, just as I had managed to prepare the chicken in the village. Heavy-duty butchery tools would have been redundant in the rainforest.[3] I had seen how quick and easy it was to make a good, sharp bamboo knife, using just a simple stone flake. Bamboo technology seemed as if it would have been a sensible, expedient adaptation to rainforest environments.

Although bamboo itself may be missing from the archaeological record, microscopic use-wear analysis of stone tools and examination of cut marks

young bamboo

on bones now provides archaeologists with an opportunity to test this theory. It may not be long before we know for sure whether bamboo use really can explain the simplicity of stone tools in East Asia.

However, we are still left with a problem: there is no clear difference between the archaeology associated with archaic humans and early modern humans in East Asia. Surely we should find some kind of 'signal of modernity' as soon as modern humans arrive on the scene? Shouldn't modern humans carry with them some badge of superior intelligence, a mark of technological sophistication that has been lacking from previous human incarnations?

But are we once again falling into the trap of divorcing stone tools from their environmental context? We should not even start to think about a tool assemblage without first firmly placing our toolmakers in their environment. The suggested bamboo explanation for the basic stone tools of East Asia could work for archaic human populations just as it does for modern ones: an intelligent solution to tool-making in a rainforest environment, where bamboo abounds.

So why *do* the tools change at around 30,000 years? If it's not a different toolmaker, what is it? What is changing at that time? The answer is: climate. And at 30,000 years ago, East Asia was starting to feel the cold and dryness of the approaching LGM. And now we really do see the signature of modern human behaviour: when the environment changed around them, the modern humans in East Asia adapted, by inventing new technologies.

East Asian Genes to the Rescue: Shanghai, China

I wanted to find out just what Chinese genes revealed about regional continuity versus a recent African origin. And I wanted to speak to a Chinese geneticist.

It was time to leave Beijing and move on to China's commercial capital, Shanghai. Beijing had seemed grey and somehow very stolid. It felt like a place where dogma would have very deep foundations. On first impressions alone, Shanghai felt more progressive, cosmopolitan and open. The centre of the city was an architectural palimpsest, with pre-war art deco hotels standing next to concrete high-rises, with ugly flyovers and an elegant museum shaped like an ancient bronze cauldron. On the main streets, international brands jostled for room alongside home-grown shops, and gigantic screens mounted high on buildings displayed streams of advertisements. There was even a huge screen being carried up and down the Huangpu River on a boat. From the Bund, I watched the Pudong district lighting up at dusk, as though it was, itself, a huge advertisement for commercialism and capitalism. China was changing.

I drove out to Fudan University, to the Institute of Genetics, to meet Professor Jin Li. He showed me around labs where teams of enthusiastic post-doctoral researchers were busy with pipettes and centrifuges. It was out of these labs, some seven years earlier, that compelling genetic evidence about the origin of the Chinese people had emerged.

Li's research group had undertaken a massive project designed to test the competing hypotheses about the origins of East Asians.

'I wanted to see if I could find evidence for regional continuity in the genes of Chinese,' he explained. 'I decided to look at the Y chromosome, and I started off using a marker of recent African origin, which would allow me to filter out those individuals and leave me with other lineages that may have survived locally through regional continuity. We took thousands of DNA samples from people all over China.'

So Li had started off wanting to prove the patriotic theory that the modern Chinese had a heritage in China which stretched back, unbroken, to *Homo erectus*, a million years ago. The genetic marker he had used was a mutation at a site on the Y chromosome called M168, a swap of a cytosine base to thymine, which previous studies had suggested was present in all non-African populations. But previous studies had taken only limited samples from Asia. Li's group had collected DNA samples from more than 12,000 men from South-East Asia, Oceania, East Asia, Siberia and Central Asia, with the idea that, somewhere among them, there would be much more ancient, non-African Y chromosomes.[1]

'We knew that the old type of M168 mutated to the new type about 80,000 years ago, in Africa. So if the Out of Africa hypothesis was true, we would expect that everybody in China would be carrying the new type of M186. But if there had been an independent origin of modern humans in China, we should be able to see that at least some people were carrying the old type.'

'And what did you find?'

'We did not see any old type in the Chinese population. In fact, we had a very large sample covering almost every corner of East Asia, and *everybody* was carrying the new type of M168.'

The ubiquity of the M168 mutation in Chinese DNA showed that modern humans emerging from Africa had completely replaced earlier East Asian populations.[1] Peking Man had no descendants alive today.

'And what did that result mean to you?' I asked.

'Well, as a Chinese, of course I wanted to find evidence that we have ancient roots in China. That was my education,' said Jin. 'It is what we are all taught. But, as a scientist, I have to accept the evidence. And the evidence showed that the recent Out of Africa hypothesis is right. Regional continuity can't be true.'

'Do you think, on balance, that other genetic evidence supports Out of Africa?' I asked him.

He was categorical in his reply: 'I would make a stronger statement: it exclusively supports the Out of Africa hypothesis.'

Genetics also offered some insight into *how* the East had become colonised by modern humans. A greater diversity of Y chromosomes in the south, including among Thais and Cambodians, tells the tale of initial colonisation of South-East Asia, followed by a northwards migration. And the Y chromosome tree suggests a very broad range for the initial date of entry into East Asia, at some time between 25,000 and 60,000 years ago.[2] Mitochondrial DNA analyses also show greater diversity in the south, and support the main theme of a south-to-north colonisation of the Far East. The four main haplogroups in East Asia (B, M7, F and R) are all about 50,000 years old.[3]

It is quite extraordinary that, despite the thousands of years since the initial colonisation of the East, and all the population movements that have occurred since then, it is still possible to look into the genes of living East Asians and find the clues to where they first came from. Like a piece of parchment that has been written over and over, the faint traces of the original story are still there. Analysis of complete mtDNA sequences shows a distinction between northern and southern East Asians, but more than that – a geographic structure can be discerned in the boughs, branches and twigs of the East Asian mtDNA tree.[4]

But as well as the mitochondrial genetic evidence for a general migration from south to north, there are lineages in northern East Asia – particularly C and Z – that are missing in the south. So where have these come from? Stephen Oppenheimer[5] traces these lineages back to India, to early Asians who had skirted the western end of the Himalayas to reach the Russian Altai and populate Siberia between 40,000 and 50,000 years ago, leaving archaeological traces at places like Kara-Bom. The Y chromosome evidence mirrors the mtDNA phylogeography, with the North Asian founder population splitting east and west. Some headed west to Europe, while the eastward-bound colonisers continued following the mammoth steppe all the way into what is now northern China.

Some remarkable details emerge from the complicated, over-written palimpsest: among the Ainu of Japan, a mitochondrial haplogroup called Y1 seems to record a specific migration from north-east Siberia into the northern Japanese islands.[6] And the north-east Asian populations share lineages, like C, with Native Americans – but that is the subject of another chapter entirely.

So it seemed that the East Asians were indeed descendants of the South-East Asian beachcombers, except for a few lineages that had made their way east from Central and northern Asia. Also on the Y chromosome, the M130 marker seems to record a migration along the South-East Asian coast, turning north to Japan, and there are a scattering of archaeological sites in Korea and Japan dating to around 37,000 to 40,000 years ago that may record this wave of colonisation.[7]

I wondered if Jin Li had any thoughts on the origin of East Asian features, and whether genetics could yet shed any light on the development of these specific facial characteristics. Although he thought they may have first arisen in South-East Asians, perhaps around the LGM, he was also sceptical about any association with cold adaptation. Another hypothesis puts the origin and spread of East Asian features down to expansion of Neolithic populations with the advent of rice farming, but that didn't seem to fit with the suggested timing of the mutations either.

Li Jin wanted to pin down the relationship between genes and facial morphology – something that might help him answer where and how East Asian features arose and spread.

'We don't know what features are determined by which genes. We are just starting to try to identify the genes underlying morphological variation, and then we should be able to tell when exactly these features developed as well.'

He was about to embark on what sounded like an extraordinarily ambitious project: to relate genetics to morphology by collecting anthropomorphic data from living people – effectively measuring them up – and using whole genome sequencing to look for genes or patterns of genes that seemed to be associated with particular features.

'We're looking at a thousand people, recording their morphological features – and we're in the process of doing whole genome scanning.'

This was just the sort of research that would start to fill in that vast gap in our understanding, building a bridge between genetics and morphology. And there were already some results …

'We already know which genes underlie the orientation of hair whorls,' said Jin grandly, but with a wry smile.

It was clear that Jin Li was hugely excited by the potential of genetics to delve into the deep past and tackle the questions of origins of modern humans and modern Chinese. I was incredibly impressed by his open-mindedness and objectivity: surely the mark of a true scientist. And it was clear that the scientists at Fudan University were operating in a culture of academic freedom. In a country where regional continuity was still a 'fact' taught to schoolchildren and endorsed by the state, Li had been able to publish evidence for a recent African origin of East Asians.

We walked out of the Institute of Genetics and across a garden dominated by a large statue of Chairman Mao. It seemed ironic. The poorly proportioned figure had been erected by students and Red Guards in 1966, but he was presiding over a very different cultural revolution now. Academic and individual freedom seemed to be winning back some ground.

Pottery and Rice: Guilin and Long Ji, China

Throughout prehistory, human populations have contracted and expanded, pushing into new territories and then withdrawing, largely under the influence of climate change. But three major episodes of Stone Age 'migrations' or population movements of modern humans can be discerned amid the general oscillations and milling about: the initial spread across and out of Africa, resettlement of great areas of the northern hemisphere after the end of the Ice Age, and the spread of expanding populations after the invention of farming.[1]

The 'Neolithic revolution' and the origin of agriculture in the East was independent of that in Europe. Just as in the West, the East Asian farmers were more successful than hunter-gatherers at a population level (if not really at the level of the health and longevity of individuals). Higher levels of food production supported growing populations, and agriculturalists spread out of their homeland, carrying their languages and lifestyles with them. Indeed, some archaeologists have argued that the facial – and dental – characteristics of East Asians are so recent that they reflect that dispersal of a quickly expanding population of rice farmers as the Neolithic got under way in the East.[2,3]

It seems that the 'Neolithic package' of settlement, farming and pottery didn't just suddenly spring into existence: the elements we recognise as being characteristic of the Neolithic way of life emerged in a mosaic fashion. In the East, one of the first elements to appear was pottery, pre-dating the development of farming. From the 1960s onwards, it was thought that the earliest pots in the world belonged to the Jomon culture of Japan, dating to nearly 13,000 years ago. But new archaeological evidence suggests that pottery may also have appeared at this time, and independently, in the Russian Far East and in south China.[4]

In the late Pleistocene and early Holocene of south China, between 14,000 and 9,000 years ago, a culture appears which is characterised by grinding stones, shell and bone tools – and the earliest pottery. Some archaeologists have called it 'Mesolithic', comparing it with the same period in Europe; others prefer 'pre-Neolithic'. I visited Guilin, in Guangxi Province, to see the oldest known pot in China.

The landscape around Guilin was quite striking and beautiful. Out of the fertile plains rose huge, wooded, karst outcrops. Some were cone-shaped, others more rounded. Guilin City itself was spread out on the flat ground among the karst hills, and along the Li Jiang (Willow River). I was visiting soon after Qingming (pronounced 'chingming'), the annual Chinese Festival of the Dead, when tombs are swept clean, inscriptions touched up and new pine trees planted in cemeteries. Qingming literally means 'clear and bright'. I drove past graveyards with cairn-like tombs adorned with bright strips of red paper – prayers for protection – and in one village a funeral procession was making its way down the road, with a banging of cymbals and drums. The coffin was brightly decorated, topped off with a giant purple moth.

But I was visiting a much older graveyard: the cave site of Zengpiyan in Guilin, where I met the deputy director of the museum, Mr Wei Jun. He opened the iron gate protecting the cave and we stepped inside.

Archaeologists had dug down through river sediments and the trenches were still open. Eighteen burials had been discovered in the cave, mostly in flexed positions, and some covered in red ochre. As well as the human burials, Wei and his colleagues had found pebble tools and animal bones.

'But the most important discovery was the pottery we found here in 2001,' said Wei.

'It happened on the morning of 7 July. I recall it was raining very hard outside. There were about seven or eight archaeologists working inside the cave. Then, suddenly, one of my colleagues came upon a piece of pottery that was a different colour – much paler – than other pottery we'd found.

'Professor Fu came over and looked at it carefully, and thought it might be very old.'

From the depth at which it was found the archaeologists immediately suspected the pot of being ancient; when it was dated, from fragments of charcoal in the same layer, it turned out to be 12,000 years old.

'From the radiocarbon dating, we know that these pieces of pottery are the oldest in China.'

These pieces of pottery were also among the oldest in the world. There was no evidence of plant or animal domestication at Guilin: the pots appeared to have been made by hunter-gatherers.[5]

The following day, I met Professor Fu Xianguo, who had directed the excavations at the cave. The earliest Zengpiyan pots were made with local clay, apparently deliberately tempered with quartz particles, and fired low, at less than 250 degrees C. They were thick, wide-mouthed and almost hemispherical in shape. On a sunny day, in a field near Guilin, 12,000 years after pottery was first made there, we recreated a pre-Neolithic pot. Fu's chief pottery technician, Mr Wang Hao Tian, collected reddish clay, and we mixed it with smashed-up quartz particles. Wang dug a round pit and we pushed pieces of clay into it, to form the hemispherical shape. Then we fired the pot, along with other pots that Wang had been busy making that week, on an open fire. Two brothers, Mr Liu Cheng Jie and Mr Liu Cheng Yi, traditional potters from nearby Jing Xi (pronounced 'Jing see') on the Vietnamese border, came to help with the firing. The Liu brothers laid logs across large stones to form a rack on which to place the pots, then pushed burning sheaves of straw under the rack, moving the flaming bundle around with a long stick to sear the pots. After about an hour of this gentle firing, the Lius built up a bonfire over the rack, with piles of straw and branches, and our pots were hidden inside the blaze. Another hour later and the potters started to dismantle the still-smoking fire, hooking the pots out on long sticks and setting them down on the ground, where they made a quiet tinkling noise as they cooled. Most of the pots had survived the firing – including our experimental pre-Neolithic cauldron, now dark grey but still speckled with quartz.

Fu had his own particular theory about why the foragers around Guilin had started to make pottery.

Newly fired 'Neolithic' pots.

'In north China, the origin of pottery is thought to be very much related to the development of agriculture,' explained Fu. 'But from our study, we think the origin of the pottery in south China is related to boiling snails.'

I was more than a little sceptical about this idea. Certainly, there were plenty of snail shells in Zengpiyan Cave, but there was no real evidence that they had been deposited there by humans, rather than washed in with the river sediments, or even that they had been cooked.

But it was time to subject our pot to a test. Most of the early pots were round-bottomed cooking cauldrons, like the hemispherical pot we had made, and seemed well designed for boiling water. After our pot had cooled, we subjected it to another test: filling it with water and bringing it to the boil over an open fire. It didn't break. In fact, there's no evidence that the Guilin pots had been heated again after their initial firing, although our experiment had at least demonstrated that they *could* survive being heated.

It may be possible to find out what the pre-Neolithic potters of Guilin were putting in their pots, using residue analysis – which can be applied to fragments. Until then, any theories about what the pots were used for, including snail-cooking, must remain speculative. Some archaeologists have argued that the early pre-Neolithic pots were used to cook wild grains, and although there is no direct evidence for this either the pots are from a time when wild grasses start to form a more important part of diets.[6]

During the LGM, East Asia became colder and drier: deciduous trees retreated south of the Yangtze River and vast areas of what is now China became grassland. After the Ice Age, the global climate became warmer and wetter, and there was more carbon dioxide in the atmosphere, which may have resulted in grasses becoming up to 50 per cent more productive. It is around this time that the archaeological record shows people in China, as well as in south-west Asia and Europe, beginning to focus on collecting wild grasses.[7] But then, around 11,000 years ago, there was a cold, dry spell, comparable to the Younger Dryas in Europe. Perhaps it was this deterioration in climate that provoked the foragers to start cultivating grasses – whose seeds could be stored through the winter.[7, 8]

Although China is now dominated by rice agriculture, millet was just as important to the early cereal growers. Genetic studies of modern cultivated and wild grasses have suggested that domesticated rice (*Oryza sativa*) may derive from the Asian wild rices *O. rufipogon* and *O. nivara*. There are two subspecies of domestic rice which may relate to two, separate centres of origin of rice farming: in East Asia and in South Asia. Foxtail millet (*Setaria italica*) may come from the wild green foxtail (*S. viridis*), while broomcorn millet may come from a wild grass of the same name: *Panicum miliaceum*. Generation upon generation, selection of more seedy plants made the domesticated varieties more productive than their wild counterparts. I had never really thought of rice as 'grass'. Then I visited the terraced paddy fields of Long Ji and saw Liao Jongpu, whose family had farmed there for generations, with a handful of grass-like seedlings, picking out sprigs of three or four stems at a time, to sow into his submerged fields. It looked like grass, but it was going to be food.

As the climate warmed back up again after 10,000 years, cereal cultivation intensified – and was there to stay. Today, rice in south China has a higher genetic diversity than that in the north, suggesting that the origins of domestication lay in the south. But the rice of the Yangtzi Basin, while less genetically diverse, *looks* more ancient and closer to its wild counterpart.[8] Climatic changes and human manipulation of the environment make it really difficult to predict the origins of plant strains from their modern-day distributions.[7]

The earliest archaeological evidence of agriculture in China comes from the area around the Yangtzi River Valley. Grinding slabs and wild rice husks have been found in Upper Palaeolithic cave and rockshelter sites, dating to beyond 10,000 years ago, showing that people were collecting and processing

The rice paddies of Long Ji

wild grasses. It's important to remember that the grass seeds were just part of a much wider diet: wild rice and millet were not productive enough to have been a staple food, and would have been just one part of a broad-based subsistence strategy. And, in fact, although we tend to focus on rice because of its importance today, the earliest domesticated plants may not have been cereals: they could have been starchy roots and tubers like yams and taro, or even non-food plants like gourds or jute. The first farmers are likely to have been cultivating a range of crops.[7]

In the 1970s, evidence for cultivation started to appear as Neolithic villages were discovered, dating to around 7000 years ago. Since then, the earliest dates for farming in the region have been pushed back to about 10,000 years ago, and even earlier. In 2001, archaeologists uncovered a Neolithic site in Shangshan, in Zhejiang Province. The site was the remains of an early village: post-holes and trenches marked the outlines of dwellings, while stone tools, large grinding stones, pebble pestles and red pottery provided clues about life in the village. Many of the stone tools were basic chipped pebbles and flakes, just as are found throughout the Palaeolithic in China, but some are something new and different entirely – stone axes and adzes – suggesting an increased reliance on cultivation. These people were working the land.[9]

The pottery at Shangshan is still similar to the earlier forms – simple, hand-formed or slab-built pots, fired at low temperatures – but it also contains important clues to the new way of life in Shangshan village. The clay was tempered with bits of plant – the first example of this – and some of those plant remains are rice husks, shorter and fatter than wild grains, suggesting these may be the remains of an early domesticated variety. Charcoal embedded in the pottery has been radiocarbon dated to around 10,000 years ago.

Earlier sites with pottery are caves, like Zengpiyan. But the Neolithic village of Shangshan is in the middle of a river basin. It represents the beginnings of a new, more sedentary lifestyle. Instead of moving around the landscape, setting up temporary camps or using natural 'homes' like rockshelters and caves, a place was chosen for its suitability for growing crops, and permanent houses were built.[9]

The transition to farming and a settled way of life was gradual and patchy. It is likely that the early cultivators were semi-nomadic 'collectors', using wild foods supplemented with cultivated varieties. Caring for crops would have increased productivity but would also have tied farmers to their fields, so this is perhaps why they stopped being nomads and settled down in villages like Shangshan. But it is also important not to imagine that hunting and gathering was completely abandoned as agriculturalism was taken up. Even in recent, historic times, farmers continued to collect wild plants and hunt wild animals.[8]

Some archaeologists have suggested that population pressure was a motivating factor in the origin of farming and a settled way of life. But settlement of large communities doesn't happen until after 9000 years ago in China. Early settlements are small-scale and most of the artefacts found in them are the tools of daily use, rather than 'luxury' items like beautiful pots or jewellery. So prehistoric Chinese society between 13,000 and 9000 years ago seems quite egalitarian; the proposition that agriculture may have arisen there to support a stratified society and accumulation of wealth – or that early pottery was developed by aggrandising individuals as a sign of prestige – seems unfounded. Although I rather like the idea of 'competitive feasting', there's no evidence that this drove the origins of agriculture or pottery. Climate seems to have been a major factor, but the precise environmental and social factors that led to the gradual adoption of agriculturalism and a settled way of life in the East are still unclear.[8, 5]

The transition to agriculture should not be viewed as inevitable or progressive, but once it originated it spread (although, in a few areas, including parts of Polynesia, New Zealand and Borneo, farmers faced with unsuitable environments reverted to foraging).[1]

So how did the spread of agriculturalism occur? Did the farmers' population expand and replace the hunter-gatherers and their culture, or did the culture of agriculturalism spread among existing populations of foragers? Ammerman and Cavalli-Sforza proposed a 'wave-of-advance' model in Europe, where the farmers' expansion can be seen as a spreading ripple of intermarriage and gene dilution, until the populations furthest from the cultural epicentre are genetically almost entirely derived from the original hunter-gatherers. Thus the population of western Ireland is 99 per cent genetically from the original foragers, with just 1 per cent of lineages traceable back to Anatolian farmers.[1]

Although we don't yet understand the relationship between genetics and morphology, we can at least assume that people's faces reflect their genetic make-up. So are East Asian features an indication that Neolithic farmers spread throughout eastern Asia, replacing earlier populations, or are those faces from more ancient lineages, before the advent of agriculturalism?

It would be nice to have a range of skulls that showed the emergence of East Asian features and allowed us to date the morphological changes, but skeletal evidence from the Far East is rare, especially before 11,000 years ago. The Niah skull, at around 40,000 years old, does not look 'East Asian', although it does bear similarities to the Ainu, the descendants of the original Jomon population in Japan. But there is a skull with recognisably East Asian features, from Java, dated to 7000 years ago, and thus pre-dating the beginning of rice cultivation in Indonesia.[3]

Some geneticists have argued that Y chromosome variants suggest that millet and rice farmers did indeed spread out from China, largely replacing earlier populations throughout East and South-East Asia,[10] but Oppenheimer[3] argues that the balance of the genetic evidence suggests a spread of people, with what we now recognise as East Asian features, much earlier than the Neolithic – around the LGM. He suggests that, at this time, people would have been retreating back towards the warmer coastline, populating the expanded coastal plains left by the drop in sea level. Oppenheimer suggests that, before the LGM, the inhabitants of East Asia looked like the original beachcombers. Then, during the LGM, 'East Asian'-looking people from Central Asia moved outwards to the coasts of East Asia.

If Oppenheimer's theory is right, it would seem that, as in Europe, most of the inhabitants of East Asia today are descended from the original beachcombers and the coastward-bound populations of the LGM, rather than from a wave of Neolithic farmers spreading through the region. Culture and language are much more labile, transportable commodities. Our genes betray a much more ancient heritage.

So, even in Shanghai, with all its skyscrapers and technology, the people may look very much like the hunter-gatherers who inhabited that coastal plain 20,000 years ago.

4. The Wild West: The Colonisation of Europe

Cave art
in Cougnac

On the Way to Europe: Modern Humans in the Levant and Turkey

Considering that Europe is geographically so close to Africa, it seems remarkable that modern humans made it all the way to Australia some 20,000 years before we find any evidence of their presence in Europe. Why did it take so long? The answer is likely to be complex, involving geographical and environmental barriers, and, perhaps, the presence of other humans already occupying Europe. Because, whereas in most of Asia (with the notable exception of Flores) earlier humans had vanished long before moderns arrived on the scene, Europe was the domain of the Neanderthals.

There is a huge gap between the appearance of the first anatomically modern humans in the Near East – in the Skhul and Qafzeh caves in Israel, some 90,000–120,000 years ago – and the first evidence of modern humans in Europe – at around 45,000 years ago. After Skhul and Qafzeh, modern humans disappear from the Levant for around 50,000 years, although it seems that, during this time, modern humans were making their way eastwards along the coast of the Indian Ocean.

From Arabia and the Indian subcontinent, it may seem that the colonisers should have been able to spread north into Europe with ease. Stephen Oppenheimer[1] suggests that, just as deserts may have blocked the northern route out of Africa for much of the last 100,000 years, the way from the Indian subcontinent and Arabian Peninsula to the Levant was also sealed off by geographical barriers: by the Zagros Mountains, and the Syrian and Arabian deserts. While the beachcombers surged eastwards, their northwards expansion into Europe was blocked. But around 50,000 years ago, the climate warmed up briefly, for a few thousand years. Oppenheimer argues that this warming opened up a green corridor from the Arabian Gulf to Syria, a gate into Europe.

The colonisers could then have spread north-east, skirting the Zagros Mountains, up the coast of what is now Pakistan and Iran, and up the River Euphrates into modern-day Iraq and Syria, making their way from the Persian Gulf to the Mediterranean coast. Some archaeologists argue that Upper Palaeolithic sites along the Zagros Mountains support this route, and indeed suggest that Upper Palaeolithic technology may have originated around the Zagros Mountains, with some dates in excess of

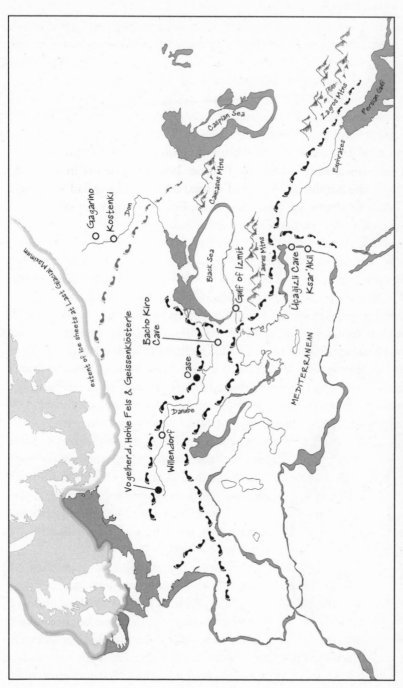

Routes into Europe. The black footprints represent the first modern humans to reach Europe, bringing with them the Aurignacian culture, starting around 45,000 years ago; the grey footprints represent the later incursion of the Gravettian people, beginning around 30,000 years ago.

40,000 years.[2] However, these dates need to be treated with caution: firstly, these are dates at the very extreme of what is considered reliable for radiocarbon dating, and, secondly, the dates were reported in the 1960s, long before the new sampling and calibration techniques were applied. However, the types of tools found at the Zagros sites are similar to the earliest Upper Palaeolithic tools found around the eastern Mediterranean, known as the 'Levantine Aurignacian'.

The route from the Persian Gulf to the Mediterranean, following an earlier southern dispersal from Africa, seems a reasonable suggestion, but other researchers still favour a simple northern route out of Africa, from Egypt. But whichever of these routes was taken, we would expect to find part of the archaeological trail in the Levant and in Turkey; in other words, in those countries that border the eastern Mediterranean.

There is a growing body of archaeological evidence for the earliest modern humans in the Levant and Turkey, in the form of Upper Palaeolithic stone tools, and ornaments – and some bones.

In the 1940s, archaeologists began to excavate down through 19m of deposits at the site of Ksar 'Akil, near Beirut in Lebanon. They found twenty-five layers containing Upper Palaeolithic archaeology. In the deepest layers, they found Levallois-type technology (stone tools made from prepared cores), typical of the Middle Palaeolithic, alongside classic, reshaped Upper Palaeolithic tools, like end-scrapers and burins. In later layers, the Levallois cores are replaced by cone-shaped prismatic cores, from blade manufacture – a hallmark of the Upper Palaeolithic. Dating of layers above and below the earliest Upper Palaeolithic stratum at Ksar 'Akil

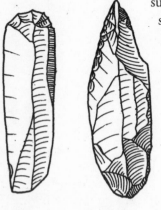

suggests that those tools were being made there somewhere between 43,000 and 50,000 years ago.[3] And the discovery of a skeleton at Ksar 'Akil confirmed that it was modern humans who were making those tools.[4] Dating of the Upper Palaeolithic at Kebara also suggests the presence of modern humans in the area by 43,000 years ago.[5]

Tracing the northwards expansion of people into Turkey has been problematic. The Palaeolithic of Turkey has long been in the shadows, mostly because comparatively little archaeological research has been carried out there.[6] Many Palaeolithic sites in Turkey are just 'findspots', where stone tools have

Upper Palaeolithic artefacts from Üçağızlı Cave

been spotted on the surface. Relatively few
have been excavated, but in the last twenty
years archaeologist have been striving to plug
this gap, and some interesting finds have
emerged, which help us to trace the journey
of those early European colonisers.

The archaeological site of Üçağizli lies on
the rocky south-west coast of Turkey, about
150km north of Ksar 'Akil, and near the city
of Antakh (ancient Antioch). The partially
collapsed cave was first discovered in the

Pierced Nassarius shells from Üçağizli Cave

1980s, and excavations began there in earnest in the 1990s, leading to the
discovery of Upper Palaeolithic artefacts in red clay sediments. The tool
manufacture at Üçağizli seems to follow a very similar pattern to Ksar
'Akil: the oldest Upper Palaeolithic layers in fact contain a mixture of
'Middle Palaeolithic' technology alongside more classic, retouched tools.
Just as at Ksar 'Akil, the Levallois cores disappear in the higher layers, to
be replaced by prismatic cores, where blades have been knocked off by
soft-hammer or indirect percussion. But at Üçağizli there are also tools
made of bone and antler. The earliest Upper Palaeolithic layers date to
between 41,000 and 44,000 years.[3]

Throughout the Upper Palaeolithic layers at both Ksar 'Akil
and Üçağizli, the archaeologists discovered classic signs of the Upper
Palaeolithic – and of modern humans: ornaments. The vast majority are
small seashells, pierced through to be used as beads or pendants. More than
five hundred shell beads have been found at Üçağizli alone. It's possible
to see how the holes have been created in the shells – some by scratching
away, while others seem to have been punched through with a pointed
tool – and it's quite clear that these holes were made by humans and not
by other, natural processes. There were shells from marine snails *Nassarius
gibbosula*, *Columbella rustica* and *Theodoxus jordani*, as well as the pretty,
ridged bivalve *Glycymeris*. From the large collection of shells at Üçağizli,
it looks as if those hunter-gatherers also enjoyed the taste of seafood:
bigger shells, from limpets (*Patella*) and the edible snail *Monodonta* also
appear, unpierced, in the archaeological strata. The archaeologists were
sure that these shells had been collected for food as they were not wave-
worn like the empty seashells that wash up on beaches, and, in addition,
many of them were burnt.

The shell beads from Üçağizli are not – by a long stretch – the
earliest ornaments: the pierced shells from Skhul in Israel date to between

100,000 and 135,000 years ago.[7] Shell beads were also found in Blombos Cave, dating to around 75,000 years ago, and the evidence for ochre use goes back to beyond 160,000 years ago at Pinnacle Point. These finds suggest that art and ornamentation is probably almost as old as the human species itself. But the Üçağizli beads are useful as a marker for modern human presence. They show that modern humans were bringing with them a shared culture, and perhaps an awareness of identity and a system of communication that did not seem to have been there among archaic human populations.[3]

There is a scarcity of Upper Palaeolithic sites in Turkey. Kuhn[6] argues that the Anatolian Plateau, most of which lies more than a kilometre above sea level, would have been a cold, unwelcoming place during the late Pleistocene, and that modern humans (as well as other animals) would have gravitated towards the warmer coast. With a higher sea level today, any Pleistocene coastal sites would now be submerged. Nevertheless, Üçağizli is an extremely important site, a stepping stone into Europe, with Upper Palaeolithic artefacts going back more than 40,000 years, anticipating the spread of the classic Upper Palaeolithic culture, the 'Aurignacian', from east to west across Europe, between about 40,000 and 35,000 years ago.[8]

This all seems to fit together very neatly, but we need to be aware that these initial movements into Europe are occurring at an awkward time for radiocarbon dating, and some dates obtained during the twentieth century might need reassessment. And not only do archaeologists and anthropologists argue about the exits from Africa, there is also debate about the route *into* Europe, and where Upper Palaeolithic culture began. Kuhn[6] seems convinced by the dating of Üçağizli, but also suggests that this culture may represent a spread southwards from Europe, rather than northwards from the Zagros Mountains and the Levant. Other researchers have argued that Upper Palaeolithic culture may have arisen in the Russian Altai, north of the Zagros Mountains, with a dispersal of modern humans, carrying this technology, coming into Europe from around the Caucasus Mountains and the northern coast of the Black Sea.[9]

However, most researchers seem to agree that Üçağizli and Ksar 'Akil fit very well with a model where modern humans, bearing an Upper Palaeolithic, pre-Aurignacian toolkit, arrive in the Levant, spread north into Turkey, and then westwards through Europe.[8]

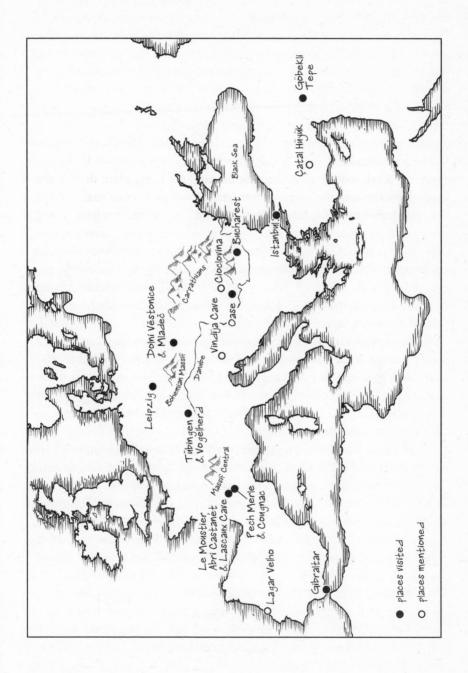

Black Sea

Göbekli Tepe

Çatal Hüyük

Istanbul

Bucharest

Cioclovina

Carpathians

Oase

Vindija Cave

Dolní Věstonice & Mladeč

Bohemian Massif

Danube

Leipzig

Tübingen & Vogelherd

Massif Central

Pech Merle & Cougnac

Le Moustier
Abri Castanét
& Lascaux Cave

Lagar Velho

Gibraltar

● places visited
○ places mentioned

Crossing the Water into Europe: the Bosphorus, Turkey

Making my own way up through the Asian part of Turkey, I reached a watery barrier: the Bosphorus. This narrow strait connects the Black Sea in the north to the Sea of Marmara, and at its southern end the Sea of Marmara narrows down again to form the Dardanelles, connecting through to the Aegean Sea.

In Istanbul, I took the ferry to cross the glittering Bosphorus, from the Asian to the European side. I thought about the early colonisers reaching this waterway, and Bulbeck's ideas about coastal and estuarine adaptations, and the possible use of watercraft in the Palaeolithic. It didn't seem to me that the modern Bosphorus and the Dardanelles would have constituted much of a barrier to the early pioneers.

In fact, when I looked into it, it turned out that the Bosphorus was dry during the Pleistocene. It wasn't until after the Ice Age that the sea level rose sufficiently to flood the Bosphorus and connect them. Boreholes drilled down through the sediments at the bottom of the strait show how and when the connection between the two bodies of water was established: the Sea of Marmara spread northwards and eventually opened into an estuary at the southern end of the Black Sea, around 5300 years ago, and the Bosphorus was formed. Interestingly, it looks like there was another, intermittently open, connection between the Sea of Marmara and the Black Sea during the Pleistocene, to the east, from what is now the Gulf of Izmit.[1] So even if the Bosphorus wasn't there to cross, there may have been times when the colonisers *would* have got their feet wet crossing from Asia to Europe.

From the Bosphorus region colonisers could have headed northwards to the coast of the Black Sea, or continue beachcombing westwards along the Mediterranean coast, and it appears that they did both. There are a scattering of sites with Upper Palaeolithic stone tools, on or close to the Mediterranean coast, in Italy, France and northern Spain, and there are also sites dotted along both the European and Asian shorelines of the Black Sea. An important early Upper Palaeolithic site is Bacho Kiro, in Bulgaria, where a 'pre-Aurignacian' toolkit – with lots of blades – has been found, dating to 43,000 years ago. Moving northwards up the west coast of the Black Sea, the colonisers would have reached the great delta of the Danube, in modern-day Romania. They could have then used the waterway as a superhighway into the heart of Europe: there are many Upper Palaeolithic sites along the Danube and its tributaries. The conventional radiocarbon dates for this westwards spread across Europe suggested it took place between 45,000 and 35,000 years ago. New, calibrated radiocarbon dates for these sites suggest that this dispersal into central and western Europe happened fairly rapidly, between 46,000 and 41,000 years ago. This speedy spread across Europe may have been helped by another episode of global warming, the Hengelo interstadial.[2, 3, 4, 5]

Face to Face with the First Modern European: Oase Cave, Romania

My next destination was a site along that Danube corridor. I travelled to Romania, to meet geologist and speleologist (cave expert) Silviu Constantin, who was going to introduce me to the cave where the earliest known modern human fossil in Europe had been discovered.

From Bucharest, we drove east, following the Danube fairly closely, like our predecessors 40,000 years before. We drove and drove, and eventually reached a small village in the south-western Carpathians, the name of which I cannot disclose because the exact locality of the cave was secret. A couple of stray dogs chased our car up the hill, excitedly barking, but when we stopped and got out they ran away. We were staying in the 7 Brazi (fir trees) Pensiune, high up on the hill overlooking the village.

The following morning, we met up with the team of cavers who were accompanying us – Mihai Bacin (leading the team), Virgil Dragusin and Alexandra Hillebrand – and sorted out caving gear for the trip before heading off to the cave itself. We drove through wooded valleys and past abandoned factories, and eventually turned off on to a dirt track. About a quarter of a mile later, we reached a point where the road had collapsed. We all got out to take a look, and, after hauling some large stones into the hole, decided that it was passable, and cautiously drove the cars over the roughly stopped-up breach. It held. We pulled up just around the corner from the collapsed track. The cave itself was below us in the steep-sided valley, and we scrambled down the wooded slope, lugging our equipment down to the stream bank at the bottom. Once we were down, I could see the tall, slit-like entrance of the cave, with the stream emerging from it.

This was Peştera cu Oase, the Cave of Bones. It was incredibly exciting to be standing there, in front of a place that I had read so much about.

Romanian beetle

A colleague from Bristol University, João Zilhao, had been part of the team that excavated there in 2003–5, so I had heard a lot about the cave from him. Unfortunately, he was in Portugal investigating another cave, but that left me in Silviu's capable hands – he had also been one of the excavating team, and he had dated the finds from the cave.

On 16 February 2002, a group of intrepid cave divers were exploring the cave. The cavers had made their way into the depths, past a duck-under and through a longer, underwater section, and up a steep ramp to an area littered with animal bones. 'They found a human mandible, right on top of the flowstone. It was probably dug out recently by an animal. It was just sitting there, waiting for someone to discover it,' said Silviu.

When the mandible was radiocarbon dated, it was found to be 35,000 years old, the oldest known remains of a modern human in Europe.

'You must have been pretty excited to get that date back,' I said.

'I remember everyone was excited,' said Silviu. 'The oldest human was here, in Romania. We were proud that it was in one of our caves.'

The cave was an exciting proposition for scientists studying this period of prehistory. In 2003, an international team of archaeologists visited the cave, and discovered masses of bones – mostly cave bear, but also some other human material: fragments of a skull. These pieces were found further down the slope than the mandible, in part of the cave that was subsequently, and evocatively, named Panta Strămoşilor (the Ramp of the Ancestors). The skull and mandible were from two different individuals.

The place where the bones were found is hard to get to now – and involves that dive.

'The archaeologists were trying to figure out how such a massive bone deposit could come into the cave,' said Silviu. As a geologist and a caver, he had joined the team to answer this question. He could also help with dating – using uranium series dating on stalagmites. From his investigation, it seems that Peştera cu Oase once had another entrance,

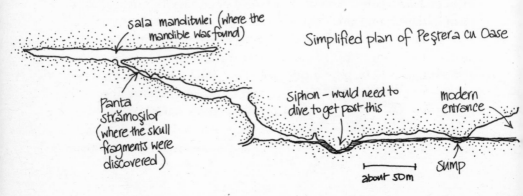

sala mandibulei (where the mandible was found)

Simplified plan of Peştera cu Oase

Panta strămoşilor (where the skull fragments were discovered)

Siphon – would need to dive to get past this

modern entrance

about 50m

sump

allowing the cave bears access. In fact, the two galleries that lead off from the Ramp of the Ancestors both seem to have had openings in the past. Today, these entrances have collapsed, and are practically blocked off, although some small animals – such as rodents – still fall down into the cave through the sinkholes.[1]

The majority of the bones in Oase belonged to cave bears. There were also bones of other cave-dwellers such as wolf and cave lion, which had presumably made their dens in there at various times. However, also found in the cave were skeletal remains of distinctly non-cave-dwellers such as ibex and red deer. These could have been brought into the cave by people, but there were no signs of such habitation in the cave, nor any cut marks on the animal bones to suggest that they had been eaten by humans. That left geological processes and carnivores to explain the accumulation of bones in the cave. In 2005, the archaeologists returned to excavate the cave; they found many more bones, but also important clues as to how the skeletal remains had ended up there. Underneath a covering of stalagmite, they found a 30cm-thick layer of cave bear and other animal bones, many bearing the tooth marks of bears and wolves. This looked like a collection of bones of animals eaten by carnivores denning in the cave, as well as the cave-dwellers themselves. Under that was a layer of bones mixed with sand, gravel and cobbles, and these were sorted by size – larger bones at the top of the slope, smaller ones at the bottom, and many of the bones in this layer had rounded-off edges. These bones had been swept into this part of the cave by flooding. So it seemed that the animal bones in Oase had ended up there through a combination of cave-dwelling carnivores bringing their dinner home, and flooding.

'So how do you think the human bones ended up in the cave?' I asked Silviu.

'Most probably they had been washed in.'

There were no gnaw marks on these bones, so perhaps a person, or even a buried skeleton, had fallen in through a sinkhole and then the bones had been washed further down the cave by floods.[1] Nevertheless, it seems rather strange that there are only cranial remains – a skull and a mandible, and no other parts of the skeleton – but there are still a lot of bones in Oase. Perhaps the rest of the skeleton will yet be found.

I followed Silviu into the cave – wading along the stream and into the first great gallery, with Mihai, Virgil and Alexandra following behind. The roof was very high, and hung with enormous stalactites. Towards the back of this gallery were a series of pools with stalagmite edges. We followed the stream around to the left, where the roof plunged down.

We were lucky that day: there hadn't been much rain the previous weeks and days, so the water level was low enough to enable us *just* to keep our heads above water. In a wetter period this would have been a duck-under. Still, it was an awkward manoeuvre: under the water, the cave floor sloped away from the gap I was aiming for, and I couldn't get a foothold. As my feet slipped, I gave up trying to walk through the gap, and swam through it instead. On the other side, I was wading shoulder-deep, hanging on to a rope to keep me close to the shallower right-hand wall. The floor of the tunnel gradually rose up until the water was just knee-high.

We all made our way along this narrow, tunnel-like part of the cave. Sometimes the floor dipped down and we would plunge in up to our chests again. Eventually, we reached a wider part of the tunnel, but the roof sloped down and down until it touched the water: this was the siphon, and the end of the line for me. I would have had to dive to get to the Ramp of the Ancestors. It was frustrating in a way, but I had known that I would only be able to get this far. I still felt privileged to have come so close, and to have explored the cave where Europe's earliest modern human had been found. Silviu and I sat down on a convenient stalagmite and I asked him about excavating in the cave. It sounded arduous and difficult – getting to the site was one thing, taking in tools and bringing back bags of sediment and bones, through the siphon and through the duck-under another altogether – but the results had been worth it. When the skull had been dated, it was even older than the mandible: around 40,000 years old. And there were some things about the skull and mandible that seemed a little odd.

Leaving Oase, Silviu and I drove back to Bucharest, where I would be able to see the skull and mandible. We had travelled close to the Danube on the way out to Oase, but on the way back we took a route through the beautiful wooded gorges of the Carpathians. Where the gorges opened out into valleys, fields were visible in which people were cutting hay and heaping it on to three-legged wooden frames to make stacks. The haystacks varied from field to field and village to village. Some were tall and thin, others squat and conical. We slowed down to pass a heavily laden hay cart pulled by two horses.

The centres of the villages were often lined with low, terraced cottages, but on the outskirts there were more often than not massive, ugly tower blocks. These buildings seemed so incongruous among the fields. 'Why build blocks of flats in villages?' I asked Silviu. 'They were built by Ceauşescu,' he said. 'He wanted to create more land for farming.' 'But it seems like there's plenty of land,' I suggested. 'Yes. There is. Ceauşescu wanted to destroy the villages.' It was strange to think that it really wasn't that long ago – 1989 – that Ceauşescu had been removed from power. Romania was a country in recovery.

Silviu told me about how he had travelled around the countryside as a young man. He recounted how, if he'd got stuck for somewhere to stay overnight, he would find a barn and sleep in the hay. He usually looked for someone to ask beforehand, and had often been invited in to dinner. 'People treated the occasional backpacker as a traveller: to be given hospitality,' he said. 'Now they want money. They think tourists have money and they want some of it.' Silviu was worried about the people and the countryside being spoilt by tourism. He liked the wilderness.

'What would this landscape have looked like 40,000 years ago?' I asked Silviu.

'I don't know,' he said, and paused for thought. 'It would have been [Oxygen Isotope] Stage 3. So – colder than now. And wetter summers. Like – perhaps Norway today, at the coast; maybe like Bergen.'

I tried to imagine the hunter-gatherers in the foothills of the Carpathians, dressed warmly against the cold, in a country with red deer, ibex, wolves and cave bears. Although cave bears are long gone, there are plenty of brown bears roaming Romania today: nearly half of all European bears are in the Carpathians. Some hunting of bears goes on today; indeed, one restaurant in Bucharest even had 'bear paw' on the menu. There were times when I was truly glad to be vegetarian.

The following day, I visited Silviu in his natural habitat at the Emil Racovita Institut de Speologie in Bucharest. He brought a series of cardboard boxes into a lab next to his office and we carefully unpacked some of the Oase bones. There were pieces of animal bone embedded in speleothem – a real gift for a geologist like Silviu who could date the stalagmite, and therefore the bone. He also brought out a huge and formidable looking cave bear skull. And then there were the human remains: the mandible and the skull.

Again, these were quite clearly modern human. The skull was globular in shape, without any of the more obvious

The Oase skull

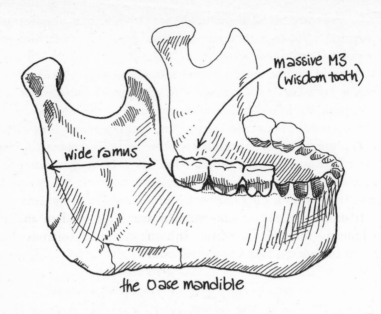

the Oase mandible

features of an archaic skull, like big browridges, or a protruding occiput at the back of the head. The jaw was quite gracile and modern-looking too, with a definite chin. But there were some oddities – particularly in the mandible or jawbone. The chin was very straight, the ramus (the part of the mandible ascending up to the jaw joint) was very wide, and the mandibular foramen (the hole on the inside of the ramus where the nerve supplying the lower teeth enters) was a little strange. Then there were the teeth, set in a very wide arc in the mandible, and with absolutely enormous wisdom teeth. These molars are usually smaller than the ones in front, but in the Oase jaw they were huge.[1,2] I had read about these odd features in the scientific articles announcing the discovery and dating of the Oase bones, but it was something entirely different to hold the actual bones in my hands and look at them myself.

Now, everyone is unique and we all have 'anatomical traits' – little variations on a theme – in our bodies and our bones. In our skulls, some of us may have one hole for a particular nerve, where others have two or three. We might have little islands of bone, or 'ossicles', embedded in the zigzag joins or sutures where the plates of the skull come together. Some of these traits are genetic, whereas others appear during our lifetimes, and might be related to diet – or other things we do to our bodies. For instance, lumpy bits of bone inside the 'external auditory meatus' – the bony part of the ear canal – can be caused by swimming in very cold water. But, like skull shape and size generally, we're not sure about how or why some of these traits occur, and the various influences of genes and environment. It's all part of that great question in developmental biology: how are our bodies shaped by our genes and our environment?

Having said all that, however, there are traits that do at least seem to hark back to an 'earlier' form. If you were being unkind, you might call them 'throwbacks'. If you were a bit nicer about it, you might think of these characteristics as echoes of evolutionary history, glimpses of where we've come from. And those odd traits in the Oase jaw seem to fall into this category: archaic features. But could this be more than just an echo of a more distant evolutionary past in these 40,000-year-old bones? Is it possible that, instead, the Oase bones show a mixture of modern and archaic traits because that person *was* a mixture of a modern and an archaic human – some sort of hybrid? This isn't such a preposterous idea, because, when modern humans started making their way into Europe, someone else was already there: the Neanderthals.

As we have seen from the archaeology and fossil record of East Asia, ours was not the first human foray out of Africa. A series of archaic human species made the leap before us.

In a review article that appeared in *Science* in 2003, Ann Gibbons wrote: 'The long-legged, relatively big-brained hominin called *Homo erectus* has long been considered the Moses of the human family – the species that led the first exodus out of Africa more than 1.5 million years ago.'[3]

The biblical analogy is great. I can imagine that striding, big-browed man leading his people out of Africa, across the Red Sea, but it's a sleight of pen in two ways. Firstly, there's a wry and unwritten 'but of course it wasn't like that', as we all have this tendency to promote our ancestors to heroes and imagine their lives as epic struggles against adversity, winning through so that we could be alive today. And secondly, Ann Gibbons goes on to write about *Homo georgicus*, one of the recent and somewhat cheeky surprises in European palaeoanthropology. The three paradigm-nudging and diminutive fossil skulls were recently discovered in Dmanisi in Georgia, and dated to 1.75 million years ago. Then another small skull was found in Kenya, dating to about 1.5 million years ago – perhaps belonging to a particularly small population of *Homo erectus*, perhaps linked with the Georgian hominins. The Kenyan skulls were the same age as another famous fossil: Turkana Boy. This young man, with a largish brain, is sometimes classified as *Homo erectus*, sometimes as *Homo ergaster*. (Remember that the world of palaeoanthropology is interpreted differently by 'lumpers' and 'splitters' – see page 3.)

Fossils can be extraordinarily slippery when you're trying to pin a species name on them. The Kenyan and Dmanisi skulls are no exception. They all look a bit like a small *Homo erectus* without browridges, but also bear similarities to another, earlier hominin species, *Homo habilis*. Although the discoverers of the Dmanisi skulls claim that they warrant a new species name, *Homo georgicus*, most researchers place the skulls in *Homo erectus*. So maybe *erectus* was the first hominin to get out of Africa after all.[3]

Georgia lies west of the Caspian Sea, but it is hard to know whether it really counts as part of Europe or Asia. Certainly, Dmanisi was sidelined when another, intriguingly ancient fossil, this time from the Sierra de Atapuerca, in northern Spain, was reported as 'The first hominin of Europe'. It dated to 1.2 million years ago, and its discoverers suggested that it was *Homo antecessor* (a category that some lump into *Homo heidelbergensis*).[4]

By about 300,000 years ago, *Homo heidelbergensis* in Europe had morphed into Neanderthals. And when modern humans got to Europe, these other hominins, their distant cousins, were still around in the landscape.[5, 6]

The first fossil of these ancient Europeans was found in 1848, in Gibraltar, but nobody paid it much attention. The bones that gave the species its name were found in 1856 in Germany, near Düsseldorf – in the Neander Valley, or Neanderthal. The valley was being quarried for limestone, and workmen clearing out mud from caves in the cliffs prior to quarrying found what they thought were cave bear bones. But a local teacher recognised that they were human and collected them up.[7]

The following year, rather bravely for the time, Professor Schaafhausen of Bonn University published a report on the skull and bones from the Neander Valley, saying that they were normal – non-pathological – but seemed to be from an ancient inhabitant of Europe as the remains were found alongside bones of extinct animals. This interpretation was challenged by Professor Mayer, also at Bonn, who said the bones were probably much more recent, probably those of a Russian Cossack dying from rickets who had crawled into the cave, with great browridges from frowning in agony. But a few years later the find had been widely published, and there was a growing consensus that the bones were very ancient. The Irish anatomist William King proposed that the skeleton should be given a new species name: *Homo neanderthalensis*. It was the first known species of a fossil human.[7, 8, 9]

Since the discovery of the first Neanderthal fossil more than 150 years ago, several thousand bones have been found from more than seventy different sites. And there are over three hundred sites where Neanderthal stone tools have been found.

Neanderthal characteristics start to appear in European *Homo heidelbergensis*. For instance, the Sima de los Huesos specimens from Atapuerca, which are over 350,000 years old, already have some 'Neanderthal' features such as protruding faces and gaps behind their wisdom teeth, as well as a characteristic shape of the browridge, and a ridge across the back of the head: the 'occipital torus'.[10] By about 130,000 years ago, 'classic' Neanderthals, with full-blown features, lived right across Europe – and beyond.[11] Their territory extended from Portugal in the west to Siberia[12] in the east, from Wales in the north to Israel in the south. And they persisted in some parts of Europe and western Asia until less than 30,000 years ago – *after* modern humans had arrived in Europe.

Given that we now know so much about Neanderthals, there are still many questions about their disappearance. Although there is no evidence of Neanderthals and modern humans actually living in precisely the same places at the same time, there was certainly a period when both species were present in Europe. It used to be thought that this period of overlap lasted around 10,000 years, but new calibrated radiocarbon dates suggest that the overlap was shorter: about 6000 years in north and central Europe, and perhaps only one or two thousand years in western France.[13] But why did the Neanderthals disappear? Did we kill them off or out-compete them? Or perhaps they are actually still around – could Neanderthals have been assimilated into the expanding modern human population as it flowed westwards across Europe?

There are certainly some researchers who think so. They put forward specimens – mostly skulls – like Oase and Cioclovina from Romania, the Mladeč fossils from the Czech Republic and the Lagar Velho skeleton from Portugal[1] – as physical evidence for interbreeding between Neanderthals and modern humans. Palaeanthropologists João Zilhao and Erik Trinkaus suggest that archaic traits in these fossils are not just 'throwbacks': they may be evidence of Neanderthal genes in the early modern human populations of Europe.

Neanderthal Skulls and Genes: Leipzig, Germany

So I made my way to Germany, not to the Neander Valley, but to the Max Planck Institute for Evolutionary Anthropology in Leipzig, where I had arranged to meet Dr Katerina Harvati, who had recently analysed the

Cioclovina skull. Katerina met me on the other side of the revolving door at Max Planck, and we walked into an enormous space, some three floors of atrium with light streaming in through glass walls on two sides. Stairs and ramps to the upper floors seemed to float in the air. Katerina led me up to the labs on the second floor, where she was going to show me CT scans of the Cioclovina skull.

But my attention was first drawn to a composite skeleton, put together from casts of fossil bones from different sites, standing in the corner of the lab. It was the first time I had laid eyes on a complete, assembled Neanderthal skeleton, and it was interesting to see just how stocky he looked. The ribcage flared out at the bottom, quite different from the modern human chest shape. Individual bones were generally quite similar to modern human bones, but nonetheless very rugged.

'We can tell from their body form and proportions that Neanderthals were showing some level of cold adaptation: they were stocky, with short limbs,' said Katerina.

But how much of an advantage would this have given them, compared with modern humans?

'It has been calculated that the advantage would be – perhaps not as great as we originally thought – maybe the equivalent of one business suit.'

It didn't sound that impressive. Cultural adaptations, like clothing and use of fire, must have been more important to the Neanderthals' survival in Ice Age Europe.

But the most 'different' part of all the Neanderthal skeleton was the skull. Neanderthals have very long, low skulls, whereas modern human human crania are much rounder. Neanderthal faces are big: they have massive browridges, large, goggly orbits (eye sockets), large nasal openings and projecting, prognathic jaws.

So what about this Cioclovina skull that had been suggested to be a Neanderthal/modern human hybrid? The skull itself had been discovered in the cave that gives it its name in southern Romania, in 1941 – during phosphate mining – and had recently been radiocarbon dated to about 29,000 years old. The skull was really just a braincase: most of the face was missing. Although its general shape was definitely modern, some researchers had suggested that the shape of the browridge and the back of the skull were Neanderthal-like.[1]

Obviously, before you can confirm or reject a claim that a skull represents a hybrid, you have to have an idea of what a hybrid might look like. Is it likely to have an even mix of features from each parent?

Or might it be mostly like one parent with just a few features from the other? Katerina had looked into the features of hybrids in other primate groups and she found that a common feature of hybrids seemed to be a size change – either bigger or smaller – than would be expected from the parent populations. Some hybrids – like a gibbon–siamang cross in Atlanta Zoo, and hybrids from different macaque and baboon species – looked anatomically like a mixture of the two species they came from. It also seemed that hybrid populations tended to be more variable than the parent species, and also had rare anomalies popping up more often than usual.[1]

So Katerina had analysed the Cioclovina skull to see if it showed any of these signs of being a hybrid: an appreciable size difference, a mixture of features, a high level of variability, or any strange anomalies. But she had also measured the skull so that she could compare its size and shape with those of other modern human and Neanderthal skulls. Describing features in skulls, even measuring them, is fraught with problems, as I had seen so vividly in China, but Katerina had also used a technically sophisticated and perhaps more objective approach to the problem of comparing skull shape and size.

The first step was to convert a real skull into a mathematical model, a cloud of points in 3D space that described the shape and size of the skull, using features or 'landmarks' that could be recognised on any skull. Rather than measuring a skull with calipers to get distances and angles, Katerina showed me how she had captured the 3D shape and size of skulls in two ways: using an electronic digitiser and CT scans. The digitiser was an elegant piece of equipment – an articulated arm ending in a stylus that could be placed on the surface of a skull – and points could be captured in 3D space, with x, y, z coordinates. It was a piece of apparatus that was widely used in design and engineering – and was now beginning to be applied to the study of old bones. 3D coordinates could also be taken from detailed CT scans of skulls, which would allow points on the inside as well as the outside of the skull to be recorded.

Having captured and quantified all that information, Katerina could then compare different skulls, and she did this in the context of variation among different primates.

'The difference between Neanderthals and modern humans is not similar at all to the differences that you would find between subspecies of primates living today,' she said.

'It is much more similar to the distances you'd find between closely related species.'

'So you can be absolutely sure that Neanderthals are a separate species?' I asked.

'Yes, that is what I'd say. They are too different to be another population or even a subspecies of modern human. They were our sister species. Closely related – but a different species.'

This seemed to refute the multiregionalist idea that all species since *Homo erectus* have essentially been one.

'So what about Cioclovina?' I asked. Katerina showed me a 3D computer model of the Cioclovina skull, based on CT scans that had been done at a local hospital. She spun the model round on the screen, and pointed out the relevant features. The browridge was big, but it was broken in the middle, unlike the uninterrupted 'monobrow' of Neanderthals. The occipital bone at the back of the skull did bulge out, and the nuchal line where neck muscles attached was well marked, but not really Neanderthal-looking. There didn't seem to be anything unusual about its size, nor were there any odd anomalies in the skull.

So what about the results of the shape analysis? Katerina had compared the 3D 'landmark configuration' of the Cioclovina skull to Neanderthal and modern human (including Upper Palaeolithic) skulls. Using different sets of statistical analyses to make the comparisons, Cioclovina always came out closer to modern humans.[1]

'From my analysis, I wasn't able to see any resemblance to Neanderthals,' Katerina told me. 'There is no evidence to support the claim that this is hybrid. It actually turns out to be very typically modern human in its anatomy.'

It was clear that Katerina couldn't wait to look at the other proposed hybrid specimens, like Oase. She was open-minded about what her results meant, and what she still might find.

'Of course this doesn't mean that hybridisation didn't happen. It could have happened and we just haven't found the hybrids yet. Or, some of the other proposed hybrids that I haven't examined yet might fit the criteria. Or it could be that it was so rare that it hasn't left a trace in the fossil record. And the genetic evidence to date suggests that *if* admixture happened, it was so low that it was really not significant in an evolutionary sense.'

Indeed, it wasn't just the shape and size of Neanderthal bones that was being studied in Leipzig, it was genes as well. In 1997, a team of scientists led by Svante Pääbo of the Max Planck Institute published the first analysis of DNA from an extinct human. They had managed to extract mtDNA from one of the original fossils from the Neander

Valley. Pääbo chose to look for a non-coding, fast-mutating section of mitochondrial DNA that had already proved useful in studies into evolutionary relationships between living species.

Getting DNA out of an ancient bone was always going to be a huge challenge – DNA starts to fall apart after death – but Pääbo and his team had hoped that some tiny fragments might still be there. The extraction was done in a sterile room, to try to reduce the possibility of contamination with modern DNA. The bone sample was ground into powder and then the sample was treated to amplify up any DNA – by getting any fragments to make copies of themselves. Then the sequencing could start, and the results were quite stunning: when they compared the Neanderthal sequence with the equivalent mitochondrial DNA sequence from nearly 1000 modern humans, they found that it was distinctly different. The modern human mtDNA sequences differed from each other by an average of eight different base pairs out of almost four hundred. But the Neanderthal sequence had an average of twenty-six base pair differences compared with the modern human samples. This difference suggested that Neanderthal and modern human mtDNA had been evolving along separate pathways for about 600,000 years. Although this seems a very long time ago, compared with the dates of the earliest known Neanderthal (about 300,000 years ago) and the earliest known modern human (about 200,000 years ago) it still makes sense, as the lineages would have started to diverge within an ancestral population of *Homo heidelbergensis*.[2]

This result seems to support the theory of a recent African origin of modern humans, and a replacement of any earlier human populations. In contrast, the multiregional hypothesis suggests that archaic populations in Africa, Europe and Asia developed into modern human populations. A halfway house theory has modern humans originating in Africa, then spreading into Europe and Asia and interbreeding with existing archaic humans.

Pääbo's findings suggested that the mitochondrial DNA lineages, at least, had separated (and stayed separate) hundreds of thousands of years before modern humans appeared in Europe. Even if you ignore the timings, then the multiregional model with hybridisation suggests that Neanderthals should be genetically closest to modern Europeans, but there was no evidence of this in the mitochondrial DNA: the Neanderthal sequence was equally different from all modern humans across the globe. Another study compared ancient DNA extracted from two 25,000-year-old European modern human fossils, and found that the Cro-Magnon mtDNA fell in the modern human range of variation, and was very different from the Neanderthal sequences.[3]

Looking at mtDNA variation as well as modelling the population expansion of modern humans in Europe, researchers in Switzerland came up with a *maximum* interbreeding rate between the two populations of less than 0.1 per cent. Statistically, this is so low as to be practically non-existent, and the Swiss scientists go as far as to say that this suggests the two species were biologically separate – and could not produce fertile offspring even if they had seized upon the chance to have sex with each other.[4]

So do these mtDNA results represent definitive evidence that the Neanderthals could not be counted as among the ancestors of modern Europeans? Well, they certainly seem to point in that direction, but, actually, it's impossible to completely rule out *any* hybridisation between modern and archaic populations. Neanderthal genes could have entered the human gene pool, but those lineages might have died out, leaving no trace of them today. And what if only Neanderthal men, not women, had interbred with the incoming modern humans? That wouldn't show up in the mitochondrial DNA – which is inherited only from the mother. So although these Neanderthal mtDNA studies are amazing, and suggest that hybridisation didn't happen, they can't rule it out. So would it be possible to probe further, to go after more Neanderthal DNA – perhaps nuclear DNA?

When Svante Pääbo was interviewed for *Science* magazine after the publication of the Neanderthal mtDNA paper in 1997, he was very pessimistic about the chances of anyone ever managing to recover and sequence nuclear DNA from Neanderthal bones.[5] But just over a decade later, I was visiting his lab at the Max Planck Institute – and they were doing just that.

The genetics labs were just along the (very beautiful, sky-lit, gently curving) corridor from the bone lab. The Institute felt like a modern monastery, with an all-pervading calm and scholarly atmosphere. But instead of monks painstakingly copying out biblical passages, scientists were locked away in their high-tech scriptoria, sequencing the Neanderthal genome.

I met up with Ed Green, one of the geneticists hard at work on the Neanderthal Genome Project. Ed had brought along some casts of the original fossils from which DNA had been extracted.

'How do you go about trying to extract DNA from these fossils?' I asked Ed.

'Well, the first thing is to find the fossil that has ancient DNA that can be extracted. Then the way it's done is to simply to take a dentist's drill, drill a bit, get some bone powder, and then use a standard extraction method where you bind DNA to silica beads.

'Then the really fun part begins – trying to sequence this DNA, and see what is there. Is this DNA from the individual that owned this bone originally? Or DNA from bugs that have crawled into the bone since then?'

'And presumably there's quite a lot of modern human DNA knocking around as well – from the archaeologists who excavated them,' I suggested. Ed agreed. He was very keen to encourage archaeologists to excavate fossils in a 'sterile' way today, but there were many bones that had been discovered decades ago, and handled by scores of archaeologists and curators.

The team had looked at more than seventy Neanderthal fossil bones, and tested them first to see if they were *likely* to contain any usable DNA by checking the condition of other organic molecules: amino acids. Six of the specimens had good levels of these protein building blocks, so there was a good(ish) chance that some DNA might be in them as well. They went on to extract DNA, but, always aware that this genetic material could come from modern people, they checked for contamination before going any further.

A sample from a fragment of Neanderthal bone from Vindija Cave in Croatia looked particularly promising. 'Luckily for us, this shard of bone was not interesting enough morphologically to have been handled and looked at a lot – so this guy is nearly free from contamination by modern humans,' said Ed.

So the geneticists chose to try out DNA sequencing procedure on the extract from the Vindija fossil. This technology is advancing at an astonishing rate. Inside insignificant looking white boxes in genetics labs there are small trays holding hundreds of wells of DNA fragments. And the genetic material they are dealing with is *very* fragmentary: over time, long stretches of DNA that start off with millions of base pairs become broken and broken again into short sections of just a few hundred or tens of base pairs each. So the process involved sequencing those fragments and then virtually sticking them back together. New technology meant that many different fragments could be sequenced at the same time. 'The throughput for DNA sequencing is hundreds of times more than it was just three or four years ago,' Ed told me.

He explained the sequencing method in a very visual way (considering you can't actually open up the box and watch it in action). In each well, there were many copies of one strand of DNA, and the machine worked out the sequence by 'asking' each strand what nucleotide base (A, C, T or G) was next. It did this by flowing a solution over the wells containing each base in turn. If the 'next' base was T, the solutions of A, C and G would flow over uneventfully. When the solution containing T was introduced, enzymes would grab the base and at the same time emit a flash of light. This is called 'pyrosequencing'. 'Every flow, you've got different wells lighting up, like a firework display,' said Ed. Every time a nucleotide solution passed through, some of the wells would answer 'yes' by emitting a flash. The machine cycled on and on, until all the strands in all the wells had been sequenced. This technique can read segments of 100–200 nucleotides in length: perfect when you're looking at tiny fragments of an ancient genome.

Many of the sequences had turned out to be bacterial, but that's exactly what the geneticists expected. But comparing the sequenced fragments with human, chimpanzee and mouse genomes, a good percentage of them looked primate. Then came the work of assembling those sequenced fragments into longer pieces. Eventually, if they managed to extract enough fragments, the geneticists would be able to sequence the entire Neanderthal genome.[6]

Analysis of Neanderthal DNA should be able to cast light on many areas of enquiry, not only the question of hybridisation. By comparing the differences between Neanderthal and modern human DNA, the geneticists can estimate the time of the 'split' between the lineages. At the moment, in Leipzig, that's looking as though it happened some time around 516,000 years ago. This is older than the split suggested by fossils, at about 400,000 years ago – but that's unsurprising. The genetic split would have happened in a population that was still 'together'.

This is ground-breaking science, so it's not surprising that there are still problems that need to be ironed out. And probably the most tricky one is that problem of contamination with modern DNA, which could skew results. Pääbo's Leipzig lab isn't the only place where Neanderthal genome sequencing is going on. A team led by Edward Rubin, in California, are also at it – and they published their first chunk of Neanderthal sequence in the same week as Pääbo's team. But they came up with different results and a different – even earlier – prediction for the divergence of Neanderthals and modern humans, of around 706,000 years ago.[7] So it seems that, even with all that careful screening, some contamination may have crept in, explaining the earlier dates coming out of the Leipzig lab.[8] With each lab acting as a check on the other, though, the scientists hope that they will be able to overcome these teething problems.[9] The Californian dates may seem very early indeed, but it's important to remember that this is the predicted date of divergence of the mtDNA lineages, not of the actual populations. Based on this genetic data, Rubin's team estimated that the population split happened about 370,000 years ago, which is quite a good match with the fossil data.

Another potential application for ancient DNA is in identifying bone fragments that are too small to characterise on the basis of size and shape. In fact, this has already been applied to fossils from at least two sites. A child skeleton from Teshik Tash in Uzbekistan has often been held up as the most easterly example of a Neanderthal, but some have disputed its credentials. Even further east, bones and teeth from Okladnikov Cave in Siberia, found alongside Mousterian tools, were too broken up for it to be decided if they were modern human or Neanderthal. Genetics to the rescue, then. Scientists working in labs in Leipzig and in Lyons independently extracted and analysed the mtDNA from the bones from both sites. The results showed that the Teshik Tash child had Neanderthal mtDNA, and so did two of the bone fragments from Okladnikov.[10] This study was very significant: it hugely extended the known range of Neanderthals to the east, right into Central Asia. Maybe they even got to Mongolia and China. Genetic analysis is clearly an exciting addition to the toolkit of the Palaeolithic archaeologist.

There is also exciting potential for finding out – at some point in the distant future, when we know a lot more about the functions of genes in us and other animals – more about Neanderthal biology.[6] But even now we know that at least some Neanderthals possessed a version of a gene that probably gave them red hair. The gene in question is *melanocortin 1 receptor* (or 'mc1r'). In modern humans today, mutations that impair the function of this receptor gene produce red hair and pale skin. A team of geneticists managed to extract DNA – including part of the mc1r gene, from two Neanderthal fossils, one from Spain and another from Italy. Both fossils contained a mutated version of the mc1r gene, different from any of the variants seen in modern humans. To see what effect this gene would have, scientists inserted it into cells in the lab and found that it had a partial loss of function – like the other variations in the mc1r gene that produce red hair in humans today.[11] It is important to note that this is a *different* mutation from that in modern human redheads. It doesn't imply any genetic mixing between Neanderthals and modern humans, and it certainly doesn't suggest that the redheads among us are Neanderthals!

Another particular gene that has been identified in Neanderthals is FOXP2. This is a gene that has two specific differences in humans compared with other living primates. People missing out on those human-specific changes to FOXP2 have problems in both producing and understanding speech. Analysis of FOXP2 in living people suggested that it appeared and swept through the human population about 200,000 years ago, which seemed to fit quite well with the appearance of modern humans in Africa. It suggests that 'modern' language and symbolic behaviour are uniquely human attributes, with a biological basis. Eric Trinkaus took issue with this interpretation. He argued that there was evidence for symbolic behaviour in the Neanderthal archaeological record, with intentional burial, for instance. And he found it hard to imagine how complex subsistence strategies would have appeared – from around 800,000 years ago – without complex social communication. And yet the 'human' version of FOXP2 was initially estimated to have arisen well after the split between modern human and Neanderthal lineages.[12] But a recent DNA study of two Spanish Neanderthal fossils showed that they both carried the 'human' form of FOXP2.[13] For Trinkaus, this showed that the 'much maligned Neanderthals' had a degree of human behaviour that was reflected in the archaeological record but that he felt had often been played down. But how can we explain the same version of FOXP2 existing in both modern humans and Neanderthals? Either it is much older than the earlier studies suggested, and was present in the ancestors of modern humans and Neanderthals, or it has passed from one population to the other by gene flow. The latter seems very unlikely as no other genetic studies to date had produced any evidence of gene flow.[13]

But what about the ambitious Neanderthal Genome Project? Was there any evidence for hybridisation emerging from the nuclear DNA? The key to looking for evidence of hybridisation was to concentrate on genes or other bits of chromosomes that are specific to modern Europeans (and this is a tall order as most genetic differences are shared between populations across the globe rather than being specific to one area), keeping an eye out for these sequences in the Neanderthal genome. If any European-specific DNA sequences were found in Neanderthals, this would strongly imply that there had been some sharing of genes between Neanderthals and modern humans in Europe.

When I visited the Max Planck Institute in the early summer of 2008 Ed told me that they had managed to sequence about 5 per cent of the Neanderthal genome. I asked him a difficult question, considering that the Neanderthal Genome Project was still such a long way from completion: 'If chimpanzees are about 1.3 per cent different from us, in terms of the sequence of DNA, do you have a feeling for how different the Neanderthal genome is going to be from ours?'

'Yes, we do,' he replied. 'It's looking about ten times closer than the chimpanzee. But Neanderthals are so closely related to us, it's hard to speak in terms of percent differences. It really depends on which Neanderthal and which human you're talking about.'

'And have you seen any suggestion at all of hybridisation with modern humans?' I asked.

'No. There's no evidence to date of any hybridisation between modern humans and Neanderthals,' he replied. 'But by the end of the summer we should have 65 per cent of the Neanderthal genome, so we'll be able to give a much more definitive answer then.'

This question of what happened when modern humans walked into Neanderthal territory was fascinating. I asked Ed what he would have done if he'd met one of our cousins.

'If I came face to face with a Neanderthal, the first thing I would do is ask for a DNA sample,' said Ed, ever the scientist.

So far, then, Neanderthal genetics has shed light on how far this ancient species ranged across Europe and Asia, has shown that they possessed the same 'language gene' as modern humans (although it must be stressed that the development of language cannot be linked to just one gene), and that some of them had red hair. And, bearing in mind that there was still a lot of genome left to sequence, there was no evidence – yet – for any mixing between Neanderthals and modern humans in Europe. (Nearly a year after I visited Leipzig, Svante Pääbo announced the completion of the first draft of the Neanderthal genome – 63 per cent of it, over three billion bases – at the annual meeting of the American Association for the Advancement of Science in Chicago. There was still no sign of interbreeding with modern humans.)[14]

But it's also important to remember that the conclusions from genetic studies like this can never rule out *any* hybridisation. Perhaps it's just that Neanderthal lineages have not survived to the present day, and maybe some Neanderthals had modern human genes – just not the ones whose genomes were being sequenced.

Does this make the whole endeavour futile? Far from it. If there is no evidence of mixing, then we can at least say that hybridisation didn't happen at a level that we could consider to be significant, and so it cannot explain the apparent disappearance of Neanderthals from the fossil and archaeological record: they cannot have been absorbed and assimilated into 'modern' populations.

Thus far, all the genetic studies suggest that any hybridisation was, at the most, insignificant. And, actually, when you take a closer look at when and where Neanderthals and modern humans were living in Ice Age Europe, this makes some sense. There are only two areas where the dates for modern humans and Neanderthals actually coincide: in southern France and in south-west Iberia, in the period between 25,000 and 35,000 years ago.[15] Even then, they could have missed each other by hundreds or thousands of years, so the opportunities for inter-species sex would have been extremely few and far between anyway. So it's not really surprising that no 'Neanderthal' genes have been found in the modern gene pool – or vice versa.

So that means that the Neanderthals – whether or not rare liaisons led to hybrids whose existence has now been expunged from the modern gene pool – really did disappear. But why did the Neanderthals, who had been living in Europe for hundreds of thousands of years, fade away when modern humans arrived on the scene?

I needed to look more closely at the archaeological evidence: was there any difference in the way modern humans and Neanderthals were subsisting in their environment? Was there anything that could have given modern humans 'the edge' in Europe?

Treasures of the Swabian Aurignacian: Vogelherd, Germany

In a complete contrast to the ultra-modern Institute in Leipzig, I next visited the medieval university town of Tübingen. I walked up cobbled roads to a castle where I passed through a great arch into a courtyard, then on past a fountain and up stone steps, then turned a corner to enter the Department of Early Prehistory and Quaternary Ecology. At the end of a corridor plastered with posters of wonderful carved animals and birds, I found Professor Nick Conard in his office.

Nick's office was lined with red cupboards on one side, dark wooden bookshelves on another, and wooden filing cabinets. There were two desks, each piled high with papers and books, and in one corner was a large grey safe with a map of the Swabian Jura hanging on it. Nick had spent years excavating sites around Tübingen, where he had discovered evidence of the earliest modern humans in Europe. But it wasn't just stone tools that he'd found: there had been some rather wonderful pieces of art and musical instruments. And he had some of them in the safe. I had to look away while he found the key and then started bringing small cardboard boxes over to a low table, where we sat down to open the boxes of treasures.

The first object Nick took out, dating to around 35,000 years ago, was an ivory flute. It was discovered in 2004, at a cave site called Vogelherd, lying beneath two other flutes that had been made from hollow swan bones. The ivory flute had taken much more craftsmanship, though: it had been carved out of a mammoth tusk, then split to hollow out the inside, and joined back together with something like birch pitch. There was a row of incised notches down each side, crossing the join, perhaps made to help when putting the two halves back together.

The ivory flute had been smashed up into fragments, which archaeologists had found and carefully pieced back together; the notches had also helped the archaeologists when it came to reconstructing the flute. Nick explained that, using mammoth ivory, the instrument-maker wouldn't have been constrained by the dimensions of a hollow bird bone and so could make a much larger, longer instrument. But it also seemed to be an exhibition instrument – designed to show off the technical skill of the instrument-maker. Nick had been completely taken by surprise by this discovery. They had found mammoth ivory carvings in Vogelherd before, but this was the first indication of music that had emerged from the site. The three small flutes represented the first real evidence of music – anywhere in the world. Nick had a replica of one of the swan-bone flutes, which I tried to play with less than impressive results, not being any sort of musician. But I could at least get a series of notes out of it. More accomplished musicians have tried and produced music that sounds quite harmonious to the modern ear, with tones comparable to modern flutes or whistles.

Opening the other boxes, Nick brought out some finds from the 2006 digging season at Vogelherd, and from the nearby cave of Hohle Fels – beautiful things nestled into cut-to-fit shapes in foam inside each box. Nick lifted out a tiny ivory mammoth, just 3cm long. It was carved in the round, with naturalistic detail, its trunk hanging down and curving over to the right, and there was a tiny spike of a tail. The hind legs were shorter than the front. It seemed perfectly

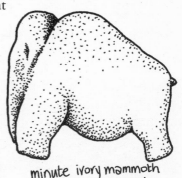

minute ivory mammoth from Vogelherd

tiny ivory bird from Vogelherd

proportioned. The bottom surfaces of the feet were scratched in a criss-cross pattern.

Then there was a lion carved in relief, again in ivory, with hatching along its back. It had a long body, and its hackles were raised. And a tiny, beautiful bird. The body of the bird had been discovered in earlier digs, and there had been much speculation about it. Was it a human torso? But then the archaeologists had discovered the head and neck – a minute fragment that could so easily have been passed over. But it fitted the body, and, suddenly, there was a bird, perhaps a duck or a cormorant, with its neck outstretched. Finally, from another small box, Nick carefully lifted out a minute lion-man. Standing just over 2cm tall, he looked like a miniature version of the famous lion-man from Hohlenstein-Stadel, near Ulm – around the corner from Hohle Fels. All of these objects dated to more than 30,000 years ago.[1]

But there was one more surprise. Nick opened a long box, and inside it was a long, smooth piece of stone, unmistakably carved into the shape of a penis, with the foreskin and glans carved into it at one end. We contemplated this bizarre object. Was it a hammer stone, carved in a phallic shape as a joke? Or could it be that this stone had a functional use more related to its shape? Nick was quietly amused by the find. It suggested that the people of the Swabian Aurignacian had, at the very least, a healthy sense of humour, and perhaps an even healthier sexual appetite.

The art of the Swabian Jura was fascinating, and this really is the earliest evidence of something that we can properly appreciate as art. I had seen pierced shells and ochre 'crayons', leaving us guessing what was drawn with them, but here were carefully executed carvings of animals, and strange therianthropic beasts – men with heads of lions. Nick said that the styles of these Aurignacian carvings were similar across different sites in the Swabian Jura, although there were many different themes. It seemed to be a time of some artistic experimentation. But recurring imagery like the lion-men from Hohlenstein-Stadel and Hohle Fels also suggested very strongly that they were made by people from the same cultural group in the Lone Valley. Many different ideas have been put forward about the meaning and function of these artefacts: some have suggested that they indicate hunting magic, and the therianthropic figures in particular have been linked to shamanism. For Nick, the discovery of the tiny waterfowl carving challenged previous interpretations of Aurignacian carvings from the Swabian Jura as representing fast and dangerous animals, with whom Palaeolithic hunters may have identified.[1]

'I think the combination of these symbolic artefacts, ornaments, figurative representations and musical instruments, shows us these people have the mental sophistication of ourselves, the same creativity that we have,' he said. 'And we can even get insights into the system of beliefs. For instance, the examples of human depictions combined with lion features show that, at least in their iconography, they were engaging in transformation: people having a connection with the animal world, being depicted as mixed animal/human figures.'

But how could these small ivory carvings hold any clue to the survival of modern humans – and the demise of the Neanderthals? Well, certainly the Neanderthals, however intelligent and whether or not they had language like us, never produced anything like the objects found at Vogelherd. I asked Nick about the differences between modern humans and Neanderthals, and it became clear that he thought culture had played a key role in the expansion of modern human, and contraction of Neanderthal, populations, during the late Pleistocene.

On their own in Europe, the Neanderthals seemed to have been getting along just fine.

'The Neanderthals were the indigenous people of the area. They had very sophisticated technology, certainly command of fire, and knew how to get along in their environment. They had everything 100 per cent under control, and they were doing very well,' said Nick.

'So if they were so good at surviving in Ice Age Europe, why did they disappear?' I countered.

'Well, I would approach that question from an ecological point of view. If you have one organism occupying a niche, it's going to stay there until something drives it out of its niche: either environmental change that makes it impossible to occupy the area, or another organism coming in and competing for resources.'

'So you're saying that modern humans were that competing organism?'

'Well, yes. It's very clear that Neanderthals and modern humans were really occupying the same niche. We see that unambiguously in the archaeological sites: the diet consists of the same foods – especially reindeer, horse, rhinos and mammoth.'

'But why did we modern humans survive and not Neanderthals?'

'Well, there's no question the Neanderthals were very effective hunters, and really were at the top of the food chain. But we do see some differences in technology. I think that the innovations that modern humans developed in Europe, the Upper Palaeolithic toolkit, organic artefacts, but also figurative art, ornaments and musical instruments – these are all things that seemed to help give them an edge against the Neanderthals.'

I found it hard to imagine why art and music might have given modern humans an advantage.

'Well, think of the lion-man,' said Nick. 'There's a lion-man from this valley, and a lion-man from the Auch Valley. It's the same iconography, the same system of beliefs, the same mythical structure, and they're the same people. And we don't see those kinds of symbolic artefacts with Neanderthals, so it seems that their social networks were much smaller than those of modern humans.

'And from my point of view,' he continued, 'the evidence even at this time, 35,000 years ago, is completely unambiguous: music was a really key part of human life. It's not entirely clear how that would give you a major biological advantage over the Neanderthals, but it seems to fit into this complex of symbolic representation, larger social networks. Perhaps music helped to form the glue that held these people together.

'When the competitor arrived, the Neanderthal way of doing things wasn't as effective in the face of people who had new ways of doing things, new technology, new culture and social networks,' explained Nick.

Whereas competition for an ecological niche seemed to have spurred modern humans on to develop wider social networks, the Neanderthals appeared to be 'culturally locked in'. It was a competition that modern humans would eventually win. Nick explained that, while their respective territories probably shifted back and forth over the centuries and millennia, Neanderthals were, on average, retreating while modern humans expanded.

'In regions like the Levant, we have good evidence for movement back and forth of the two populations. It's certainly not the case that modern humans always immediately expanded at the cost of Neanderthals; there are some good examples of Neanderthals displacing early modern humans, too.

'When the new people came in, resources got tight, and modern humans were able to develop new technologies and new solutions quicker than the Neanderthals. In a sense there was a continual cultural arms race going on. And here, in this setting, it seems like a lot of innovations took place that gave the modern humans a bit of an edge. But it wasn't a sudden, blanket devastation of the Neanderthals: there was a lot of give and take, but, ultimately, they were pinched out demographically.'

'So do you think modern humans and Neanderthals were actually in contact with each other?' I asked.

'Well, in some areas, there were fairly dense populations of Neanderthals. And I think they did meet. And I think they would have been checking each other out from a distance, often avoiding each other. That was probably the most common scenario, but there may have been times when they came together, in peaceful co-existence, and times when there was quite a bit of conflict.'

'What do you think about the question of interbreeding?'

'Any place where people come together, interbreeding is the most normal thing in the world. So I think there were occasionally encounters where interbreeding took place, but not very often, and so it didn't contribute very much to our genetic make-up or our anatomy.'

While it's difficult – even impossible – to summarise the interactions that may have occurred between the two populations over so many thousands of years, the question 'Why did we survive to the present while Neanderthals disappeared?' is still relevant. Even if members of the two species never came face to face, they were in competition with each other in the landscape. And there was archaeological evidence for different subsistence strategies, which, for modern humans, included different and possibly more flexible technology, as well as culture and complex social networks, which may ultimately explain why we are here today and the Neanderthals aren't.

Later that day, Nick took me to Vogelherd itself, in the lush Lone Valley, where excavations were ongoing. A team of archaeological technicians and students were busy digging down through the spoilheap (or 'backdirt') of the original excavation, finding plenty of evidence that had been discarded by the first archaeologists who dug there.

'The site was first dug in 1931, and all of the material was dumped outside the cave,' explained Nick as we walked past the cave entrance. 'We're systematically digging through it all to find out what they missed.'

It looked like a very pleasant place to be digging. The cave was set on a hill above an idyllic, lush, green valley. I asked Nick what it would have been like 35,000 years ago.

'If you're talking about the Ice Age, you think of ice: white, stark, and inhospitable. That's wrong. I mean, it was cold in the winter, but in the spring and summer it would be more like it is today: lots of grass, greenery, really abundant fodder for the animals. Just think about a mammoth: the archetypal animal of the Ice Age. A mammoth eats about one 150 kilos of grass every day to stay alive. The mammoth steppe was a very rich environment, and at these sites in the Lone Valley we see abundant remains of woolly rhino, mammoth, reindeer, horses, all kinds of animals.'

'But it must have been very cold during the winter here?' I suggested.

'Well, yes. But humans – and Neanderthals – can live almost anywhere as long as there's something to eat, and materials, particularly hides, to make clothing out of, and controlled use of fire.'

Nick wandered around the site, visiting trenches to see what finds had emerged that day. These included fragments of flint blades, quite typical of Aurignacian toolkits. The sediment was being bagged up as it was removed, and would be sieved. It was only through this careful sifting of the soil that Nick's team had found the fragments of ivory that made up the ivory flute, and the head of the bird carving.

But Vogelherd had contained disappointments as well as revelations. On first excavating it in 1931, archaeologists dug out some 300m³ of sediment from inside the cave, finding Middle and Upper Palaeolithic artefacts in distinct layers. The latter included a rich collection of Aurignacian tools and artefacts, which have since been radiocarbon dated to 30,000 to 36,000 years ago. The archaeologists also found modern human remains including two crania and a mandible, embedded in the Aurignacian layer. These bones, in association with the Aurignacian tools, seemed to provide conclusive evidence that modern humans were the makers of this technology. The findings from Vogelherd tallied with the discoveries at the Cro-Magnon rockshelter in France, where modern human skeletal remains – 'Cro-Magnon Man' – had been found associated with Aurignacian tools.[2] This link between modern humans and a particular technology meant that archaeologists could assume the presence of modern humans, in the absence of skeletal remains, when they found Aurignacian tools and artefacts. At other sites, Neanderthal remains had been found associated with Mousterian (Middle Palaeolithic) tools. So it seemed that each of these populations had a clear 'signature' that archaeologists could use to map out their sites and territories in Europe.

In 2004, Nick Conard and his colleagues published radiocarbon dates for the skeletal remains from Vogelherd: they dated to a mere 4000 to 5000 years ago. It looked as if they were intrusions from late Neolithic burials near the cave entrance. The 'association' with the Aurignacian layers in the cave was incidental. This disappointing result had wide implications. Vogelherd had been a key site for demonstrating that modern humans made the Aurignacian. And in 2002, radiocarbon dates had been published for the Cro-Magnon skeletal remains as well, showing them to be about 28,000 years old: too young for the Aurignacian tools in the rockshelter, although nowhere near as young as the Vogelherd bones had turned out to be.[2] The identification of modern human sites through Aurignacian tools alone was starting to look decidedly shaky.

Vogelherd, along with a generous scattering of other Aurignacian sites, including Hohle Fels and Geissenklösterle in Germany, and Willendorf in Austria, had also been used to support the theory that the 'Danube Corridor' provided a route for the early modern human colonisation of central Europe.[3] But the radiocarbon dates of the Vogelherd and Cro-Magnon bones meant that archaeologists could no longer assume that the spread of Aurignacian technology represented the ingress of modern humans in Europe. As shocking a suggestion as it may seem to Palaeolithic archaeologists who have relied on Aurignacian tools as signs of modern humans, there was now nothing to suggest that this was a reasonable assumption to make. Indeed, there was nothing now to refute the hypothesis that Neanderthals might have made those tools and even those beautiful ivory carvings and flute from Vogelherd.[2]

When I interviewed Nick, however, he did not seem like a man about to consign a lifetime's work to the dustbin of prehistory. There was still reason to think that it was modern humans who made the Aurignacian. The Oase skull and mandible showed that modern humans were *in* Europe, close to the Danube, by 40,000 years ago. The Aurignacian appears suddenly in Swabia, and is always 'on top of', i.e. later than, Mousterian (Middle Palaeolithic) archaeology, and in some places where both Middle and Upper Palaeolithic artefacts have been found, there has been a distinct gap, an 'occupation hiatus' between them.[1] It seems too much of a coincidence to think that the existing, Neanderthal population of Europe would have started manufacturing a completely new-looking toolkit just as modern humans arrived on the scene: it seems more likely that it was the moderns who brought that technology in.

And there were still some sites that seemed to hold up that association between modern humans and the Aurignacian. The site of Ksar 'Akil in Lebanon is important in this respect, as it produced a modern human skeleton alongside that 'transitional' half-Middle, half-Upper Palaeolithic 'pre-Aurignacian' industry. The burial dated to 40,000–45,000 years ago, and the layers overlying the skeleton were full of classic Levantine Aurignacian tools.[4, 5]

The site of Mladeč in the Czech Republic, which was first excavated in the nineteenth century, has produced over one hundred modern human fossils, in association with classic Aurignacian tools, including bone points. In 2002, a date was published for the calcite overlying the skeletal remains: 34,000–35,000 years ago.[6] This looked like a good candidate for reaffirming the link between moderns and the Aurignacian, but a direct date on the bones themselves was needed to clinch it. A few years later and a radiocarbon date for the fossils themselves was published, placing them at around 31,000 (uncalibrated radiocarbon) years old.[7] Although less well dated than Mladeč, there are also several French sites where modern human remains and Aurignacian tools have been found together: Les Rois and La Quina in the Charente, and Brassempouy in the Pyrenees.[5]

Palaeolithic archaeologists everywhere must have breathed a huge sigh of relief. Once again – and actually even more surely than before the Cro-Magnon and Vogelherd redating upset, now that it was based on direct dating of human remains – the assumption that the Aurignacian – everywhere – was made by modern humans seemed well founded.

'We don't have Neanderthal bones, or modern human bones, here,' said Nick. 'We have dates that correspond to the period of the last Neanderthals and the first modern humans – in theory, it could be either. But the most plausible explanation is that it's modern humans.'

———

Having seen the beautiful artefacts from Vogelherd, I left the dig and Nick Conard, and drove a mile or so down the valley to meet experimental archaeologist Wulf Hein for a practical lesson in the differences between Middle and Upper Palaeolithic technology. I wanted to get to grips with the Mousterian and the Aurignacian – literally.

Wulf Hein was an expert flint knapper and aficionado of all things Palaeolithic. His car was full of crates of flint, various spears, spear-throwers, bows and arrows. The finished objects were beautiful, but I really wanted to know how the flint tools were made, and to see the Levallois and prismatic core techniques in action.

We sat in a field in the idyllic Lone Valley, having spread out a blanket to catch any stray bits of flint. In this archaeologically rich area, Wulf was very concerned that he didn't confuse any archaeologists in the future by adding new, twenty-first century flint tools to the archaeological record. He took a Levallois core out of a crate; it was flattish and he had made it into the shape of a tortoise shell by knocking flakes off the periphery – making it into a 'prepared core'. He struck it with a pebble – expertly – and a large flake detached itself from the middle of the disc-like core. Middle Palaeolithic technology starts off with simple flakes struck from cobbles. The Mousterian industry of the Middle Palaeolithic takes this a stage further, with flakes produced from a prepared (Levallois) core, just as Wulf Hein had demonstrated. It is named after the site of Le Moustier in the Dordogne, where Neanderthal fossils were found alongside their characteristic tools.

Wulf then took a stone that had been carefully prepared into a cone shape out of the crate, and handed it to me: I was to make a blade from this prismatic core. Under his expert guidance, I gripped the core between my knees, held the tip of a piece of antler close to the edge of the stone, and struck the antler with a pebble. A long, thin and extremely sharp blade detached itself from the side of the core and fell to the ground.

'Oh, nice one!' exclaimed Wulf.

'Are you proud of my work?'

'Yes, I am. I'm astonished. It took me forty years.'

I think my success had more to do with beginner's luck – and having a good teacher.

There it lay, a blade, the hallmark of the Upper Palaeolithic, the foundation of the classic Aurignacian industry. This technology was named after the site of Aurignac in the lower Pyrenees, excavated in 1860. The sort of tools that characterise the Aurignacian include long, thin slivers of flint called *lames Aurignaciennes* (Aurignacian blades) – like my blade but retouched all around the edges, as well as end-scrapers and burins, carinate (keeled) scrapers and tiny bladelets. The functions of these tools are being debated – especially the carinate 'scrapers' and the bladelets that are made from them. Are the unretouched bladelets waste products from making such a scraper? Or are the bladelets really what the maker was after, and the 'scraper' is actually just a small core – not a tool in itself?[5] Even though the precise functions of all these pieces of flint aren't yet understood, the shapes are very characteristic.

Archaeologists have long believed that knapping blades from a prismatic core was much more efficient than the old-fashioned Levallois technique: the manufacturer can produce many blades from one prismatic core. In comparison, the Middle Palaeolithic Levallois technique just gives you a few flakes from one tortoise-shaped core. However, I later met up with experimental archaeologist Metin Eren, in Exeter, who had spent years making and comparing Middle and Upper Palaeolithic flakes and blades. His results had been surprising: the initial preparation of the prismatic cores actually produced more waste than discoidal cores, and thin blades didn't last as long as flakes. So, in terms of producing usable cutting surfaces, it seems that there actually wasn't much to choose between discoidal and prismatic cores, flakes and blades, in terms of efficiency.[8]

Now I had seen very early evidence of (small) blade production, in Africa, Europe and Arabia, going right back into the Middle Palaeolithic. But there's more to the Upper Palaeolithic than prismatic cores and long blades. Bone and antler tools are also seen to be characteristic of this culture. These materials do (rarely) pop up in other Middle Palaeolithic contexts – like Blombos Cave and Howiesons Poort in South Africa. Perhaps this isn't surprising for those are modern human sites.[9] (Just to throw a spanner in the works, there's some evidence of Neanderthals making bone points as well.) Other elements that are considered particularly characteristic of the Upper Palaeolithic are grinding and pounding stones, suggesting more processing of vegetables was going on, and widespread use of body decorations like shell, tooth and ivory beads. (Now, I had seen evidence of much earlier use of ochre and ornamentation, but before the Upper Palaeolithic it is very patchy.) There appears to have been much more long-distance transport of raw materials, sometimes over hundreds of kilometres, compared with generally shorter distances in the Middle Palaeolithic. Carved figurines – as I had seen from Vogelherd and Hohle Fels – appear in the Upper Palaeolithic, along with cave painting (more of which later). It's not straightforward, but, as a package, the Upper Palaeolithic does seem to be something special: it is still a useful category.

Moving beyond the Aurignacian, improved hunting tools appear, like spear-throwers in the Gravettian, and, eventually, bows and arrows and boomerangs.[9] Wulf had brought some of these along with him. The atlatl – or spear-thrower – was a very simple tool, essentially a stick, about half a metre in length, with a hooked end that fitted into a recess in the end of the spear.

'This spear-thrower has a beautiful carved antler end. We don't know what the original shaft looked like, but this is a reconstruction. It's interpretation, but it works.'

'I'd like to see how far you can throw these slender spears, with and without a spear-thrower,' I challenged him.

'OK, and you will be astonished.' Wulf took up the gauntlet.

He threw a spear without the spear-thrower first. 'That was 220 grams thrown by hand,' he said. Then he got ready to throw another spear, fitting it into the hook and laying it above the spear-thrower, gripping them both with one hand. 'And this is 220 grams thrown with a spear-thrower – with the same power.'

He threw it – rotating the spear-thrower so that it became like an extension of his arm – and the spear flew off ...

It was extraordinary: it had double the range. I had a go as well and amazed myself by how much further I could throw the spear with the atlatl. It was such a simple but impressive piece of kit.

'If you're trained you can get it even further. The record is 180 metres,' said Wulf.

Wulf had more examples of Stone Age projectile weaponry with him.

'At the end of the line stands the bow and arrow,' said Wulf. He had a beautiful replica of a Mesolithic bow with him. It was found in Denmark, preserved in a bog called Holmegaard, near Copenhagen. 'The original is about 8,600 years old. It's the oldest bow we've ever found.'

Even earlier arrows have been found, dating to more than 11,000 years ago, and some archaeologists argue that there is evidence, albeit fragmentary, going even further back, into the Upper Palaeolithic. It seems that the development of bows and arrows was related to ecological changes: they appear as the world was warming up after the Ice Age, and Europe was becoming wooded. 'If you're hunting in the woods, a bow and arrow is much more effective than a spear-thrower. And it's easier to aim,' said Wulf.

I enjoyed playing with Wulf's 'Mesolithic' bow and flint-tipped arrows, and I am very sorry to say that two were left behind – somewhere in the long grass by the stream in that field; after so much time had been spent looking for them, we had to accept that the arrows were gone. Perhaps they will be found by archaeologists one day.

Neanderthals, as far as we know, never made spear-throwers or bows and arrows, although they did progress from using thrusting spears to throwing spears. But does that really indicate an innate technological superiority of modern humans? Stone Age weapons expert John Shea has argued that the development of true projectile technology was key to our species' ecological success, giving our ancestors an advantage in hunting and even providing them with long-range weapons that could be used to eliminate rivals, of our own species and perhaps others as well.[10] But there's no direct evidence to suggest that spear-throwers were ever used against Neanderthals. And the first sign of the bow and arrow is well after the LGM, when Europe was warming up and woodland was returning – long after Neanderthals had disappeared from the landscape.

But I'm getting diverted here by later developments of the Upper Palaeolithic. Sticking with the Aurignacian, which was contemporary with the Mousterian, there is still a distinct difference between Middle and Upper Palaeolithic. Bar-Yosef[9] argued that the characteristic elements of the Upper Palaeolithic are 'evidence for rapid technological changes, emergence of self-awareness and group identity, increased social diversification, formation of long distance alliances, [and] the ability to symbolically record information'.

Across Europe, then, the change from Middle to Upper Palaeolithic cultures, between 40,000 and 30,000 years ago, is taken to represent the replacement of Neanderthals – bearing Middle Palaeolithic technology – with modern humans, carrying with them Upper Palaeolithic tools and artefacts. It has been called the 'Upper Palaeolithic Revolution', but this is a problematic label as it suggests that the late Pleistocene, in Europe, is the time and place of some kind of emergence of 'fully modern' behaviour. This is difficult to argue, as this pits modern humans in Europe not only against the Neanderthals but also against modern humans elsewhere. Were the people in Africa and Asia, still making Middle Palaeolithic stone tools, cognitively inferior to the Europeans? This is obviously a very divisive and Eurocentric viewpoint, and recalls that comment by Movius about Eastern Asia being an area of 'cultural retardation'. It seems much more likely that modern behaviour was essentially born at the same time as our species, and that modern humans came out of Africa, as Oppenheimer puts it 'painting, talking, singing, and dancing'.[11] And, as we have seen, the African evidence supports this idea.

But there is no doubt that the Aurignacian represents a new sort of culture. So how can we explain it? Well, culture is something that represents human interaction with the environment, and with other humans, so changes in culture can be seen as being driven by changes in climate and environment, as well as changes in society – and rather than a biological change. The Aurignacian didn't require a new brain, or a few new genes. It was a new product of a brain that was already well equipped to develop behavioural solutions to environmental challenges. We have seen that the hallmark of modern humans elsewhere was ingenuity, adaptability and inventiveness. So it seems utterly reasonable to suggest that, in Europe, the new, modern human culture of the Aurignacian represented adaptations to a new environment – perhaps including the existence of a competitor in the landscape.

The Aurignacian objects that have been passed down to us from our earliest ancestors in Europe show us that much more was going on on top of the change to a new method of tool manufacture. They show us that those people were flexible and adaptable, and forming complex social networks. They seem to indicate that these people felt *part of* something which extended far beyond their immediate families and familiar landscapes. Archaeologists argue endlessly about classification and naming of toolkits. The arguments can seem – especially to someone more schooled in bones than stones – incredibly complicated and esoteric. But the arguing itself means that this was an interesting time: several things are happening all at once: new stone tools are appearing in Europe, and at the same time there are new social structures emerging.[9] Maybe the new styles of stone tools are part of a badge of identity, a fashion, for those early modern humans in Europe.[8]

The Aurignacians spread inexorably across Europe. Perhaps at times the Neanderthals regained territory, but, ultimately, they were to disappear from the landscape. Their traces gradually contract and disappear, and the last Neanderthals seem to have occupied a lonely outpost in the very south-west corner of Europe: Gibraltar.

Tracking Down the Last Neanderthals: Gibraltar

The first Neanderthal fossil to be found was in Gibraltar – a skull blasted out of Forbes Quarry, in 1848[1] – but no one recognised it at the time, so the German find, six years later, had the honour of giving the species its name.

The Gibraltar skull, probably that of a female, remains one of the best-preserved Neanderthal fossils. In 1926, another Neanderthal fossil – parts of the skull of a four-year-old child[1] – was excavated from the Devil's Tower site on Gibraltar. So Neanderthals certainly lived on the Rock – and more recent excavations have revealed quite challenging details about their way of life, and even why they might have died out.

As I flew into Gibraltar, it was much smaller than I expected it to be: a 6km-long rocky headland sticking off the south-west corner of Spain into the Mediterranean. The runway of the airport was crossed by the main road into Gibraltar. From the plane window I could see the white cliffs of the limestone outcrop, dotted with square caves. The Rock

is riddled with caves and tunnels, some
natural, but many man-made. A great deal
of them date from the Second World War, when
civilians were evacuated and Gibraltar became a
military fortress. The military presence on Gibraltar
is still very evident today, but the Rock is also home
to nearly 30,000 people. Sheer cliffs, more than
400m tall, rise out of the sea on its east side, but
on the west the houses and hotels of
the town cluster at the foot of a more
gentle slope.

   I had travelled to Gibraltar to see
Clive and Gerry Finlayson, and we met
up on the terrace of the Rock Hotel, looking out
over the harbour. They had studied the Gibraltar
Neanderthals for years, and were keen to set the record
straight when it came to perceptions of these ancient
people as brutish, cold-adapted savages. In fact,
the evidence emerging from Gibraltar seemed
to indicate people much more like ourselves.

gull over the Rock

   Clive's background as a zoologist and ecologist informed his approach
to Palaeolithic archaeology. He wasn't into constructing theories based on
stone tool typologies, but favoured a landscape-first view where humans
(modern and Neanderthals) were firmly placed in their environmental
context. He saw differences in technology as very much driven by
ecological and social changes – not the other way around. From this
perspective, technological and cultural differences represented quantitative
rather than qualitative differences between the people making them. He
was clearly not a supporter of the Upper Palaeolithic revolution, and
argued that there was no sudden appearance of 'modern behaviour'. He
wasn't suggesting that there were no differences between modern humans
and Neanderthals, but rather that those differences were shaped by the
environments in which cultures emerged, and by the social structures
adopted by each population.[2] The Finlaysons' research in Gibraltar had
provided evidence not only for late survival of the ancients in this corner
of Europe, but for something which I had, perhaps naively, previously
thought of as being exclusively 'modern' behaviour: the Neanderthals
had beachcombed.

   Early the following morning, I met up with Clive down at the
marina, and we took a boat out, heading around the point to the cliffs on

the east. As we set out, the sea was as smooth as glass, and we seemed to glide along, the Rock bathed in golden sunlight and looking magnificent. There were natural caves, half in, half out of the water, all along the east side of the Rock.

'There are over a hundred and forty caves in the whole of the Rock, and there's probably twenty or thirty just along this cliff,' Clive told me. 'We get the feeling this was almost like a "Neanderthal city". On Gibraltar, we've got ten sites, two with fossils and the other eight with tools: occupation sites. That's probably the highest density that's been recorded anywhere. And bear in mind, that's just the sites where the archaeology has been preserved to today. It's difficult to say how many people were living here, but I imagine the Rock would have had maybe a hundred Neanderthals living on it.'

Clive wanted to show me one site in particular where he had been excavating in search of Neanderthals for several years – with very successful results: Gorham's Cave.

'Most of the evidence is submerged or has been lost. So we're lucky to have Gorham's Cave: the sea hasn't washed the sediments away there. And we've found a lot of evidence of Neanderthal occupation: stone tools, animals that they've butchered, hearths – Neanderthal barbecues, if you like.'

The site had first been excavated in the 1950s, then had lain forgotten for decades. Clive had started excavations again in 1991.

'We come every year and excavate the site. It's huge. There's eighteen metres' depth of archaeology in there, and there are new results coming out every year.'

The results of radiocarbon dating on charcoal from the cave showed that it was occupied until 28,000, and perhaps even as late as 24,000 years ago. It made this cave the last known Neanderthal outpost.[3]

Some archaeologists have argued that these late dates for the uppermost Mousterian level in Gorham's Cave are due to contamination with charcoal from even higher, Upper Palaeolithic layers. But Clive has argued that the stratigraphy inside the cave is reliable, and that indeed, after the last Middle Palaeolithic occupation, the cave seemed to have lain abandoned for 5000 years. After 19,000 years ago, the cave was reoccupied, by modern humans bearing a later Upper Palaeolithic toolkit called the Solutrean.[3]

Clive didn't like the idea of Gibraltar as a 'refugium' for Neanderthals. 'When we talk about a "refuge", the impression is almost that this is a place that Neanderthals came because there was nowhere else to go,' he said. 'The reality is that this was a good place to be. And for a period of 100,000 years, there were Neanderthals living here.'

Even when Europe was beginning to feel the full chill of the LGM, between 28,000 and 24,000 years ago, south-west Iberia was experiencing a mild but still balmy Mediterranean climate, with mean annual temperatures of around 13–17 degrees C – in fact, very similar to today.[3, 4]

The large caves that now lie right on the rocky coast would have been set back from the seashore 25,000 years ago. Offering protection from the elements and from predators, I imagined the caves would have made quite sought-after Palaeolithic homes. Hearths had been lit deep inside Gorham's Cave – but it was tall enough for smoke to rise to the ceiling and find its way out without getting into the eyes and throats of the cave-dwellers. And there were plenty of resources just on the doorstep. The sea would have been much lower than it is today, with a mosaic of habitats including woodlands and wetlands between the foot of the cliffs and the coast. The sandy coastal plain would have been dotted with sand pines and junipers, and liberally watered with streams. 'The vegetation also indicates seasonal pools,' explained Clive, 'and that's also borne out by animal remains that we've found, including newts and frogs, and waterfowl like ducks and coots.'

We had hoped to access the cave from the sea, but just an hour or so after we had set out on a millpond, the winds and currents had conspired to create a sizeable swell that would make any landing attempt dangerous. We continued northwards in the boat, Clive pointing out other caves that had been home to Neanderthals. But, of course, with the change in sea level, it was likely that a lot of evidence was under the waves. And Clive and Gerry were looking there, too.

Gorham's Cave, Gibraltar

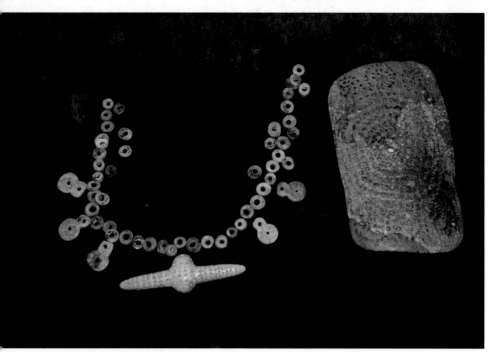

Beads and ivory plaque from Mal'ta, Siberia: beautiful objects from the time of the Last Glacial Maximum (the peak of the last Ice Age), now kept in the Hermitage collection

People and reindeer assemble for the festival on the frozen river at Olenek, Siberia

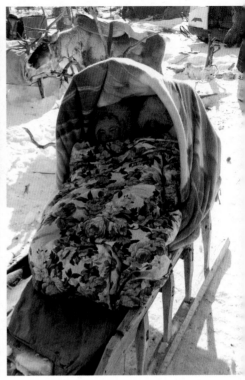

A little Evenki girl riding her reindeer

Evenki babies tucked up ready for the journey

The welcome committee at Olenek, Siberia

The view from Zhoukoudian, China

Bamboo forest above the village of Zhujiatun. Our ancestors may well have used local supplies of bamboo to make tools

Bamboo is used today for a huge variety of purposes in China, from building houses to making baskets

The village of Zhujiatun

One of the Liu brothers hooking pots out of the ashes, near Guilin, China

Long Ji rice paddies: rice was just one of the crops grown by the first farmers of South Asia, but today, rice farming dominates the landscape, and provides China's staple food

Europe

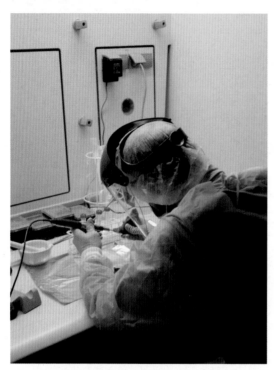

The Neanderthal Genome Project in action: a researcher drills out a sample of Neanderthal bone ready for DNA extraction and sequencing

The Max Planck Institute for Evolutionary Anthropology, Leipzig, Germany

Wulf Hein flint-knapping in a field near Vogelherd, Germany

Rounding the southern point of Gibraltar, on my way to see Gorham's Cave – the last known place inhabited by Neanderthals

A cast of the skull of Gibraltar woman alongside a selection of finds from Vanguard Cave, including the dolphin vertebra (immediately below skull)

Michel Lorblanchet – with a beard full of charcoal – making a twenty-first-century hand stencil

Ice Age art: the horses painted on the cave walls at Pech Merle, France. Cave art expert Michel Lorblanchet believes that the same person created the horses and then made hand stencils around them – perhaps as a signature

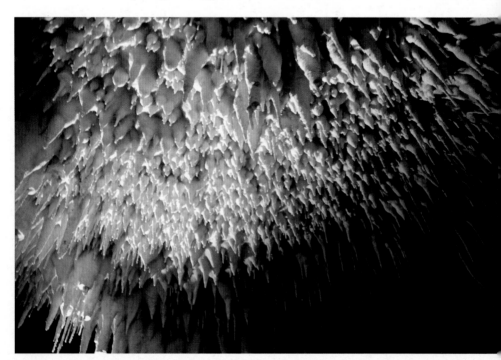

Stalactites dripping from the roof of Cougnac Cave, France

Turning theories on their head: one of the monumental standing stones, with carved reliefs of animals, at Göbekli Tepe, Turkey

A half-excavated standing stone – T-shaped and adorned with mysterious animal images

One of the circles of huge standing stones at Göbekli Tepe. Amazingly, it seems that these massive temples were built by hunter-gatherers

Meeting Chief Big Plume, resplendent in eagle feathers and weasel skins

Echoes of the Ice Age: the Ipsoot glacier, Canada

Filming in La Brea tar pits, Los Angeles, where skeletons of Ice Age beasts have been preserved for thousands of years

Models of mammoths at La Brea tar pits

The skull of Luzia, dated to 13,000 years old, and one of the earliest known Americans

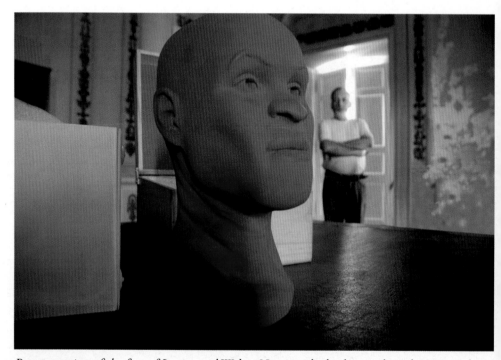

Reconstruction of the face of Luzia, and Walter Neves in the background, at the National Museum in Rio, Brazil

Observing life on the Amazon today: a fisherman throws a weighted net into the water

Geologist Mario Piño, one of the team who excavated the extraordinary site of Monte Verde, Chile

Chinchihuapi Creek and the site of Monte Verde

Turning around, we headed back around Europa Point and into the Bay of Gibraltar, to rendezvous with the dive boat. Already on board were Gerry and marine ecologist Darren Fa. As we transferred on to the boat, I learnt that work had already started: somewhere beneath us, two underwater archaeologists were busy trying to move a large rock, using air-filled lift bags. Gerry was hoping that removing this rock would reveal an undisturbed layer of sediment. Surveys of the seabed had revealed a large reef, some 20 to 40m beneath the waves. Caves in this reef would have been on the coast – rather than underwater – during the time of the Neanderthals. The archaeologists were interested in collecting samples which would provide even more information about the palaeoenvironment, but they were also hoping that they might find signs of ancient occupation hidden beneath the seabed. Mounting an archaeological investigation of this type requires expertise in diving, marine ecology, underwater archaeology – and patience.

'The logistics involved are really difficult,' said Gerry. 'It's expensive and very time-consuming. If you're excavating in a cave and your pencil breaks, for example, you can just go to your box of tools and get another one. If you're on the bottom of the sea and your pencil goes, and you don't have another, that's the end of your dive. We take two of as many things as possible. We plan very carefully. But sometimes things go wrong, and you have to come back to the surface.'

Each diver could stay down for only an hour at most – so each day's digging was about meticulous planning, and small but effective steps towards collecting data. They worked in shifts; Gerry and Darren got into their wetsuits and dive gear ready to step in as the first team resurfaced.

Out on the boat, I could clearly see the mountains of Morocco in the distance: the Rock of Gibraltar lies just 21km across that narrow strait from Morocco. Once again, I found myself surprised that there was no evidence that this had ever been a crossing point in Palaeolithic times. Modern humans appeared to have stayed firmly put on the African side, and there's nothing to suggest that Neanderthals ever made the journey over to Morocco either.

When Gerry and Darren came back up, I asked them what they'd found.

'We actually managed to move the rock using the lift bags,' said Darren. 'And we got some sediment,' said Gerry. She was very pleased that they had made steady progress and managed to safely shift the rock, opening up the previously covered sediments for excavation. I had sort of hoped that one of the divers might bring up a nice Mousterian tool from the seabed on the day I was there, but that was a bit too much to hope for. It was slow and painstaking work – but essential if we are to learn more about the ancient environments that our ancestors and Neanderthals inhabited.

Back on dry land, Clive was keen to show me some of the animal bones that had been found in Vanguard Cave, the neighbouring cave to Gorham's, occupied by Neanderthals over 50,000 years ago.

I picked up a metapodial, a foot bone. It looked a bit like a sheep bone to me. 'That one,' said Clive, 'well, either it's their favourite, or it's easier to catch, or there's a lot more of them than other herbivores. It's a wild goat: Iberian ibex. Eighty per cent of the large mammal bones belong to this species.'

Red deer bones were also fairly common. 'They were quite partial to those.' There was also a bone from an aurochs: a massive, ancestral cow. Clive thought that bringing down an animal of this size, with its huge horns, would have required a certain degree of courage, planning and cooperative hunting. 'It's probably not remarkable that we don't find that many of them,' he said.

'Perhaps surprisingly, given what has been said about Neanderthals, we're also finding smaller animals. Nearly 90 per cent of all the mammal bones are rabbit.'

I did find this surprising. It didn't fit with the traditional view of Neanderthals as big-game hunters.

'And it's not just land resources that they're using,' said Clive, rather proudly. 'They're eating limpets and mussels. And just look at this,' he picked up a jaw with sharp teeth. 'It's a monk seal. And this isn't an isolated case. A lot of the bones have cut marks. I think there's too many of them to be stranded animals; I think they're hunting them. And it gets even more complicated when you look at things like this ...'

He handed me a dolphin's vertebra. I could just make out some cut marks on one of the bony levers – the transverse processes – sticking out from the body of the bone. These weren't trowel marks from a clumsy modern archaeologist: they were thin cuts from a flint tool. It appeared that we were looking at the remains of a Neanderthal's dinner.[5] So it looked like Neanderthals were just as capable of adapting to a coastal way of life as our ancestors had been. There were bird bones as well – but could Clive be sure that these had been eaten?

'There's very little evidence for other predator action. We don't have cut marks on these bones, but it seems that there may be tooth marks – from Neanderthals. The dominant species are partridge, quail, ducks:

the kinds of birds that you or I might find palatable. Clearly, birds are on the menu.

'We've looked so much at Neanderthals in the north where large mammals were available, and it's created a slightly biased impression,' suggested Clive. 'Here, big game was in the diet,' he continued, 'but I see the Neanderthals more as beachcombers, collectors of plants, hunters of birds and rabbits, who occasionally caught a goat, more occasionally a deer, and even more occasionally got a large, dangerous animal. Like human societies today, I'm sure the Neanderthals had cultural and geographic diversity. They were exploiting whatever was available on their doorstep.'

Other sites also challenge the idea of the Neanderthals as behaviourally inflexible. Arcy-sur-Cure in south-west France has produced stone tools that are rather weird – almost sitting on the definitional fence between Middle and Upper Palaeolithic – combining Mousterian-like flake-based tools with 'Upper Palaeolithic' blades and bone tools. While it appeared to have developed out of the Mousterian tradition, it was seen as marking the beginning of the Upper Palaeolithic in Europe, and used to be called the 'Early Aurignacian'. Along with the 'transitional' toolkit, there were also pierced fox canines, apparently evidence of personal ornamentation. The site is not unique: the tools from Arcy are classified as *Chatelperronian*, after the cave at Chatelperron in south-west France where this technology was first described. Examples of this technology have been discovered throughout central and southern France, and in northern Spain. But who made it? Fossils had been found in association with Chatelperronian tools at Arcy-sur-Cure, but they were so fragmentary that it seemed impossible to tell if they came from modern humans or Neanderthals. They dated to 34,000 years ago – when both populations were in Europe.

But then, in 1996, an international team including French archaeologist Jean-Jacques Hublin and Fred Spoor from UCL showed that the anatomy *inside* the temporal bone – the bone at the side of the skull that contains the workings of the ear including the semicircular canals – was characteristically Neanderthal.[6] So this suggests that it was Neanderthals who made the Chatelperronian. But did they invent it independently or copy it from modern humans, or is this a change in technology that shows behavioural adaptation to a changing environment? Clive put it to me that it represented a change from ambush hunting to projectile technology. It seems pertinent that the Chatelperronian appears only after modern humans have come into close proximity with Neanderthals – but this was also a time of major climate upheavals.

In fact, both possibilities – copying or innovation – suggest a certain behavioural flexibility and intelligence that we have not always credited the Neanderthals with in the past. 'I think we may have underestimated the Neanderthals,' said Clive.

The Chatelperronian can be seen as rather problematic as it blurs the distinction between what archaeologists would consider to be a Neanderthal and modern human technologies. It challenges our pre-conceptions. Archaeological evidence like this makes it seem less inevitable that modern humans survived where Neanderthals didn't – and shows that the Neanderthals were much closer to us than we've perhaps liked to think in the past. But they did disappear. On a continental scale, climate changes and the presence of a competitor in the landscape probably played their roles in the demise of our sister species. Their population gradually contracted, until, it seems, just a few of them were left in Gibraltar, happily living a coastal lifestyle in this idyllic corner of Europe, and probably completely unaware that they were the last of a long European lineage. So what happened to the last Neanderthals?

From Clive and Gerry's work on Gibraltar there doesn't seem to have been a 'last stand' with modern humans taking the Rock from their distant cousins. There's a distinct gap between the last evidence of Neanderthals, at 24,000 years ago, and the first evidence of modern humans – at about 18,000 years ago. 'There's a gap of five thousand years when there's nobody living in these caves,' said Clive. So it seems that competition between the two populations could not have played a role in this particular location.

Clive suggested that, in the end, it may have just been a numbers game. If the Neanderthal population was very small, then they could have easily died out – just think of threatened species today. 'These last Neanderthal populations would have been very small and very vulnerable,' he said. Inbreeding and increased rates of congenital disease could also have contributed. But Clive also thought that climate played a crucial role in the demise of those late-surviving Neanderthals. 'We've looked at deep-sea cores offshore here, and the most severe climatic conditions of the previous quarter of a million years hit precisely at the point that Neanderthals disappear.'

Although Gibraltar had enjoyed a mild climate up to 24,000 years ago, there then seemed to be a sudden and severe downturn in conditions. Evidence from marine cores shows a drop in temperatures at this time, known to palaeoclimatologists as the 'Heinrich 2 Event'. These Heinrich Events are moments of 'ice-rafting': icebergs breaking off the northern ice sheets then drifting south in the Atlantic caused a distinct chill in the ocean. The Neanderthals had survived cold before, but during this 'event', the sea surface temperature was the coldest it had been for a quarter of a million years. For Clive, this sudden cold and dry period could explain the demise of the last Neanderthals. 'So it could be that climate change was the final nail in the coffin for these last Neanderthals.'

For Clive, the *final* extinction of the Neanderthals and the expansion of modern humans are separate events. Though he accepts that contact may have happened elsewhere, the evidence on Gibraltar shows that those late-surviving Neanderthals disappeared thousands of years before modern humans arrived on their patch.[3]

'But I think it's a long, drawn-out process [across Europe]; there's not one event that causes their extinction.'

In contrast to the traditional view of Neanderthals as cold-adapted people of Ice Age Europe, Clive Finlayson viewed them as warm-loving humans, who survived late in Gibraltar because Mediterranean conditions lasted longer there. Even in winter, days were long, and the range of environments around the coast of Gibraltar allowed people with a diversified subsistence to survive in rough times – at least until the Heinrich 2 Event. In contrast, Clive saw the *modern humans* as the more cold-adapted bunch of people. He may be right. Although Neanderthals are stocky and short-limbed – suggesting they were biologically better adapted to cold climates than rangy, long-limbed modern humans, they may not have had the cultural adaptations to beat the chill of the Ice Age. And their physical adaptations may even have held them back. Contrastingly, if modern humans were physiologically more vulnerable to cold, this may have spurred them on to develop advanced clothing – which later meant that they had the wherewithal to survive through the LGM.[7]

Certainly, at around 25,000 years ago, we see a new culture emerging in Europe, apparently brought in by a second wave of modern humans from the north-east: people who appeared to have been quite comfortable in temperatures that are more reminiscent of Siberia today.

A Cultural Revolution: Dolní Věstonice, Czech Republic

The period between 30,000 and 20,000 years ago was one of global climatic instability, leading up to the LGM.[1] During this time, a new culture and technology spread across Europe – the Gravettian, named after the site of La Gravette in the Dordogne, where the characteristic stone points of this technology were first recognised.

This culture appears to have started in north-eastern Europe about 33,000 years ago, with sites like Kostenki, on the Don River. It comes along as the world was cooling down, in the run-up to the peak of the last Ice Age, and many archaeologists see the Gravettian as an adaptation to colder, periglacial environments. Perhaps not surprisingly, then, although this culture is a European phenomenon, there are similarities with the middle Upper Palaeolithic culture of Siberia. The Gravettian is the 'technology of the steppes': these people were reindeer and mammoth hunters, who thrived in cold climates, and spread into Europe as the temperature dropped off.

Innovations in the Gravettian include better shelters – such as the 25,000-year-old semi-subterranean dwelling at Gagarino on the Don. The hearth in this Ice Age house contained burnt bone: it appears that the Gravettians were turning to alternative sources of fuel. Stone lamps appear in the archaeological record, and, at Kostenki, lamps seem to have been made from the heads of mammoth thigh bones. Just as in Siberia at this time, eyed needles indicate that people were capable of making clothes. The Gravettians also seem to have invented cold storage – digging pits (using mammoth-tusk mattocks) which may have been used to store meat or bones for fuel. The hunting technology of the Gravettians included a range of innovations: bevelled stone points, ivory boomerangs and even woven nets – perhaps for hunting small game. Compared with the Aurignacian, stone blades were narrower and lighter, often retouched into very sharp points. Some assemblages include tanged or shouldered points.[2] There also appear to have been changes in society. Massive, complex occupation sites have been found that suggest people were 'getting together' on a large scale. Whether these are gatherings for communal hunts, feasts or other social occasions is unclear, but this certainly implies that social networks were enlarging and society was getting more complex for these Ice Age hunter-gatherers.[3] Perhaps the most intriguing material evidence passed down to us from the Gravettian, though, are objects that have no obvious function: the so-called 'Venus figurines'.

I headed to the Czech Republic, taking the train to Brno, then heading south to the small town of Dolní Věstonice (pronounced 'dolny vyestonitseh'), close to where one of these mysterious female figures was found. There I met up with Jiri (pronounced 'Yurjy') Svoboda at the museum in Dolní Věstonice. Upstairs in the museum, he brought out shallow boxes full of the finds from the archaeological site. There were numerous animals carved from mammoth ivory, including a tiny, beautifully observed head of a lion. There were also strange bone spatulas, about the length and shape of shoehorns, but flat.

'What on earth', I asked Jiri, 'are these?'

'If you go to ethnography, you can find a whole variety of uses for pieces like that. You might use it for cutting snow, for example. Eskimos have tools like that. But I have also seen a similar tool in Tierra del Fuego for taking bark from trees. And the Maoris of New Zealand have similar pieces as prestigious, symbolic weapons. Unfortunately it's very difficult to find any use-wear on the edges, so we can only guess.'

And then Jiri brought out the Dolní Věstonice 'Venus'. She was a strange little figure, just over 10cm tall, made from fired clay. She was very stylised, with an odd, almost neckless head; the face was a mere suggestion, with two slanting grooves for 'eyes'. She had globular, pendulous breasts, and wide hips. Her legs were separated by a groove, and another groove ran around her hips, as though indicating some kind of girdle. On her back, pairs of diagonal grooves marked out her lower ribs. Compared with the stick-thin and often clothed female figurines I had seen in the Hermitage, from the Siberian site of Mal'ta, this 'Venus' was more rounded and buxom, and splendidly naked.

I asked Jiri what he thought she represented. He was cautious: the real meaning of these mysterious prehistoric objects is lost, and we can really only guess what they symbolised. Was she a deity? Did she represent some kind of archetypal female, or did she perhaps combine male and female sexuality in one? Jiri covered up her upper half, and the legs of the figurine, with that deep groove between them, could certainly be taken to represent a vulva. Then he covered up the legs, and her head and breasts were transformed into male genitalia.

'So perhaps she is a combination of the two symbols: male and female.'

'Do you think she could be a deity, a goddess?' I asked.

Jiri laughed. 'Well, maybe a deity. But it depends on how you define that. I think there is some kind of personification, some symbolism here, that is depicted in

Dolní
Věstonice
Venus.

the shape of a female body. There is certainly some meaning, but it's very difficult to know exactly what.'

There were other objects in the collection from Dolní Věstonice that seemed to represent sexuality in some way, including a small ivory stick with a pair of protruding bulges – which could be seen as breasts or testicles. Jiri certainly saw some connection between the archaeological signs of widening social networks and the development of symbolism represented by objects like the 'Venus'.[1]

Whatever the Dolní Věstonice 'Venus' meant to the society who made her, she was special in another way – because she was made of clay. She is among the first ceramic objects in the world – and one of some 10,000 pieces from Dolní Věstonice and the nearby site of Pavlov. At 26,000 years old, these ceramic objects pre-date any evidence of utilitarian pottery, i.e. ceramic containers, by some 14,000 years.[4] Many of the fired clay pieces were just irregular pellets, but among them were works of art – more than seventy, nearly complete, clay animals, and the Venus figurine. But there were also thousands of fragments of figurines, many of which had been found in what seemed to be purpose-built kilns up the hill from the occupation area. Analysis of the kilns suggested that they produced temperatures of up to 700 degrees C. The preponderance of ceramic fragments has led some archaeologists to formulate a rather bizarre theory: that the figurine-makers were pyromaniacs, deliberately exploding their creations in kilns – and the smashed figurines were the relics of a strange prehistoric form of performance art. Pottery specialists argued that the pattern of breaks in the clay fragments was commensurate with heat-fracturing, but I remain quite sceptical about the deliberate explosion theory. We were looking at the earliest ceramics in the world – presumably there was a fair amount of experimenting going on – so was it really that surprising that many of the pottery creations had exploded? And it seems reasonable to imagine that successfully fired figurines would have been removed from the kilns, while leaving behind fragments of exploded pieces. Not only that, but the exploded fragments could feasibly have been incorporated into the structure of the kilns; this is something that has been seen, admittedly thousands and thousands of years later, in clay pipe kilns, where broken pipes are included in the kiln.[5]

Dolní Věstonice is also famous for a strange burial, where three individuals were placed in the ground at the same time. There were two male skeletons on either side of a skeleton whose sex was difficult to determine, but which was definitely pathological. The skeletons lay in unusual positions – the individual on the left lay with his arms stretched down and out towards the person in the middle, his hands lying over the pelvis of his grave mate, while the male skeleton on the right was buried face-down. Red ochre covered the heads of all three, as well as the pubic area of the middle skeleton. Wolf and fox canines and ivory beads were found around the heads of the three skeletons.[6]

It seems that these three individuals were indeed buried together, at the same time, and this in itself is very unusual. It is possible that they were related: certainly, they shared unusual anatomical traits, including absence of the right frontal sinus (the space in the skull bone above the eyes), and impacted wisdom teeth.[7] But why did they end up in the same grave? They are all quite young: one was a teenager, the other two were in their early twenties. Some archaeologists have suggested their youth indicates a particular adverse circumstance befalling the Gravettian population of Dolní Věstonice, but, from the other human remains at the site, death in early adulthood does not seem to have been particularly unusual.[8]

The deformed leg bones and spine of the middle skeleton have been variously ascribed to rickets, paralysis or congenital anomalies, but it is difficult to be certain about what caused the bony abnormalities in this individual. Although only in their early twenties, he (or she) had already developed osteoarthritis in the right shoulder. Despite the fact that this individual was young when he or she died, there is no indication that the deformities caused this person's death. What is certain, though, is that this individual would have had a very obvious pathology; some archaeologists have argued that this could have been part of the reason that he (or she) was accorded respect, and selected for what seems to have been a special burial.[6]

There are certainly unique things about the Dolní Věstonice burial – but there also aspects of it, in particular the use of ochre and the ivory ornaments, that indicate it was part of a culture that stretched across Europe – and I mean *right* across Europe – in the Gravettian. Around the same time as the Dolní Věstonice burial – about 27,000 years ago – a man was buried at Paviland Cave on the Gower in South Wales, with ochre and ivory rods in his grave. And at Sunghir, some 200km north-east of Moscow, about 24,000 years ago, a man and two children were buried along with ochre, fox pendants and thousands of ivory beads, apparently sewn on to clothing.[9]

Jiri Svoboda and I left the museum and drove up the road to the archaeological site of Dolní Věstonice – which was now, like most of undulating lower slopes of the Pavlov Hills, covered in grapevines. We climbed the hill to the top of the vineyard, where Jiri pointed out the locations of the two main sites, on slightly raised ridges perpendicular to the slope.

'To our left is the first site that was excavated in this area, in the twenties,' Jiri explained. 'There was a priest going to Dolní Věstonice church, from Pavlov, and he noticed, in the cut of the road, bones and charcoal coming out. When it was excavated, that is where the Venus and the other clay figurines were found.'

The second site had been discovered during commercial quarrying in the mid-eighties.[10]

Jiri pointed along the ridges to our right. 'The triple burial was on the next one – at the site of Dolní Věstonice 2. It looks like people may have been burying their dead in the settlement, inside a hut. And probably other people didn't go in any more, and the hut collapsed, and that would be the burial.'

Jiri explained that Dolní Věstonice was just one of a series of settlements along this escarpment, and he also placed it in its wider geographical – and chronological – context. Where we were standing, on that hillside in Moravia, formed part of a corridor that also led through southern Poland and lower Austria, a low-lying passage between the Carpathian Mountains in the east and the Bohemian Massif in the west. It allowed fauna – including humans – to move from south-west to north-east on the European plain.

At the end of the Pleistocene, between 30,000 and 20,000 years ago, the Moravian landscape would at times have been partly wooded, mainly with conifers, but also with oak, beech and yew. These species make it sound as if the climate was rather more pleasant than it actually was; even when there were trees around, snails indicate very cold temperatures, more like subarctic tundra. The climate was fluctuating, and there were even colder, drier periods when the landscape would have been transformed into a treeless steppe.[1]

'It's difficult to find a present-day analogy,' said Jiri. 'Siberia has a zonality of its own today – it's different in the south and the north – but it's possible to imagine what it would have been like here. Mean annual temperatures were very low. But although the winters were much colder than today, some of the summers could have been quite hot.'

Whereas Aurignacian sites often occupied higher ground, the Gravettian sites of Austria, Moravia and southern Poland are clustered along the mid-slopes of river valleys. The hilltops would have been too cold. The large mammals hunted by the Gravettians would have passed through the valleys, and so the camps were well positioned to intercept the herds.[1]

'We can imagine a forested landscape below, and the slopes covered by steppe, but again with some conifers. The sites control the valley, and the game was in that valley. This is the place where we can imagine mammoth herds.'

These Gravettian sites seemed to represent more than just temporary hunting camps: they appear to have been occupied through the year. The intensity of the occupation layers, the richness of artefacts – including things which were both delicate and time-consuming to make, the stability of the house structures – all point to a less nomadic and more sedentary existence.

'These big sites were almost long-term settlements,' explained Jiri, as we stood looking down over the vineyard and the gently rolling ridges. 'But at the same time people were quite mobile. So probably they combined the two: some people stayed at the campsite – here – and others went off to find raw materials and to hunt.'

It reminded me of the Evenki, with their villages and satellite hunting camps.

'It always depends on local conditions,' said Jiri. 'Because normally, of course, hunter-gatherers are mobile people, but there were time periods and specific environments and strategies that enabled more sedentary ways of life.'

The new culture that swept across Europe seems indeed to have been a movement of people and genes – and not just ideas (see map on page 207). Analysis of European mtDNA has revealed two lineages: the haplogroup H (the most common in Europe) and pre-V, which appear to have originated in the east, around the Caucasus Mountains, between the Black and the Caspian seas, and spread across Europe between 30,000 and 20,000 years ago.[11, 12, 13]

But just as this second wave of Europeans spread from east to west, Europe was becoming colder. As the LGM approached and the ice sheets descended, northern Europe – as well as northern Siberia – was all but abandoned. Even the cold-adapted, fur-clad, reindeer-hunting Gravettians couldn't survive in those truly Arctic conditions. Archaeology and genetics – European mtDNA and Y chromosome lineages – record the contraction of the population into refugia in the south-west corner of Europe.

Sheltering from the Cold: Abri Castanet, France

So it was that I made my way to south-west Europe – to the Périgord region, practically synonymous with the Dordogne *département*, whose caves and rockshelters contain an incredible record of Ice Age life.

Heading north from Toulouse, I drove through up through wooded gorges so typical of the Dordogne, where the large rivers running west from the Massif Central to the Atlantic have etched the limestone bedrock. I eventually hit the Vézère Valley and followed it west, reaching the town of Les Eyzies, famous for its many rockshelters. The valley was wide and plunging, framed by limestone cliffs which were incised with a deep, horizontal groove. These grooves in the cliffs – high enough to stand up in – were used as rockshelters by modern humans, or, as these ancestors are known in France, in honour of the first fossil found, Cro-Magnon man.

I continued along the Vézère Valley, through the small hamlet of Le Moustier – famous as the place where Mousterian tools were first discovered, though in my journey I had now left the Neanderthals behind. Turning off the main river valley and following a road leading up one of its tributaries, through the village of Sergeac, and into a narrow wooded valley, the Vallon de Castel-Merle, I eventually reached the site of Abri (rockshelter) Castanet. As I pulled up, American archaeologist Randall White emerged to greet me.

'When does the occupation here date to?' I asked him.

'We have good radiocarbon dates of 33,000 years ago,' Randall replied.

As an early Aurignacian site, Abri Castanet represents evidence of that first wave of modern human colonisers into western France. At that time, during the Würm interstadial of 40,000 to 30,000 years ago, the climate would certainly have been cold, though not fully glacial.[1] During early Aurignacian times, the valleys of the Vézère and its tributaries would have been covered in grassy steppe, with woodland on south-facing slopes and in sheltered valleys.

'We know the people were living here at Castanet in mid-winter,' said Randall, 'when it was probably about 35 degrees below zero outside.'

That sounded cold enough – almost as cold as it had been in Olenek – but Randall thought that people would have been able to keep warm in the deep rockshelters.

'There are holes in the rock where we think they were running lines to drop animal skins down to close off the interior space,' explained

Randall. 'And then are fireplaces inside. I think they're making their world a fairly comfortable place.'

A team of American students were busy trowelling away in the rockshelter. They had come down into the Ice Age floor surface, where, among natural stones pressed into the ground, there were stone tools and flakes lying around. There was a dense black layer in one part of the rockshelter – the remains of an ancient hearth. As sediment was excavated, it was bagged up and carefully wet-sieved. The sievings were retained and dried, then Randall and his team painstakingly picked through the material, looking for minute fragments of flint and bone. There were clues here as to how the Aurignacians had survived the long, hard winters.

'The bottoms of our sieves are absolutely full of burned animal bone. A large percentage of the animal bone they're bringing in from hunting is being consumed as fuel,' explained Randall. But bone fires would not have been straightforward. 'Bone is a terribly difficult thing to burn. There's relatively little wood charcoal here, but they seem to be adding wood to bone fires to keep the temperature above a certain threshold. It must have been a constant focus of their attention, keeping these fires going. They're collecting dung for fuel as well. We take fire and heat for granted, but I don't think they did at all.

'This was a very diverse environment,' Randall told me. 'Reindeer dominate at Castanet, but we have the remains of nine large herbivore species as well as a lot of birds and some fish. This was a pretty good environment for hunters and gatherers, even though it was cold.'

The fauna roaming this landscape would have included reindeer, horse, bison, ibex, as well as animals more suited to forest environments, like boar, roe and red deer.[2]

Excavations at Abri Castanet had also turned up hundreds of stone beads. Most of the beads from Castanet were quite tiny, the majority less than half a centimetre across. They were shaped like tiny baskets, and had been carved out of soapstone. Without wet-sieving of the sediments from the rockshelter, to wash out the dirt from the holes in the beads, they would just have looked like tiny stones and could easily have been disregarded.

'It's interesting to imagine people here, in winter, making all these thousands of beads. Probably done around these fireplaces, like the way we do embroidery or knitting, like a craft. It occupied time on those long nights,' mused Randall.

Like Nick Conard, Randall White was interested in what we could learn about Ice Age society from its art and ornaments – in fact, particularly from personal ornaments, which he believed were much more than just trinkets: to him they were important representations of belief, values and social identity.[3] While many facets of personal adornment – clothing, body painting and organic ornaments – don't usually survive in the archaeological record, those little stone beads had stood the test of time at Abri Castanet.

Randall and his team had discovered beads in various stages of manufacture, so it had been possible to work out how the beads had been made: starting with a rod of stone, which was then scratched round and round until a bead blank could be snapped off. The blank would be thinned and flattened at one end, then pierced by gouging on both sides, and finally trimmed down into the classic basket shape. Although made of soapstone, the method – of taking a baton and dividing it into blanks, then gouging out perforations – was similar to that in German Aurignacian sites, like Geissenklösterle, where ivory beads had been made.[4, 5] The beads had also been polished to a high lustre. Thousands of years before metals were discovered, these people were clearly after the same qualities we enjoy in jewellery today.

There are no burials from the Aurignacian (did they bury their dead at all, or perhaps leave them out – like the Siberian sky burials?) so it is difficult for archaeologists to determine how these beads were used. However, experimental work with beads combined with electron microscopy suggests that the Aurignacian basket beads were sewn on to something – presumably clothing.[4] But why was Randall so fascinated by Stone Age beads?

'Twenty years ago everybody laughed at me when I started working on beads. But it's about what beads *say* about the societies that these

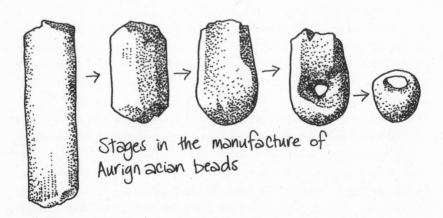

Stages in the manufacture of Aurignacian beads

people were making for themselves. The moment you can begin to construct identities by ornamenting yourself differently, by clothing yourself differently, you can provide a more complicated and effective organisation within a group, but you can also create identities across landscapes,' he explained. 'It may well be that the people in the Basque country felt themselves to be part of the same cultural entity as the people here in France – it's a very large area. Not many people would say that Neanderthals had those sorts of societies. I think the ability to organise large numbers of people across large territories would have been an enormous advantage, and I personally think that's part of the reason why Neanderthals wilted away.'

As the Last Glacial Maximum approached, northern Europe became virtually uninhabited, with ice sheets and permafrost blighting the ground. But in south-west Europe, modern humans clung on. The seasons there were moderated – as they are today – by proximity to the Atlantic: summers were cooler and, more significantly, winters were warmer than in central Europe. But although Iberia and southern France were south of the permafrost zone, the ground was still often frozen. In the Vézère Valley, the frozen uplands would have become uninhabitable, but the protected valleys still supported the hunters of the steppe. The Vallon de Castel-Merle and other valleys were not abandoned. It seemed remarkable.

But despite these harsh climatic conditions, the steppe-tundra grasslands of south-west Europe were fairly teeming with game. Mellars describes south-western France as being almost like a 'last-glacial Serengeti game reserve'.[5] Reindeer, horse and bison – all migratory, herd animals – remained common in southern France, along with ibex, chamois, red and roe deer, saiga antelope, and the occasional mammoth and woolly rhino.[6] But perhaps this makes it sound too idyllic ...

'At the Last Glacial Maximum, the reindeer here reduce in size. Even for reindeer this was an incredibly cold environment: even the animals were stressed,' explained Randall. 'I know that when I was growing up in Canada, we had some winters in which we had five or six weeks when the temperature never got above zero. It starts to act on your head. I can't imagine what it must have been like spending three months in a rockshelter in those kinds of conditions.'

Those hunters were under considerable stress, but, necessity being the mother of invention, they changed their subsistence patterns and broadened their food base: they were still hunting large animals like horse, reindeer and red deer, but there was an increasing reliance on smaller mammals, fish and birds. This intensified subsistence was accompanied by a change in technology: the hunters started to make finely chipped points with concave bases, characterising a whole new industry, called the Solutrean.[1, 7]

As the grip of the Ice Age tightened, we might perhaps expect to see arts and crafts gradually diminishing, as life became harder. More Solutrean points to make, less time for art. But, rather interestingly, we see exactly the opposite. In the Vézère Valley the rockshelters not only continued to be occupied into the LGM, but the hardy Ice Age people of south-west France made their way deep into the caves that riddle the limestone hills – to paint.

Visiting the Painted Caves: Lascaux, Pech Merle and Cougnac, France

Ornaments, portable art and cave art are classic features of the Upper Palaeolithic in Europe, but they did not suddenly appear everywhere; rather, they popped up at different times in different places. As I'd seen, early examples of portable art appeared in Germany more than 30,000 years ago, as part of the Swabian Aurignacian. The ceramic models of animals and people from Moravia were much later, dating to 26,000 years ago. Pendants and beads – like those at Abri Castanet – appeared in the early Aurignacian, and even in the Chatelperronian, in France – but were not found in other parts of Europe until much later.

Cave art is concentrated in western Europe, in south-west France and northern Spain. The limestone formations in this area certainly provided the perfect canvas, but there are plenty of limestone caves in other parts of Europe and the world. To find out why cave art happened in south-west Europe in particular, we need to look at the environmental and social context of the paintings. To start with, we need some dates.

There are some very early dates for cave art in France and Spain, going back to perhaps 30,000 years ago, although some of these dates need to be treated with caution. Although many of these painted caves have been known since the nineteenth and early twentieth centuries, it is only very recently that archaeologists have been able to obtain meaningful dates and to fit the rock art into the wider picture, alongside other archaeological evidence. In France, archaeologist and cave art expert Michel Lorblanchet welcomed the opportunity to place the paintings in time, and wrote of the 'post-stylistic era', where cave art could be directly dated, rather than relying on style alone to establish chronologies. Where paintings included charcoal, radiocarbon dating could be used to provide a precise timing. Unfortunately, the majority of paintings do not contain charcoal or other organic remains, so they are not amenable to direct dating, in which case indirect dates from excavations within caves, as well as the style of the paintings themselves, remain important clues as to their date of inception.[1]

Sometimes it is very hard to get a decent date, even when charcoal exists in a painting. Organic residues from micro-organisms and carbonate from the cave wall itself can affect the date. Radiocarbon dates of black dots from the painted cave of Candamo in northern Spain range widely, and have been reported as being as old as 33,000 years ago – or as young as 15,000. It is difficult to decide which of these is the true age: the dots may be 33,000 years old, and may have been painted over again 15,000 years ago. Or perhaps the younger date is true – and the older date was obtained as a result of contamination with more ancient carbon.[1] Similarly, dating of a black horse from Chauvet Cave in France has produced estimations of about 21,000 years ago (Magdalenian), and around 30,000 years ago (Aurignacian); the style of the paintings in Chauvet suggest that the younger date is more likely. Dating specialists certainly hope that advances in radiocarbon dating, and the use of different labs to test the reproducibility of results, will help to clear up such discrepancies in the future. But for now, the potential Aurignacian dates for Candamo and Chauvet have to be treated with caution.[1]

Experts tentatively agree, however, that most cave art, including paintings and engravings, seems to be part of the Solutrean and subsequent Magdalenian cultures of the late Upper Palaeolithic – in other words, it was created around the time of the LGM.[2]

I wanted to see some of this cave art for myself and so I made my way to Lascaux, near the town of Montignac in the Vézère Valley – the most famous of the French painted caves. Unfortunately – but entirely understandably – I was not to see the original cave paintings: the cave has been closed up while conservators attempt to eradicate the mould that has been threatening to destroy the precious paintings. Instead, I visited Lascaux II, the reproduction of the cave that is open to visitors. I had almost visited Lascaux some years before, but then decided against when I found that only the replica cave was open to viewing. This time, however, I had to admit that Lascaux II was worth a visit. I walked down a passageway lined with photographs of the original excavations, and into the 'cave'. It was great – I was completely taken in. The cool air, the shapes and texture of the walls, and the paintings themselves seemed quite authentic.

But this was a reproduction – by a single, modern artist. The colours were true to the original – manganese black, ochre yellows and reds, a similar palette to the rock paintings I had seen in Australia. Lascaux II is a re-creation of the splendid 'Hall of the Bulls' and the passageway known as the 'Axial Gallery'.

The original Lascaux was discovered by four teenagers exploring the hills above Montignac in 1940. A pine tree had fallen, and where its roots had torn up the earth the boys found a hole in the ground. The sinkhole led straight down to what would become known as the Hall of the Bulls. The young discoverers went straight through this chamber – presumably not looking up, or they would have seen that the walls curving in above them were emblazoned with huge bulls – and on into the Axial Gallery, where they first noticed the cave paintings. I stood in the replica Hall of the Bulls, gazing up at the beasts – the so-called 'Unicorn' (strange, as he patently has *two* horns), above me and on my left, followed by great black-outlined bulls, facing each other, with smaller, antlered deer filling the space between them. On the right-hand part of the ceiling were more bulls in black, and red ochre. Making my way down into the narrower, keyhole-shaped Axial Gallery, I could see a procession of animals on the ceiling above me: a beautiful black deer or reindeer with branching antlers, more bulls, and pot-bellied horses.

With my appetite whetted, I went off in search of an original painted cave: Pech Merle, in the Lot *département*. A flight of stone stairs led down to a somewhat incongruous white-painted door, through which I passed to emerge into a limestone cave deep within the hillside. I walked through magnificent chambers with huge flowstone creations, enormous stalagmites and stalactites, some of which had met between ceiling and floor to form massive pillars. The cave opened into a great chamber, high and wide and elaborately adorned with speleothem creations. It was like walking into a gothic cathedral. What on earth would it have been like for Ice Age hunter-gatherers? For someone who had never been in a church, let alone a cathedral? If we still wonder at the natural beauty of these caves today, just imagine what it would have been like for our ancestors. It must have seemed magical, otherworldly, and sacred.

I was so distracted by the natural splendour of the cave that I almost missed the cave art. But there, on one rare, smooth part of the cave wall to my left, were two beautiful horses outlined in black, facing away from each other, their hindquarters partly superimposed. They were covered in black spots which also flowed on to the background around them, as though they were somehow camouflaged. There were red ochre spots, too, on the belly of the horse on the left, and on the flanks of the other. I noticed that the flat wall of rock had a strange contour where it ended on the left – almost like a horse's head. It was as though the artist had taken this suggestion from the natural shape of the rocky canvas and allowed it to direct his hand as he (or she) created these wonderful beasts.

The horses were stylised rather than naturalistic representations. They had great curving necks and small heads, rounded bodies and slender legs. Were they artistic representations of real horses or mythical beasts?

I imagined the artist painting them, in the darkness of the cave, with a tallow lamp flickering and lighting up small areas of the wall as they applied the black and red pigments. There were six negative handprints, stencilled on to the wall around the horses, some left, some right hands, but all matching. Were these the signature of the original artist or additions by a later cave painter?

Further along in the same chamber there was another hand stencil, this time in red ochre. I found these hands incredibly moving – it was amazing to think of that Ice Age artist, so many thousands of years ago, placing a hand on the wall and recording that moment. I felt very privileged to be seeing those images. It was like a message that had been passed down from ancient times to the present. What did it say? I know the real meaning is lost for ever, but for me those hands say 'we are people, just like you'.

Pech Merle is incredibly rich in cave art; it contains more than seven hundred images, including many black-lined pictures of mammoths, bison and horses.

My next subterranean stop-off was the cave of Cougnac, and, outside the cave, on the wooded hillside, I met Michel Lorblanchet, the man who had spent so many years studying – and re-creating – the ancient art of the French caves. I had many questions for him, but first Monsieur Lorblanchet wanted to show me how the hand stencils were made. He had studied the pigments used and how they might have been delivered on to the wall, and had concluded that the ghostly handprints had been created by spitting colour on to and around a hand laid against the stone.

hand stencil above Pech Merle horse

He demonstrated the technique for me outside Cougnac, on an exposed limestone cliff. First, he donned overalls and an artistic-looking black beret, then fetched a collection of things from the boot of his car: stones, charcoal and a bottle of water. Then he ground up some charcoal, using a pebble to crush the pieces to dust on a large flat stone. He explained that the black pigments in the cave paintings were usually manganese oxide: 'It's also black, but it could be dangerous for the experimenter. I spoke to a toxicologist in Paris and this man told me: don't use manganese oxide, you'll get poisoned. So, I prefer to use charcoal.'

It became clear why, because the next thing he did was to take a generous pinch of ground charcoal, and transfer it into his mouth. I could hear him crunching it up into an even finer consistency.

'I grind the pigment between the teeth,' he said, through blackened, gritted teeth, then he chewed some more, laid his hand against the stone and started spitting the charcoal around it. He spat quickly – 'pup pup pup pup pup' – and a fine black spray landed on the wall and on his hand with each spit. He had transformed himself into a human airbrush.

After five minutes there was a slight misting of charcoal around his hand on the wall. He would pause to put more charcoal in his mouth, chew it up, then resume the spitting, an action so fast that I worried he might hyperventilate. But Monsieur Lorblanchet was well practised in this technique. As a piece of experimental and experiential archaeology, he had re-created the entire Pech Merle spotted-horse frieze, in a cave, using his spitting technique – it had taken a whole week to execute. While I suspected there might be a quicker and more efficient way of transferring the paint on to the wall, (through a hollow tube, perhaps?) I admired his dedication. His technique was also based on ethnographic studies of Aboriginal painting in Australia, where handprints also exist. The pigments were similar, too. 'Red ochre and charcoal, and manganese oxide have been used everywhere in different parts of the world,' he said. 'There are not many solutions. Some vegetables can provide pigment, but Palaeolithic people mainly use charcoal, manganese oxide and red ochre. Not vegetable pigment.'

After half an hour, Monsieur Lorblanchet stopped and stood back from the wall. His lips were black with wet, powdered charcoal, and his white beard had assumed a black streak in the centre. He took a swig from the water bottle to clean out his mouth. There on the wall was a modern hand stencil.

'So I spit the pigment on the wall. It is the same method that was used during the Ice Age because I get a foggy image, a foggy hand stencil. And this sort of form is exactly the same as in Pech Merle.'

'How did you hit on this method?'

'Oh, it is difficult. It's necessary to do it several times to get the right methods, yes. To have a good result it's necessary to have some experimentation.'

Stencilling wasn't the only method used by the ancient rock artists. Monsieur Lorblanchet described how lines would be drawn using fingers or chewed sticks to form paintbrushes. But stencilling was an excellent solution to applying paint to a rough or crumbly wall surface: 'Cave walls are often full concretions, stalagmite and stalactite, so it's impossible to draw with a finger. And if the wall is soft sandstone, and you paint with a brush or with a finger, you destroy the surface,' he explained. 'But with this spitting technique you can paint without touching the wall.

'Pech Merle is very interesting. In this case, there are six hand stencils around the horses, sometimes right and sometimes left hands. But from the same individual, the same man. And pigment analysis showed that the pigment of the horses is exactly the same as the pigment of the hand stencils. So probably the same man did the horses and the six hand stencils, at the same time.'

'So you think those stencils are the artist's signature, a maker's mark?' I asked.

'Yes, I think the main meaning is really a signature. Like in Australia. Australians visiting burial sites, they left their hand stencils during the visit, just to say I went there; I visited my uncle or my grandmother and I left a mark of my visit near the burial.'

'But the French caves are not burial sites,' I volunteered.

'No. It is very exceptional to find a burial in the European caves. But it could also be a way to say, I went to this church and left a mark of my visit.'

I was intrigued to hear Monsieur Lorblanchet talk about the caves in religious terms.

After the hand-stencil demonstration, Monsieur Lorblanchet took me into the cave of Cougnac itself. We walked down steps again, to a door in a wall and into a dank room displaying various stone curiosities, including pieces of medieval masonry and an elaborately carved stone sarcophagus lid. Then we descended more steps into the cave itself. Just as in Pech Merle, I was taken aback by the natural beauty of the cave. Cougnac was smaller in size, and the ceiling was much lower, but it was absolutely crammed with slender stalactites. It felt as if we were entering a temple, and I asked Lorblanchet if he thought that the caves had indeed been sacred places for their Ice Age decorators. He thought that they had: he saw the caves as sanctuaries, special places to which artists kept returning.

'Yes, it is a natural temple. About 10 per cent of the caves in this area have painting in them, and, usually, it is the largest caves which have been chosen. So these paintings have been made for religious reasons, if you like. These are sacred sites, and they are not painting here just for fun.'

We turned a corner and the walls were covered – wherever there was a flattish space – with line drawings of animals: elk, horses and ibex. Lorblanchet gestured around him.

'Animals everywhere ... because these people were, of course, hunters and gatherers, so they painted their world around them, and the animal world. And for them, the animals were not only game, but also spirits.'

At the back of the cave there was a small image of a man – lying stretched out, with what looked like spears stuck into him. This was an unusual theme. I felt as though I was looking at an illustration of a myth that has long since lost its meaning. There were also abstract shapes that could be seen as the head and shoulders of humans, or perhaps as vulvas. These were ancient drawings, dating to before the LGM.

'We know now, after studying this cave for years and years, that Cougnac was used intensively during the Gravettian,' said Monsieur Lorblanchet, 'then the cave was forgotten. But then it was rediscovered by Magdalenian people, and the cave again was a sanctuary. There is a gap of 10,000 years between the oldest paintings and the most recent painting in Cougnac.'

Around the earlier paintings there were finger-daubed spots of paint, often double, and around some a mist of applied red ochre. And these had been dated much later, to around 20,000 years ago, the Magdalenian period. Monsieur Lorblanchet mused about the original meaning of these paintings, and the meaning to the later artists. Did they imagine these images had been created by their ancestors or by ancient spirits? From archaeology carried out in the cave, there seemed to have been a large patch of red ochre on the floor at the entrance to the main chamber – Lorblanchet imagined the later, Magdalenian visitors dipping their fingers in the pigment and touching the walls around the ancient symbols.

For Lorblanchet, the next, modern rediscovery of the paintings meant that they were once again being incorporated into a belief system, as we tried to understand what they meant to our ancestors, and gazed at, analysed, replicated and reproduced the ancient images. It is interesting to think about these artistic creations as *still* conveying information, communicating messages within complex social networks. How many people around the world today have seen those images and gone home with the postcards?

It is also rather wonderful to think of those hunter-gatherers, those ancient Europeans, who, even as the climate chilled around them, carried on making art. During the icy grip of the LGM, it was somehow still important to decorate spear-throwers, carve mammoth-ivory animals – and paint caves. The Solutrean is marked out not only by a change in hunting technology, but by a flowering of art and ornamentation. Although making art doesn't seem to be immediately relevant to survival in an increasingly hostile environment, many archaeologists believe that the painted caves tell us something very important about Ice Age society: that the proliferation of art indicates an increasing complexity of social networks. So – and perhaps even more tellingly than a new style of stone point – the cave art represents a social and cultural adaptation to survival in extreme environments.[2]

Perhaps these 'cave art sanctuaries' were landmarks in the Ice Age landscape, marking out territories for particular groups, and maybe they were places where people aggregated and which confirmed a feeling of group identity.[4]

'These people were nomads, of course,' said Lorblanchet, 'but they had a territory. And by painting a cave it is a way for them to say, here is our sacred place, here are our gods, our belief, *we are here*. Like the church today is in the middle of the village, the painted cave was in the centre of the tribal territory.'

Meeting up of local bands – to exchange materials, plan collective hunts or hold ceremonies, perhaps – would have provided opportunities for the exchange of information important to long-term survival. Information could also be passed down the generations. Many archaeologists view the cave art as part of an 'information system' – very similar to the place of rock art in Australian Aboriginal culture. 'Information system' is little more than a very dry term for what is, essentially, storytelling. Stories could contain useful information – about the landscape, animals or society – extending human experience beyond a single lifetime. Maybe images, like those spotted horses at Pech Merle, were used to illustrate tales. For the hunter-gatherers packed into that corner of Ice Age Europe, on the edge, art and storytelling may have been crucial to survival.[3, 4]

Lorblanchet also thought that paintings contained an expression of identity that we could still understand.

'I also believe, by their painting, they show that they were exactly the same as us, you know,' he said. 'They are good artists, excellent. They had the feeling, the sense of artistic beauty. And by this act of painting in caves they also expressed themselves as different from the neighbours. The neighbours who were the Neanderthal people.'

We can't leave our ancestors, struggling to survive the wintry chill of the LGM, without looking at them a little closer. Compared with the brown-skinned ancestors of all the Out-of-Africa lineages, it seems that Europeans were getting paler, with a handful of mutated genes reducing melanin production in skin cells. One gene, imaginatively called SLC24A5, appears to be responsible for around 30 per cent of the skin colour difference between indigenous Europeans and Africans.[5]

Northern and eastern Europeans are incredibly diverse in appearance, in the variety of hair and eye colours in particular. Hair can be black, brown, pale blond, yellow-blond, or red. Eyes may be brown, hazel, blue or green. Some have suggested that this great variability in coloration is random: a product of genetic drift, perhaps, or the result of relaxed reduced selection pressure for dark skin (along with hair and eyes), as populations spread north, leaving other 'colour genes' free to vary. But there is an interesting theory that all this diversity is due to sexual selection. What if the harshest years of the LGM meant that many young men died while hunting, leaving women outnumbering men? Polygyny might be one solution, but it would have been difficult to provide food for a harem. So competition for men may have been fierce. In that competition, something that made a woman stand out from the crowd – a strikingly different hair or eye colour, perhaps – could make the difference to whether or not she managed to pass her genes on.[6] It's a fascinating theory, but ultimately impossible to test, and doesn't explain why coloration should have become so diverse in Europe and not elsewhere.

Archaeology and genes tell us how, after the LGM, humans re-expanded across Europe, from their principal Ice Age refugium in Iberia.[7, 8, 9, 10] The changing climate was once again matched by a change in technology. The new European industries – the Magdalenian and Epigravettian – were hugely variable, but there was a general explosion in the use of antler as a material for producing tools, and the harpoon was invented.[2] Tiny bone points were fitted into antler and wooden spear points.

At 16,000 years ago, populations are re-established north of the Loire, and by 13,000 years ago Britain was reoccupied. The human recolonisation of Europe was part of a general faunal expansion northwards, but some animals were missing from the post-glacial landscape: whether because of climate, hunting or both, there were no more mammoths or woolly rhinos.

Technologies continued to evolve in the late Pleistocene, and by 11,000 years ago the first definite evidence of the bow and arrow appears. The replacement of steppe-tundra with woodland meant that the large herds of horse, bison, saiga, and that 'Ice Age larder-on-the-hoof', reindeer, disappeared.[2] It seems quite counter-intuitive that the warming of Europe actually created a significant challenge to its human (and animal) inhabitants, but the altered environment required the invention of new ways of living off the landscape. The trend towards intensified subsistence that began in the run-up to the LGM continued, with even more hunting and trapping of small animals and birds, fishing, and gathering of molluscs. Ice Age technology was gradually replaced by Mesolithic tools: bows for hunting, reaping knives and axes for felling trees, as the Europeans adapted to newly wooded landscapes, as well as to estuarine and coastal environments. Europe became populated more widely and more densely than ever before, and quite soon an innovation from the Middle East would allow even greater population numbers to be sustained on the landscape.[2]

New Age Mesopotamia: Göbekli Tepe, Turkey

The end of my European journey would take me back to where I had started, to Turkey, and to the most spectacular archaeological site I have ever seen.

I travelled down the south-east of Turkey, about thirty miles north of the Syrian border, to the ancient town of Sanli Urfa. I was in Mesopotamia, between the Tigris and Euphrates rivers. In Urfa, modern buildings clustered around Roman ruins on the slopes, but I was after much more ancient archaeology.

I travelled for around an hour, west of Urfa, then turned off the main road on to a dusty track, which wound up through a rocky valley and on to a limestone escarpment. Eventually, the track ran out and I found myself at the bottom of a conical hill, with a site hut and an empty tent. As I walked up the hill, German archaeologist Klaus Schmidt came halfway down to meet me.

'This hill is not made by nature,' Klaus explained as we walked up it. 'It's a *tell*, a man-made mound created by the ruins of these Stone Age structures. It reaches up about fifteen metres above the natural limestone plateau. I was suspicious when I first saw this site: no force of nature could make such a mound of earth in this location.'

Klaus had discovered this site while surveying the area for potential Palaeolithic sites in 1994. Local farmers had been turning up masses of stone tools in the fields on and around the hill, and occasionally hitting very large stones with their ploughs. Archaeologists had got wind of this before, but assumed that the stones were the remains of a medieval cemetery. But when Klaus investigated, he found finely worked blades, and large, rectangular stones buried in the ground – and so large that they could not be moved or lifted. When he started excavating the site in 1995, he found that these stones were something truly remarkable. They were just the tops of great, T-shaped standing stones. Some were more than two metres high, and, as the archaeologists dug deeper, they found that the stones were arranged in a circle, with two larger standing stones in the centre of the ring. But there wasn't just one stone circle at Göbekli Tepe: Klaus had excavated four so far, and from geophysical surveys of the hill he supposed that there might be twenty to twenty-five of them, still buried in the rubble of the hill.

wheelbarrows
at Göbekli Tepe

Klaus led me to the top of the hill and showed me a stone circle. I was bowled over by it. Archaeologists were busy with a small crane, moving a large fragment of one standing stone from where it had fallen. There was an engraved figure of a person on the narrow outer side of one of the standing stones, at the edge of the circle. There were other stone walls at the top of the hill, which Klaus thought might have enclosed smaller sanctuaries. Rather strangely, there was no sign of habitation, such as hearths, up on the hill: it appeared to have been exclusively a sacred site rather than a settlement.

As we walked down the other side of the hill, I suddenly saw more stone circles, even more impressive than the first one at the top of the hill. The circles were wider and the standing stones taller – and more elaborately decorated. On the sides of the stones there were beautiful, low-relief carvings of foxes, boar, birds, scorpions and spiders. On the inner edge of one stone there was a full-relief, 3D carving of an animal, perhaps a dog or a wolf. It had been carved from the stone in one piece. While I was there, Turkish archaeologists uncovered an animal head sticking out, gargoyle-like, with formidable fangs, from the wall enclosing one of the stone circles.

There seemed to be several phases of building at Göbekli Tepe. The lower, older and more impressive stone circles appeared originally to have been put up as just that: a ring of large standing stones. Then stone walls had been added, creating inner and outer circles. In some places, there were slabs laid on the standing stones as though the ring had originally been covered with a corbelled roof.

The architecture and stone sculpture at Göbekli Tepe is remarkable. But what made it even more arresting was the date. 'It's been here, buried in this hill, for 12,000 years,' said Klaus.

So Göbekli Tepe appeared to be a temple site – built by hunter-gatherers. It challenged paradigms of the origins of the Neolithic. Based on previous evidence, archaeologists have suggested that the emergence of the Neolithic involved a sequence of developments that goes something like this: population pressure led to increased need for food, led to the adoption of agriculture, led to stratified societies and new power structures, led to organised religion. What Göbekli Tepe seemed to show, though, was a complex, hierarchical society – where stonemasons could be tasked with building temples – and organised religion, in the context of a hunter-gatherer society.

It's difficult to know how to categorise Göbekli Tepe. As a site representing something somewhere between the Upper Palaeolithic

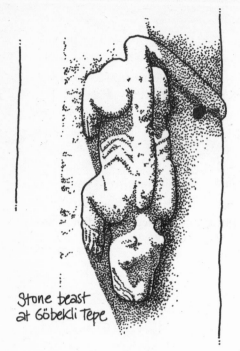

Stone beast
at Göbekli Tepe

and the Neolithic, Klaus thought that perhaps the best label would be 'Mesolithic', but then this would have been a very different Mesolithic from the nomadic hunter-gatherer lifestyle, further north in Europe. Having said that, the stone toolkit at Göbekli Tepe was similar to some of the tanged point cultures of central Europe, in the late Palaeolithic and Mesolithic. But other archaeologists, too, talk about a transition straight from the hunter-gathering Palaeolithic to the agricultural Neolithic in the Levant, missing out the Mesolithic. The first step towards this new way of life began around 14,500 years ago, with the appearance of the Natufian culture in the Levant.[1] Hunter-gatherers began to do a new thing: they settled down in villages, in which they stayed all year round. At this point, the development of agriculture seems almost inevitable. The appearance of grinding stones, mortars and pestles at archaeological sites suggests that wild cereals were important in the Natufian diet. There are also dog burials from this time: it seems that man's best friend had arrived.[1]

Klaus suggested that this may have been because gazelle hunters in Turkey required similar technology to the reindeer hunters further north, but also that perhaps there was some kind of connection or communication between the societies of Turkey and those around the Black Sea and the Crimea. However, the complex society and ritual suggested by the stone-circle temples was something else entirely. There was nothing even vaguely like it in Europe – with this scale of monumental architecture – until well into the Neolithic. So, instead, then, Göbekli Tepe is called 'early Neolithic', with the implicit understanding that many of the classic features of the Neolithic, like pottery – and in particular, farming – were yet to come.

What sort of rituals were taking place at Göbekli Tepe? The symbolism there seemed to be dominated by animals: snakes were the most common, sometimes appearing singly, or stacked up like waves.

Wild boar and foxes are also common motifs, along with leopard-like creatures and stylised aurochs heads. Birds were also depicted: perhaps geese or ducks. The images didn't seem to relate directly to animals that were being hunted, as boar and snake bones are very rare in the rubbly backfill at Göbekli Tepe. Gazelles were an important food animal, but only one carving of a gazelle has been found at the site so far. This is a bit like the Ice Age caves of France, where one of the most commonly hunted animals, reindeer, is rarely depicted.[2] Perhaps the various animals represented different 'clans' that came to the temples?

Weirdly, the large T-shaped standing stones appeared to have arms, bent at the elbows, with hands clasped at the front. They had no faces, no eyes, nose or mouth, but Klaus thought the stones represented immense, abstract human figures.

'Who are these beings made of stone?' he asked, rhetorically. 'They are the first deities depicted in history.'

He suggested that the animal motifs carved into the sides of these giant 'figures' might be guardians or protectors of the megaliths, but in some cases carefully grouped animals suggested that their meaning went further: perhaps they were representing stories or myths.[2]

'Maybe these are pre-hieroglyphic messages,' Klaus suggested.

Where the sex of the animals could be discerned, they always appeared to be male, and a small 'ithyphallic' figure with an out-of-proportion semi-erect penis had also been discovered at the site. This was very different from the famous, much later site of Çatal Hüyük in Turkey, where female imagery was common – although inside houses rather than temples. Çatal Hüyük also contained many depictions of vultures, which seemed to be associated with death. There were no pictures of vultures at Göbekli Tepe – but there were many snakes, and objects decorated with snakes have been found among grave goods at other sites of almost the same age.[2] To Klaus, the predominance of 'aggressive' animals, snakes, and absence of female/fertility symbols suggested that Göbekli Tepe might be a burial site – or at least a place dedicated to a cult of the dead.

The monumental architecture and the clustering of the stone circles certainly seemed similar to complexes of burial mounds found in the (much later) megalithic cultures of Neolithic Brittany and the UK. But Klaus hadn't, as yet, found any evidence of burials at the site. He was just about to excavate part of one of the higher stone circles where he thought that a large, flat stone might just be covering a burial.

Klaus's team had also found large quantities of stone tools in various stages of production, including nodules, half- and fully-prepared cores, and blades. Evidence of flint-working usually occurs in occupation sites, but here it was in what seemed to be a purely ritual space. Klaus thought that the flint-knapping was somehow part of the rituals that had been carried out there. But it was difficult to draw conclusions about how Göbekli Tepe fitted into the lives of the people who had constructed it, given that no settlements or any kind of camps of the same age have yet been found in the area around the hill.

However, other archaeological sites in the area give us an insight into how life was changing for the hunter-gatherer societies of the Levant. Archaeologists now see the Neolithic as emerging in stages: first, hunter-gatherers became more settled, with more complex social structures; then plant cultivation appeared, and then we see the appearance of large villages and intensive food production.[3] So it appears that social change came first – followed by agriculture. Göbekli Tepe fits into the early stage of this transition, as a pre-agricultural, pre-pottery site, demonstrating the existence of a complex society. For Klaus, the social changes may have been the impetus for the development of farming – perhaps to provide feasts to honour the gods. 'For them it would have been a logical step, to manage Nature and get more food for themselves,' he said. 'Religion created the pressure to invent agriculture.

'It's very clear we must change our ideas,' he continued. 'Hunter-gatherers don't usually work in the way we understand work.' But, from the scale of construction at Göbekli Tepe, it was clear that people here had jobs: specific things to do which weren't immediately about obtaining food, water or shelter, but were nevertheless important to society.

'They started to work in quarries. They started to have engineers to work out how to transport and erect the stones. There were specialists in stone-working, whose job was to produce sculptures and pillars from stone,' said Klaus. It appeared that the society that produced Göbekli Tepe could support both a workforce and professional artists.

Klaus also believed that the transition to agriculture might have led to the abandonment of Göbekli Tepe and its gods. His team of archaeologists were digging down through rubble that appeared to have been deliberately piled on top of the stone circles to cover them up. So, social changes may have driven the development of farming, but subsequent changes in society, and the emergence of new religions, had then been disastrous for Göbekli Tepe.

'The hunter-gatherer societies were on the threshold of inventing a new way of life, of becoming farming communities. And by the ninth millennium BC this process was successful and this new way of life was developing in this region. The old spiritual world of the hunters was without any use, and so this site was abandoned. This world was completely forgotten, completely lost, never repeated.'

I found this story, of the rise and fall of the old hunter-gatherer gods, very compelling. But in fact it is very difficult to rule out farming entirely at Göbekli Tepe, as the earliest farmers would have been planting wild foods. It's also very difficult to pinpoint that transition from collecting wild-growing plants to intentional planting. And, of course, the first planted crops would have been wild varieties: domesticated varieties would have emerged later as farmers selected particular characteristics. For hunter-gatherers for whom wild varieties of cereals and legumes were already staple foods, it may only have been a short step from gathering them to intentional planting and cultivation.[4] Klaus thought that feasting, or at least the formation of more settled communities, may have prompted this step, but there may also have been climatic reasons for the adoption of agriculture. Around 14,600 years ago, and again around 11,600 years ago, there was a period of increased temperature and rainfall, each lasting only one or two decades. During each of these warm, wet phases, cereals and legumes would have flourished – providing humans with plenty of easily exploitable resources.[3] Between these two wet, warm periods was the cold, dry snap of the Younger Dryas. The effects of this may have been over-emphasised, but for people getting used to living on plentiful cereals and legumes, a period of worsening climate may have encouraged them to start cultivating crops. This climate explanation echoes what I had discovered from reading about the origins of agriculture in China, and I find it very persuasive. Because, at almost exactly the same time, on opposite sides of the globe, people were inventing agriculture. It may not be such a coincidence: both populations were living in environments affected by global climate change.

The date of Göbekli Tepe places it slightly earlier than other archaeological sites where the transition to agriculture is clearly documented. It now seems clear that this area, between the upper reaches of the Tigris and Euphrates, was indeed the place where farming got started in the West.[5] Early farming communities – still without pottery – became established in Turkey and northern Syria between about 11,600 and 10,500 years ago.[4] At these very early sites, archaeologists have found burnt remains of wild cereals (like einkorn wheat, rye and barley) and legumes (peas, vetch and lentils). Slightly later, from around 9500 years ago (7500 BC), evidence of domesticates such as emmer wheat and barley appears – sometimes in higher (more recent) layers at the same sites.[5] Later still, herd animals are domesticated.[3] People whose recent ancestors had hunted wild animals in the foothills of the Taurus and Zagros Mountains started corralling, tending and breeding them. Along with plant and animal domestication came a new toolkit, with sickle-knives for cutting and querns for grinding cereals.[5]

Botanical and genetic studies also point to this area as the 'cradle of agriculture'. Wild varieties of the important Neolithic crops (einkorn and emmer wheat, barley, lentil, pea, bitter vetch, chickpea and flax) are found growing together in this area. The limited genetic variability of domesticated crops also lends support to the idea of a single, core area of plant domestication.[5]

Once agriculture began, it allowed populations to expand further. Food resources became more reliable. People settled down and large villages started to appear. It all sounds wonderful – until you take a closer look at what was happening to the health of individual people. Because, although much of the risk of inadequate food supply may have been shed by growing crops and keeping domesticated animals, the quality and range of the farmers' diets was not particularly good. Archaeologists have long supposed that the transition to agriculture was positive in all sorts of ways, bringing better health and nutrition, increased longevity, and more leisure time. But the truth is a little harder to bear and somewhat counter-intuitive. From the study of human skeletons from this crucial period of changeover, biological anthropologists have found that the switch from foraging to farming brought with it a *general decline in health*.

Compared with hunter-gatherers, farmers had more tooth loss and more dental caries, had restricted growth and shorter stature, and reduced life expectancy. Skeletal evidence of trauma becomes more common, indicating an increase in violence and conflict. Neolithic people also suffered more from infectious diseases than previous groups, probably because of the combined effects of a poor diet and more crowded living conditions. Anaemia was also more common.[6, 7] Traditionally, archaeologists have argued that the Neolithic brought with it improved health and reduced mortality, and so populations could expand rapidly, but the bones of our ancestors show us that this was not the case at all. The dawn of agriculture and permanent settlement brought about worse health and reduced life expectancy. But in spite of all these disadvantages for individuals, agriculture brought with it an increased birth rate that outstripped the reduced life expectancy – so the populations expanded.[6, 8] It is difficult to work out the sizes of past populations, but researchers seem to agree that population growth was extremely slow during the Palaeolithic, and that at 10,000 years the world contained around eight million humans. By AD 1800, the global population was a stonking thousand million people.[8]

Agriculture then spread out of the Levant: into central Anatolia by 8000 to 9000 years ago, then eastwards into the Zagros foothills and Indus Valley, and westwards, along the Danube and along the Mediterranean coast.[4] By 7500 years ago the first farmers appeared in Hungary, and by around 6000 years ago the culture had spread all the way to northern Spain, where the Neolithic seemed to have been adopted as a 'package': evidence of domesticated crops and livestock, pottery and megalithic monuments suddenly appears in the archaeological record.[9] So, the spread of agriculture into Europe, between 10,000 and 6000 years ago, followed similar routes to the original, Upper Palaeolithic colonisation of Europe.[4] But was this a movement of people (demic diffusion) or ideas (cultural diffusion) – or both?

When I started considering this question I thought that genetic studies might provide the answer. But in fact, over the years, various studies have turned up somewhat conflicting results. Y chromosome studies have revealed what appears to be a cluster of lineages spreading from the Levant across Europe: in other words, a movement of people.[10, 11] Some analyses of mtDNA data have also uncovered a potential Neolithic contribution to the European population,[12] but others have produced very little evidence for demic diffusion. Extraction of ancient DNA from some Neolithic skeletons from central Europe revealed that a quarter of them had a type of mtDNA that is now extremely rare in Europe, suggesting that the genetic contribution to the European gene pool from Neolithic incomers was probably minor, at least within maternal lineages.[13] The discrepancy between the maternal mtDNA and paternal Y chromosome patterns has been explained by some as indicating a spread of male farmers out from the Near East, intermarrying with indigenous women. Other researchers have suggested that the contradictory results actually warn us against coming up with too simple a model. The spread of farming across Europe would have been a complicated matter, with the degree of demic and cultural diffusion varying in different regions.[10] Broadly speaking, it seems that there was a movement of some people out of the Near East in the Neolithic, but these people mingled with the ancient populations already established in western Europe, rather than replacing them. At the moment, the picture is far from clear, but, as more work is done and more samples collected, particularly from ancient skeletons, the way that the Neolithic made its mark on Europe should become clearer.

The beginning of food production was a revolutionary event in prehistory. It paved the way for large-scale settlement – for civilisation. After hundreds of thousands of years of being nomadic hunter-gatherers, humans began to settle down and farm. It seems like a long time ago, but set against the scale of human prehistory it's a relatively recent development.

Göbekli Tepe had lain hidden under rocky fields at the top of a hill for 12,000 years, but when the ancient gods had been rediscovered they held their discoverer firmly in their grasp. When Klaus found the place, he was exhilarated, but he also knew at that moment that he had two options: either to walk away right there and then, or to spend the rest of his life there, excavating this remote hill where human society had started to change.

'I was very excited. But it was clear from the first minute that I had a choice, too: to turn back and not tell anyone about this discovery, and never come back ... or to stay and work here for the rest of my life.

'It is a dream,' he said, 'but it is also difficult to have such a site. You are captured by it. You belong to it.'

I left Göbekli Tepe very sure that I would hear more about this amazing place in the future. But, for now, my journey would take me to the lands of my final destination, to the Americas, the last continents to be reached by humans.

5. The New World: Finding the First Americans

tepee at the Powwow

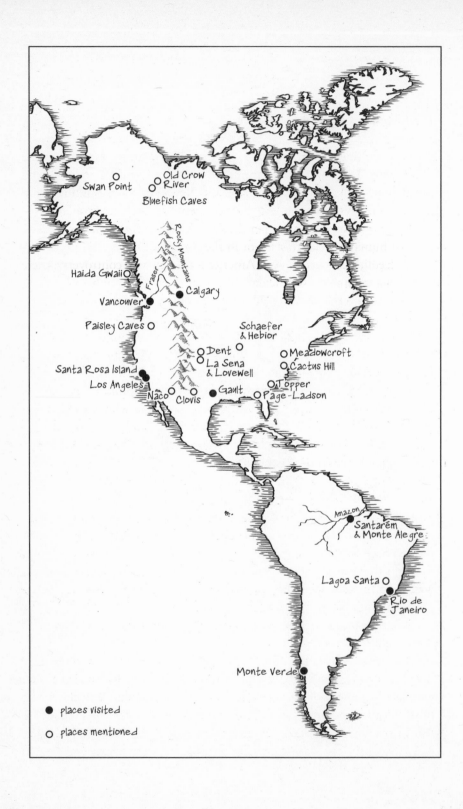

Swan Point

Old Crow River

Bluefish Caves

Rocky Mountains

Haida Gwaii

Fraser

Vancouver

Calgary

Paisley Caves

Schaefer & Hebior

Dent

Meadowcroft

La Sena & Lovewell

Cactus Hill

Santa Rosa Island

Topper

Los Angeles

Naco

Clovis

Gault

Page-Ladson

Amazon

Santarém & Monte Alegre

Lagoa Santa

Rio de Janeiro

Monte Verde

● places visited

○ places mentioned

Bridging the Continents: Beringia

This vast continent was the last of all the great landmasses to be populated by humans. For hundreds of thousands of years, while humans spread throughout Africa, Asia and Europe, the New World was unknown and unreachable. Whereas modern humans replaced archaic populations – *Homo erectus*, Neanderthals – in the Old World, *Homo sapiens* would be the first human species to set foot in the Americas.

Palaeolithic archaeology in America is riddled with controversy: there is ongoing argument over the date of colonisation, the number of ancient migrations into the New World, the origin of the incoming populations and the routes they took.

Some have suggested very ancient dates of colonisation, as early as 40,000 or 50,000 years ago, but the evidence is very shaky indeed. Most researchers agree that humans did not reach the Americas until after the LGM. Until very recently, the prevailing opinion was that the first inhabitants of America were hunter-gatherers who moved down through North America about 13,500 years ago, and who carried with them a stone toolkit called 'Clovis', named after the site in New Mexico where Clovis stone points were found in the 1930s, associated with the remains of a dozen mammoths.[1] But now, with the emergence of new archaeological sites, the redating of some previously discovered sites, as well as a growing database of genetic evidence from living Native Americans, the dates for the occupation of the Americas have been pushed back further and further.

Most archaeologists agree that the route taken into the New World was, very broadly speaking, eastwards from north-east Asia, and then down through North America and into the south.

Getting into the Americas via this northern route, during the Ice Age, required an ability to exist and subsist in extremely cold, Arctic environments. Vladimir Pitulko's Yana site in the north-east of Siberia – the place I had arranged to visit but then been thwarted by Russian airline schedules – showed that modern humans were living far up in the Arctic from at least 30,000 years ago. Some of the elements in the Yana toolkit are similar to the Upper Palaeolithic technologies of North America, including ivory foreshafts for spears, which are also found among Clovis tools.

And, unlike today's geography, where the two landmasses are separated by the Bering Strait, when people were living at Yana, there was a connection between Siberia and Alaska. So people *could* have crossed over into the Americas. This now-submerged land is sometimes referred to as the 'Bering land-bridge', although this term is somewhat misleading as the 'bridge' was about the size of Europe: a landmass in its own right. It stretched all the way from the Kolyma River in Siberia to the Mackenzie River in Canada: about 3200km wide and 1600km from north to south.[2] Many archaeologists prefer to call it Beringia.

Even during continental glaciations, it seems that Beringia remained, although very cold tundra, ice-free, and there would have been plenty of herd animals roaming the grassy tundra to tempt the Siberian hunters eastwards, including woolly mammoth, horse, steppe bison and saiga antelope.[3] As the Ice Age rolled on, and northern Siberia became less and less habitable, Beringia would have formed a refuge for the hunters of the north.

Looking at a map of this area as it would have been 20,000 years ago, the route across from Asia to Alaska may have been 'open' but the rest of North America was sealed off by a vast ice sheet. In fact, there were two great bodies of ice – that merged east of the Rockies – known to palaeoclimatologists as the Cordilleran and Laurentide ice sheets (see map on page 300).

But archaeological evidence of humans in Alaska and the Yukon dating to before the LGM is very scrappy, inconsistent – and hotly debated. The first well-dated, generally accepted evidence of modern humans in eastern Beringia – what is now central Alaska – comes from a site called Swan Point, where a collection of stone tools, including microblades and burins, has been found, similar in many ways to the Siberian lithic industries. The archaeology at Swan Point dates to around 14,000 years ago. Then there are a number of other archaeological sites in central Alaska dating to between 13,000 and 14,000 years ago.[4] These dates are well after the LGM, but there are some controversial sites in the northern Yukon that may hint at a human presence there – well before the height of the last Ice Age.

Fragments of bone – which some archaeologists think are tools – have been found at Bluefish Caves and Old Crow River sites in the northern Yukon. Canadian archaeologist Richard Morlan claims that there are definite signs of human interference with these bone fragments: there is a bison rib, dating to around 40,000 years ago, bearing a cut mark that appears to indicate that it has been sliced into with a stone tool. Other, apparently butchered, bison bones have been found at sites near the Old Crow River, as well as mammoth bones that appear to have been split to access the marrow inside, then broken further to use as tools. Richard Morlan has gone as far as to call the bone fragments 'cores and flakes', by analogy with stone tools. Similar bone 'tools' and apparently butchered bones have been discovered in the Bluefish Caves, to the south of Old Crow River.[2] These have been dated to around 28,000 years ago.[4] (There were also stone tools in these caves, but they have not been dated, and are similar to much later tools from Alaska, dating to well after the LGM.)

But other archaeologists have argued that the bones could have been broken by natural processes: through being gnawed by carnivores, trampled by large animals, being carried along and smashed up in rivers, becoming frozen and thawed, or even blown to pieces by volcanic eruptions. Even though some of the bone fragments *look* as though they have been deliberately flaked, most archaeologists find this evidence too flimsy. Flaked bone tools are not unique to these controversial sites in the Yukon: there are precedents for bone-flaking from a range of sites across Eurasia, including the Dyuktai culture of Siberia and in North America.[2] However, the peculiar absence of any other well-stratified, well-dated evidence (stone tools, human bones, indeed anything else) from these sites means that most archaeologists view them with caution.

But these very early dates (for America) are certainly quite intriguing, and they don't seem that unbelievable given that we know people were living in Yana, in north-east Siberia, 30,000 years ago. Those Cordilleran and Laurentide ice sheets expanded to seal off the rest of North America by about 24,000 years ago. Before that, if there really were people living in Alaska that early, they *could* have made their way south, down through gaps between the ice sheets.[4] The archaeologists working at the Old Crow River and Bluefish Cave sites are even now uncovering and analysing evidence, so this opening of the American chapter is still very much in draft form.[2]

At the moment, though, there is no evidence of humans in the rest of North America, or in South America, prior to the LGM, so although an early entry into the Americas, before the ice sheets spread across from coast to coast, is a possibility, it remains a hypothesis with no firm archaeological evidence to support it.

But what about genetic evidence? Can we be sure that Siberia was the ancestral home of the first Americans, and can genetic analyses offer any further insight into dating the colonisation of the Americas?

Mapping Native American Genes: Calgary, Canada

I travelled to Canada to meet up with American geneticist Tracey Pierre. We were to meet, not at a university or a research lab, but at a First Nation Powwow, near Calgary, Alberta.

Having flown over the Rockies, from Camloops to Calgary, I arrived at the powwow early in the morning, before the events of the day had begun. There was a circle of tepees around a central, circular, roofed arena and I had a moment of *déjà vu* when I saw them: they were so similar to the Evenki chums in Siberia. Some even had stoves inside them, although there was really no need for heating in Alberta in July.

People were wandering around the camp in various states of preparedness and some of the dancers were already fully decked out in beaded costumes and feather headdresses. There were, too, lots of very excited small girls, many in costumes which were largely pink. But, despite the range of colours, including bright neon pinks and greens, and the very untraditional fabrics that some costumes employed, there was a definite style that everyone had adhered to. This was a Tsuu T'ina (formerly Sarsi) get-together, and the identity of this First Nation group was stamped on their costumes and would be reiterated throughout the day, as the master of ceremonies listed again and again the families and chiefs who were present.

Walking around the powwow ground, listening to the drumming and watching the dancing, I felt a mixture of emotions. It was clear that the Tsuu T'ina people had a strong sense of identity. Their culture was under threat, and they needed to keep it alive, somehow to sustain it in the wider Canadian culture they were now part of. The powwow was like a metaphor for this struggle: inside the covered arena, Native Americans drummed and danced and honoured their leaders and ancestors. Outside, there was the ring of tepees which men entered in T-shirts and jeans and came out of dressed in their costumes, as warriors. And outside the tepees was a village of burger vans and hot-dog stalls, white Canadians selling Native American kitsch, and a fairground.

But when I met Chief Big Plume in his tepee the struggle for culture didn't seem so futile. He was a strong leader, and exuded an amazing sense of pride in his heritage and his people. He sat on a bison skin, dressed in a regal red robe with weasel skins sewn, hanging, on to the sleeves. He wore a magnificent eagle-feather headdress. Invited in, I sat down beside the Chief and asked him about stories of the origin of his people.

'As it was told down through the generations, there was an argument within the camp. A dog had knocked down a tripod with a warrior's hunting tools – his arrows, his quiver. So there was an argument and a separation of people.'

He described a small splinter group heading south. They made off, across the frozen water of the Great Slave Lake.

'There was a young child on her grandmother's back, and there was a horn sticking out of the ice. The child was asking for the horn, so they had to stop the migration. They started to chip away at that ice, to bring out the horn, and it split the ice. The clans already on the south side continued migrating south. The ones in the north stayed in the north.'

So the wanderers were cut off from their original home – and became the ancestors of the Tsuu T'ina. It was a wonderful story, and I was intrigued by the mention of ice. I had read about another Native American tale, told by the Paiute, involving ice, where a raven pecked at a great wall of ice until it cracked and allowed people to pass through. It is tempting to see these stories as memories of a time when much of North America was covered by massive ice sheets, though it's impossible to know if the oral traditions actually stretch back that far.[1]

I asked the Chief what he thought about connections with Siberia. He knew the theories and thought it very likely that his people had originally come from north-east Asia. I showed him some photographs, and he was very interested in the Evenki chums, and their similarity to tepees. He also thought the Evenki resembled some of the more northern Native Americans. Then it was time for the Chief to get back to the powwow, as the afternoon's events required his presence. I left the tepee and went off to find Tracey.

Tracey herself was a Native American, of Navajo origin. She had studied archaeology as an undergraduate, and, realising how few Native Americans there were working in the field of genetics and phylogeography, she had made it her job to plug that gap. It was not an easy task. Even for Tracey, there was much suspicion and wariness to overcome among Native American communities. And understandably so. Genetics has been much misused in the past, and, when it came to Native Americans, there had been some gross misconduct perpetrated against them.

'A lot of times in the past, there's been research conducted without the knowledge, consent or participation of the tribes who are under study. In one particular study, involving the Navajo and Apache tribes, the FBI used DNA samples that were acquired through hospitals and prisons. It's a classic example of why tribes are wary and suspicious of genetic research,' explained Tracey.

This malfeasance had been picked up by Tracey's Ph.D. supervisor, Peter Forster, who had been asked to review the paper and who had spotted the ethical blunder. I was amazed by the story. It seemed as if I was hearing about some dreadful anthropological story from the nineteenth century. But this was a twenty-first-century tale.

Tsuu T'ina man in traditional costume

To many Native Americans, genetic science is too closely tied up with US government attempts to classify racial groups and confer rights accordingly. In 1887, when the Allotment Act was brought in, Native Americans were assessed on the basis of 'blood quantum'. This meant that anyone who was less than half-blood was denied land rights: their land would be appropriated and given to 'white' settlers. The US government also requires Native Americans to carry official 'Certificates with Degree of Indian Blood'. Artists cannot sell their work as 'Indian' if they do not have federal certification. It seems a highly divisive approach in a multicultural society.[2]

Tracey was at the powwow to enjoy the dancing, but also to take some more mtDNA samples. That day, she had little success in persuading people to swab their cheeks. Gaining the trust of people who had seen their heritage trampled upon and their DNA stolen would be a long process, but Tracey believed it was worth it, and that the history she was finding in the genes belonged to the people – her people – who had given her their genetic material.

Tracey's own doctoral research focused on later migrations within established Native American populations, but I asked her what mtDNA analyses had to say in general about the colonisation of the Americas. Firstly, she was very clear that the genetics confirmed a Siberian homeland of the palaeoindian ancestors of the Native Americans.

'So far, mtDNA analyses have identified five major lineages, A, B, C, D and X, that are indicative of Native American populations today, and these can be traced back to southern Siberia.'

'So does that mean that *all* Native Americans originally came from Siberia?'

'That's what the genetic evidence is implying at this point, and the archaeological record supports it, too. For decades, scientists have been suggesting there's a link with Siberia, based on physical attributes – and I think the genetics now proves that.'

As well as those mtDNA haplogroups, there were also Y chromosome variants (Q and C) in Native Americans that could be traced back to the Russian Altai.[3, 4] Tracing the route up through northern Siberia was complicated by the fact that people had cleared out of there during the LGM, and moved back in afterwards. But the Siberian genetic origin of the Native Americans certainly supports the hypothesis of a migration westwards across Beringia.

Whereas early mtDNA studies had indicated anything up to five separate migrations into the continent, the newest data – looking at complete mitochondrial genomes – suggested that there was just one.[5] In the 1980s, genetic, dental and linguistic studies suggested that there were probably three waves of colonisation from Siberia into the New World, corresponding to three language groups among Native Americans: Amerind, Na-Dene and Aleut-Eskimo. (Although, it has to be said, this is a very conservative number; some linguists would claim there are more than 160 language families.) The first Y chromosome studies, also in the 1980s, seemed to suggest either one or two waves of colonisation. In 2004, the results of a large Y chromosome study showed that the Native American population had two major founding Y chromosome haplogroups: Q (76 per cent) and C (6 per cent). But rather than these two lineages representing separate migrations, the mixture of Y chromosome types among living Native American men was best explained by a single, polymorphic (i.e. containing more than one Y chromosome type) founding population.[4] A recent study of Native American genes from DNA inside the nucleus has also supported the idea of a single wave of migration.[6]

This also makes sense if Beringia is viewed less as a 'land-bridge', with migrating Asians beetling across it, and more as a 'staging post', receiving incomers from various parts of Asia – some of whose descendants, bearing that mix of lineages, then went on to colonise the Americas. It also means that it would be naive to look for a geographically discrete 'homeland' in Asia – people could have been coming in from lots of different regions. Instead, it is really Beringia that was the homeland of the first Americans.[7]

The genetic data also suggest a timescale for the colonisation, with modern humans spreading into Central Asia by 40,000 years ago.[8] Fossils of modern humans from Tianyuan Cave and Yamashitacho also date to around 35,000 to 40,000 years ago, hard evidence that people were in East Asia by that time.[9] Analysis of Asian and New World nuclear and mtDNA suggests a three-stage model of colonisation. First, there was a gradual population growth, and a movement of Amerind ancestors into north-east Asia, between 43,000 and 36,000 years ago. The effective founding population was small, perhaps only a couple of thousand, and this is why Native Americans have much lower rates of genetic variation than Asians: the people who went on to colonise the New World were only a small 'sample' of the gene pool in Asia. The proto-Amerind population grew and spread throughout Beringia between 36,000 and 16,000 years ago, then there was a 'population bottleneck' around 20,000 years ago – at the height of the LGM. As the world warmed up again, between 18,000 and 15,000 years ago, there was a major population expansion throughout the Americas. Based on genetic variation, the founding population could have been as small as just 5000 people.[10] The range of dates suggested by Y chromosome analysis is later, at 10,000 to 17,000 years, but the upper end of this range overlaps with the mitochondrial dates. It may seem strange that there is such discrepancy between dates worked out from different genetic trees, but there are a number of reasons for this. Variation in mutation rates could throw a spanner in the works, when scientists are basing their calculations on an assumption of a steady rate of mutation. Genetic drift within a population could also muddy the waters.[8] But that's not to say the genetic dates are of no value. And if there is some overlap, it suggests that, despite all the potential pitfalls, we can place some trust in the reconstruction of our past from our genes.

So this genetic contribution to the debate, with dates of a population expansion later than 20,000 years, makes a pre-LGM migration between the ice sheets seem very unlikely. It doesn't rule it out, though, and Stephen Oppenheimer believes that the founder mtDNA lines may have arrived in Beringia well before the LGM, becoming cut off from the founder Asian populations for at least a thousand years, and spreading throughout the Americas – before the way was sealed off by the ice sheets. He argues that the greater genetic and linguistic diversity in South America fits well with a model of early – pre-LGM – colonisation of the *entire* continent, with North America becoming largely depopulated during the LGM. The spreading ice sheets meant that the North American population dwindled, and was driven into a refugium in the far north – in modern-day Canada and Alaska. After the LGM, people could once again expand across North America.[7]

But whether the first wave of colonisation into most of North America happened after the LGM, or represents a repopulation after the initial colonisers were pushed back, it is clear that it happened long before Clovis appears in the archaeological record.

How did the first colonisers get down into North America when the way was blocked by ice? The ice-free corridor between the Laurentide and Cordilleran ice sheets didn't appear until between 14,000 and 13,500 years ago.[8] Again, the genetics seemed to indicate an answer: the ivy trail of mtDNA lay along the coast.

Exploring the Coastal Corridor: Vancouver, Canada

I therefore made my way to the coast, to the city of Vancouver, at the mouth of the Fraser River where the Coast Mountains come down to the sea. In 1808, a fur trader named Simon Fraser set out with a band of twenty-three men, in four birch-bark canoes, to explore the great river that would eventually bear his name. They paddled 800km downriver, all the way to its mouth on the west coast of Canada: they were the first non-indigenous people to see the mouth of the Fraser River. Along the way, they encountered various groups of indigenous people. Sometimes the blue-eyed, pale-skinned Fraser was taken for a supernatural being, one of the 'transformers' from the beginning of time. But when he reached the Pacific, Fraser was met by fierce Musqueam warriors armed with bows, spears and war clubs. They were right to be defensive, for these men would indeed transform their land. The trade routes Fraser mapped out facilitated the Gold Rush of 1858 and western Canada experienced its own industrial revolution as mining, fishing and forestry took off. Today, the economy has changed yet again, with information and knowledge becoming powerful commodities.[1] And up in the foothills, on Burnaby Mountain to the east of Vancouver, is the campus of Simon Fraser University, rising up the slopes like a concrete Aztec temple.

After driving around the campus a couple of times, I eventually managed to find the Department of Biological Sciences and track down Professor Rolf Mathewes in his book-lined office.

Rolf studied pollen in both ancient (palaeoecological) and modern (forensic) contexts. As well as using pollen to track down criminals, he had been investigating the possibility of ice-free refugia along the northern Pacific coast. His studies had been concentrated around the Haida Gwaii (formerly Queen Charlotte Islands) off the west coast of

British Columbia. In the early 1980s, radiocarbon dating of plant fossils had suggested that the islands were ice-free perhaps as early as 18,000 years ago – when the mainland was still covered with ice. Genetic studies of fish, birds and mammals have also suggested that the islands acted as refugia for a variety of species, soon after the LGM. With sea levels lower than today, the exposed areas of the continental shelf may have provided a 'coastal corridor' along which humans could have migrated soon after the LGM, as the edge of the continental ice sheet began to melt back. Rolf and his colleagues had taken samples from beneath the sandy seabed of Dogfish Bank, part of the now-submerged continental shelf between the Haida Gwaii Islands and the mainland, as well as a sample from the south-western coast of the islands. They had then carried out detailed pollen analysis – and radiocarbon dating – on these cores.[2]

Rolf was very excited about the acquisition of a new microscope with built-in digital screen and camera, and he used it to show me some pollen from the Dogfish Bank core.

'This is one pollen grain of a sedge,' explained Rolf, zooming in on a tiny grain. 'It's only about one-thirty-thousandth of a millimetre in size, 30 microns. More than half the pollen on this slide is sedge pollen.'

'And have you been able to date this sample?' I asked.

'Yes, and it's very exciting,' said Rolf. 'This particular sample dates to somewhere between 17,000 and 18,000 years ago.'

'So if there were sedges growing there, that meant the ice had retreated?'

'Absolutely,' replied Rolf. 'The mainland of British Columbia to the east had a huge ice sheet on it which was still kilometres thick. But these islands off the mainland were already deglaciating, much earlier than most people thought when I started this work back in the 1980s. In fact, we published the discovery of these early ice-free sites in the Queen Charlotte Islands in *Science* – it's the earliest known area for deglaciation and this early vegetation.'

'So, 17,000 years ago, the ice sheets were pulling back from the west coast of Canada?'

'That's right, but the Queen Charlotte Islands had their own little ice cap, which was much thinner than the big one. This is one of the reasons why this area deglaciated earlier. It's not like a single wall of ice that's pulling back.'

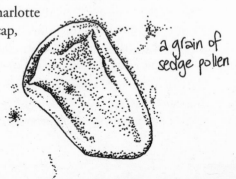

a grain of sedge pollen

The pollen in the Dogfish Bank sample showed that this coastal plain was exposed between around 17,000 and 14,500 years ago – after the glacial ice retreated and before the sea-level rise from all that melting ice flooded it. Careful analysis of the pollen in the sediments from beneath the seabed showed that Dogfish Bank had been a wetland environment, with lots of sedges, along with various grasses, horsetails, willow and the dwarf evergreen shrub, crowberry. Pollen from wetland species like bayberry, as well as fossil green algae, indicated that the land was marshy.[2]

'Do you think people could have lived here?' I asked.

'Well, based on this – absolutely,' said Rolf. 'This was dry land, with vegetation growing on it.'

'And do you think the coast could have formed a route for the first colonisation of the Americas?' I asked.

'There were a number of sites from the Gulf of Alaska right down the side of British Columbia that were free of ice, from 16,000 years ago. And, each summer, the green patches would have expanded. There were even forests starting to form on the Queen Charlotte Islands when the inland corridor was still locked up in ice,' explained Rolf.

The sample from the southern end of Haida Gwaii, from a site above sea level, showed that sedges also dominated the early plant life there, but that there were lots of other herbs and shrubs as well. Some of these plants, like ferns, prefer warm, moist habitats, whereas others, such as crowberry and bearberry, survive in cold, dry places. This mixture of plants is similar to that found in the coastal tundra of modern south-western Alaska. The concentration of pollen grains in the sample gradually increased over time, as the plants gained a firm foothold on the post-glacial landscape. By 15,600 years ago, the first trees, lodgepole pines (*Pinus contorta*), had arrived and the environment of the islands changed from tundra to pine woodlands with ferns covering the ground between them. Pine pollen can blow a long way, but the presence of this species on the islands was also confirmed by radiocarbon dating of a lodgepole pine needle to just over 14,000 years ago. Other coastal sites in British Columbia have shown pine woodland to be widespread by about 14,000 years ago. As the climate grew even warmer, the succession of plants continued: the pine woodlands started to disappear, to be replaced by spruces.[2]

'People could have travelled down the coast, probably by boat, from oasis to little oasis. The environment was certainly suitable for humans, but we'll leave it to the archaeologists to find the hard evidence that they were actually there.'

296 The Incredible Human Journey

After lunch in one of the pleasantly wooded courtyards of Simon Fraser University, I explored the campus. The original sixties buildings were designed by architects Arthur Erickson and Geoff Massey, and the modernist design principles have been maintained in later buildings, so that the whole scholarly city has a stylistic coherence and grace. I sat down to read in the calm serenity of the Academic Quadrangle, where a huge jade boulder was reflected in the still waters of a pond. The monumental buildings had a lightness and aesthetic appeal that I didn't normally associate with sixties concrete architecture. Their futuristic qualities had made them the ideal backdrop for sci-fi series: James Fraser University has featured in *The X-files*, *Stargate*, and as a city in the Cylon-occupied planet of Caprica in the film *Battlestar Galactica*.

Back in Ice Age Canada, though, and aliens aside, British Columbia held other clues to the potential coastal dispersal of its earliest human inhabitants. Bizarre as it may seem, but very fitting for Canada, one of those clues is *bears*. The Haida Gwaii Islands are full of limestone and are rained on a lot: perfect conditions for cave formation. Limestone caves are particularly good for fossil preservation, and so the Haida Gwaii caves seemed like a pretty good place to start looking for fossils. In 2000, a team of Canadian palaeontologists and archaeologists started excavating the caves – and quickly started to find animal bones. These included dog, mouse deer, duck – and bear. Radiocarbon dating showed the dog bones to be quite recent in palaeontological terms: less than 2000 years old. But the bear bones were very exciting – they were much older.[3]

I travelled north from Vancouver, taking the 'Sea to Sky' Highway along the magnificent coast of Howe Sound up to Squamish, then winding through the wooded gorges of the Cheakamus River valley to Whistler. This ski resort – in summertime – was full of holiday-makers going mountain-biking, whitewater-rafting, and shopping. But that day Whistler also hosted archaeologist Quentin Mackie, who had kindly arranged to come down from the University of Victoria especially to meet me, and he had brought along a couple of old friends: two bear skulls from excavations on Haida Gwaii.

When I arrived in Whistler, Quentin Mackie was already there, in the foyer of the hotel, with the two bear skulls on a coffee table in front of him (I wondered what the other hotel guests thought of this). The skulls were different sizes – one was a brown bear, the other a black bear – but both looked formidable beasts. They were ancient specimens, excavated from caves on Haida Gwaii.

'These are just a couple of the specimens we've discovered in the caves,' explained Quentin. 'We've found more than six thousand bear bones, from black and brown bear, and some of them date to as early as 17,000 years ago.'

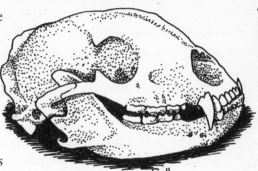

Black bear skull

So, very soon after the peak of the last Ice Age, Haida Gwaii was supporting not only sedges and grasses, but large – very large – mammals.

'We were really surprised to get this early date, because conventional wisdom suggested that, at that time, there would have been nowhere for bears to live on the north-west coast: it would have been covered by ice. There was some pollen evidence for ice-free areas, but we thought that these were small, very windswept, perhaps not very viable environments.'

'How do you think the bears got there, that long ago?' I asked.

'With these early dates we have to consider one of two possibilities. Either these bears were able to get into the coast from somewhere [north] very early – as early as 17,000 years ago, or they spent the last part of the glaciation on the coast, in some kind of refugium. Both of those are really interesting – because if bears could get in early or if bears could survive through the LGM on the coast, then it implies that humans could have done as well.'

I could see Quentin's point – that the presence of a large mammal in Haida Gwaii might have implications for human migrations[3] – but I was also a little sceptical.

'But you haven't found any evidence of *humans* that long ago?'

'Well, no. But the interesting thing about these bears is that they make a pretty good analogy for humans. They're large land mammals, they're quite territorial, and they're omnivorous. They eat berries, roots, insects, even small mammals like ground squirrel – and they may even chase down deer and caribou. They're also very good beachcombers. They will go along the strand line – and I've seen them do this many times – roll over rocks and lick up sand hoppers and crabs. But they're also quite capable of taking migrating salmon from rivers.

'Humans would have been able to eat pretty much anything that a bear would eat ... so I think that if it was a good environment for bears – which it clearly was – then it was a great environment for humans.'

Quentin had found some evidence of human activity dating back to 13,000 years ago. It seemed that people had been hunting the hibernating bears at that time, which sounded like a highly dangerous pursuit, but Quentin said there were historical accounts of people waking hibernating bears with burning braziers, then waiting by the cave mouth to spear the angry animal as it emerged.

'Bear meat is quite edible. I've eaten it myself. If you cook it well it's delicious, and very nutritious. The bears' winter pelts would also have been extremely valuable to these people, and we also have a really good record of them making tools out of bear bones, so it's a fabulous resource from nose to tail.'

However, these bear hunters weren't early enough to have been among the first colonisers of the Americas. But Quentin was not about to give up the search.

'These caves are still some of the oldest sites in Canada. And we feel fairly confident that in the next few years we'll be able to push that date back a little bit.'

The mitochondrial genetic lineages of brown bears have also been investigated – including mtDNA from bears preserved in permafrost – and they tell us something very interesting. It seems that brown bears, like humans, came through Beringia – and had persisted there throughout the LGM.[4] Not only that, the mtDNA of brown bears suggests that, having survived in refugia on islands off the western coast of Canada, these formidable animals went on to repopulate the mainland from about 13,000 years ago.

I missed out on seeing actual bears in Canada (although I did see a whole bear skeleton in a gem shop window), but I did get to visit a remnant of the Great Cordilleran ice sheet: the Ipsoot glacier, high in the mountains above Pemberton, near Whistler. A helicopter dropped me off on the glacier, along with mountaineer Jim Orava, and I spent a thoroughly enjoyable day learning to ice-climb. Equipped with ice axes and crampons, we crawled over the glacier – climbing and traversing vertical walls of ice, leaping over crevasses with ropes for security. I had rock-climbed before, but it was a very different sort of climbing on ice. In just one day, I began to understand *why* people want to climb mountains, but it also made me absolutely sure that the unbroken coast-to-coast glacier of the merged Cordilleran and Laurentide ice sheets would have been an absolute barrier to hunter-gatherers at the peak of the last Ice Age.

From one extreme to another, I also went snorkelling in the chilly coastal waters of British Columbia. Warmly wrapped in 5mm of neoprene wetsuit, I explored the shallow waters around the edges and islands of Sechelt Inlet, north of Vancouver. As well as plenty of sea stars and fish, I saw lots and lots of kelp. And, believe it or not (and at this point it does start to feel as though archaeologists are clutching at straws, or seaweed, or whatever, along the Canadian Pacific coast), kelp has also been implicated in the proposed coastal route into the Americas.

Kelp forests are incredibly rich ecosystems – among the most productive and diverse habitats on earth. They support a great wealth of marine life: fish and shellfish, marine mammals and birds. For hunter-gatherers on a kelp-forested coast, the marine resources are easily as rich as – if not richer than – those on land, provoking some archaeologists to call the potential coastal route into the Americas the 'Kelp Highway'. Bears come into the story again here: grizzly bears on the richly provided coast grow two or three times as large as their inland cousins.[5] Kelp thrives in cool nearshore waters, preferring temperatures of less than 20 degrees C, and can even survive a winter beneath sea ice. Today, the coastal kelp forests of the north Pacific stretch all the way from Japan to Baja in California. There's a break along the tropical coasts, where the sea is too warm for kelp, then it starts again along the Andean coast. The kelp forests would have been there, in the rising post-glacial waters, so any prehistoric beachcombers would have been well provided for. Today's Pacific kelp forests are home to a huge variety of fish, shellfish (such as abalones, sea urchins and mussels), seabirds and sea otters. Before Beringia disappeared beneath the waves, its bays and inlets supported large mammals like walrus and sea cow. Although the sea level around western Canada early in the post-glacial would have been about 100m lower than today, the coastline would have been almost as convoluted and fragmented as it is now. Any coastal hunter-gatherers *must* have had boats in order to get around and efficiently exploit this environment.[5] But as it has been argued that modern humans emerging from Africa – some 60,000 years before the Americas were even glimpsed – were adapted to coastal and estuarine environments and probably had watercraft, it seems quite reasonable to assume the early Americans would have used boats to navigate and forage along the coast.

Reasonable assumptions are all well and good, but ultimately all this musing about a North American coastal route is, at the moment, conjecture. The genetics may suggest a coastal route, and the conditions may have been suitable (judging from pollen, bears and kelp) for a coastal expansion from 17,000 years ago, but there is no hard evidence for people living along Canadian shores that long ago. The Palaeolithic

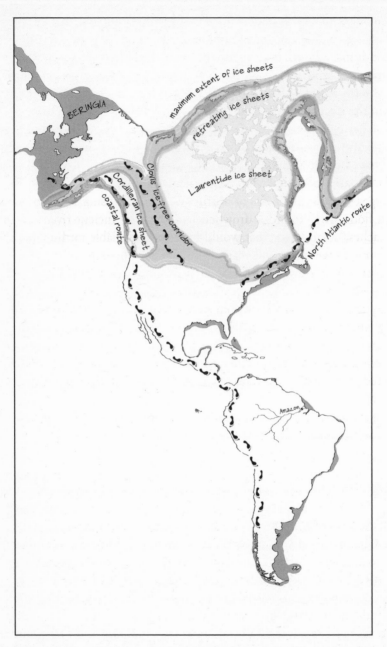

The map shows the ice sheets of North America during the Last Glacial Maximum, stretching from coast to coast, and the retreating ice sheets. The north-west coast became deglaciated about 16,000 years ago, freeing up a potential coastal route into the Americas; the interior ice-free corridor opened 14,000 to 13,500 years ago; the map also shows Bruce Bradley's suggestion of a North Atlantic route.

archaeologists of North America suffer the same problem that plagues those trying to trace ancient migrations just about everywhere else in the world: glaciation has scoured continents and destroyed a wealth of archaeological evidence in its path, and the rising sea levels produced when the ice melted have concealed the Pleistocene coasts. Presumably many of the traces of the first colonisers could be lost beneath the waves. Most of Beringia is now under water, like a Stone Age Atlantis, and the coastline of Alaska and Canada is much higher and further inland than it would have been at the end of the Pleistocene.

There are some intrepid archaeologists who have explored the submerged Pleistocene coasts of north-west Canada, using sonar to map the sea floor. They found a drowned landscape of ancient rivers, lakes and beaches,[6] a landscape that would have been available for humans to inhabit one to two thousand years earlier than the opening of the inland ice-free corridor. This ancient coast was later covered up by a relatively rapid rise in sea level, and this makes the archaeologists very optimistic that any archaeological sites would have been quickly buried under the seabed, and are therefore likely to have been well preserved. Gradual increases in sea level cause much more erosion as they occur,[7] and in fact one piece of archaeology *has* emerged from under the waves: a barnacle-encrusted stone tool was recovered from an ancient river mouth now 53m underwater.

For the first actual, datable evidence of people on the North American coast, I would have to travel south: to California.

Finding Arlington Woman: Santa Rosa Island, California

Just off the coast of Southern California, across the Santa Barbara Channel, lie the Channel Islands. These islands are rich in wildlife and have been home to humans for thousands of years. Historically, the archipelago was settled by Chumash and Tongva Indians, who used purple olivella shells as currency. Spanish explorers, missionaries and ranchers arrived on the islands in the sixteenth century, and nineteenth-century fur traders hunted the native sea otters and seals nearly to extinction. Now the islands and their diverse wildlife are protected, as part of a National Park.

I flew to Santa Rosa in a six-seater prop plane, late in the day. We took off from Camarillo airstrip and were quickly over the Santa Barbara Channel, heading west. I could just make out the Channel Islands on the horizon. We flew over Anacapa Island, then over the larger island of

Santa Cruz, just to the north of its Diablo Peak. The islands were rocky and rugged, with very little sign of human habitation.

We started descending over the sea, then came into land on a short, dusty airstrip just set back from the shore at Bechers Bay. I was met by rancher Sam Spaulding in his Hilux and we set off on a track that rose steeply into the mountainous interior of the island. Although park wardens and scientists were frequent visitors, there were no permanent residents on Santa Rosa Island. I was staying at the National Park lodges, high up in the hills. I arrived there just as the sun was setting, and I met John Johnson of the Santa Rosa Museum of Natural History, who had come out to the island to be my guide for a trip to Arlington Springs on the other side of the island.

The following morning we set off early and drove for two hours along rough tracks to the gully known as Arlington Springs. The landscape we passed through was magnificent, sculpted by water. Over the millennia, streams and rivers have gouged deep gullies and canyons in the sandstone, naturally exposing ancient sediments. And in those sediments are fossils of the various long gone inhabitants of Santa Rosa.

John and I scrambled down the steep banks of the gully to the stream, which meandered its way towards the coast. On the way there were waterfalls and plunge pools that we had to edge around to follow the water down to the sea. In the walls of the gully I could easily make out the layers and layers of sediment that had built up over millennia to then be cut back by the stream. In deep layers, there were fossilised shells: great, clam-like things and ridged bivalves. These were very ancient, from a time when this sediment had been on the bottom of the sea. As we walked downstream, we came across younger sediments, and John pointed out fossil bones sticking out of a reddish bank: a line of crumbly vertebrae and some long bones. Pygmy mammoth bones, no less.

In 1959, an archaeologist named Phil Orr had been drawn to Santa Rosa Island by its rich Pleistocene fossils. He was trying to create a new track down the side of Arlington Springs gully with a grader, to reach an excavation near the beach. His vehicle became stuck in the gully, and when he got out and looked around he spotted two long bones poking out of the side of the gully, 11m below the top of it. The bones were in the same – Pleistocene – layer as the mammoth bones he was more familiar with. But these weren't mammoth bones. Orr called on the advice of other experts who confirmed what he had suspected: the bones were human. It looked as if he had found the first evidence of ancient humans on the island. The bones were two femora (thigh bones). There was no sign of the rest of the skeleton: the femora were 'disarticulated'. They appeared to have been washed into the gully and then covered up by later layers of sediments. Orr removed the bones in a block of earth, and, back at the lab, removed fragments to send for radiocarbon dating. In 1960, he published the results: the femora were 10,000 years old.[1] Orr thought the bones to be robust and most likely male, and the find became known as 'Arlington Man'.

In 1987, John Johnson and his colleague Don Morris had found Orr's plaster block containing the Arlington Springs bones in the basement of the Santa Rosa Museum, and decided to subject the bones to some modern analyses, including DNA tests and the more reliable AMS radiocarbon dating. Anthropologist Philip Walker looked at the bones and decided that they were most likely to be from a woman: the rough line down the back of the femur where thigh muscles attach was very faint, not robust as it often is in male skeletons, and the slender diameter of the bones also suggested that they were female.[1] John showed me reconstructions of the femora, and they were indeed very lightly built. Radiocarbon dating of the human bone itself, and on an associated extinct mouse deer mandible that was buried close by, yielded a date of around 12,900 years ago[2] – meaning the skeletal remains of 'Arlington Woman' were indeed among the oldest known in North America.

Although these human remains date to several thousand years after the coastal corridor is thought to have become accessible, they do prove something about the earliest Californians: they must have used boats. At the height of the last Ice Age, the three islands of San Miguel, Santa Cruz and Santa Rosa were joined together, as a single super-island which has been called Santarosae, but it was always separate from the mainland.[3, 4]

In the 1990s, John Johnson was part of a team of archaeologists and palaeontologists who, following up Phil Orr's work, undertook a comprehensive survey of the island, looking for Pleistocene fauna in particular. They didn't find any more human bones, but they did discover plenty of mammoth fossils, all around the coast of Santa Rosa, in alluvial sediments cut through by water action – just like the bank with the mammoth fossils that John had shown me. They also uncovered an almost complete skeleton of a pygmy mammoth, and when they radiocarbon dated the bones they found them to be almost exactly the same age as Arlington Woman. In fact, there seemed to be an overlap of around two hundred years. This was an important result. Mammoths had been on the island for around 47,000 years, and now the archaeologists could be sure that there was a period when pygmy mammoths and humans were contemporary on Santa Rosa, as Orr had suspected. Orr also believed that humans had played a key role in the extinction of these diminutive mammoths. It is difficult to support such a theory on the basis of the Santa Rosa material alone, founded on dates from 'one mammoth, one human, on one island',[3] and there is no 'smoking spear' to prove human predation on mammoths. Nevertheless, it still seems a remarkable coincidence that, within two hundred years of humans arriving on Santa Rosa, the mammoths had disappeared. In many ways, the evidence from this island is tantalising but frustrating. In spite of some continued archaeological investigations on the island, the only evidence of humans there in the Pleistocene is those thigh bones. Nothing else: no other bones, no evidence of any occupation sites. What was that woman doing on Santa Rosa? If she was part of a band of hunter-gatherers living there, no other trace of these people has appeared – yet.

Humans may have killed off the miniature mammoths on the island, but for more evidence of human involvement in the extinction of Pleistocene fauna, and of the beasts themselves, I would need to return to the mainland.

Pygmy mammoth bones in the sandy bank on Santa Rosa

John and I had wandered down Arlington Springs gully all the way to the seashore, and we sat down on a sand dune to talk about his other research, in genetics and phylogeography. John had become fascinated by the language families of the Californian Native Americans, including the Chumash.

Languages can certainly suggest population origins and movements, but they are also transferable – like other aspects of culture – and so it's very difficult, even impossible, to reconstruct the deep evolutionary history of people around the world on the basis of variation in language. For John, the suggestions emerging from the linguistic studies warranted further investigation, and mtDNA analysis seemed to be the perfect tool. But it was a telephone call from a friend that had really set John off on the trail of tracing ancient migrations.

'A few years ago, a friend of mine called me up and told me there was this tooth from an ancient jaw from southern Alaska, that had DNA that was a very rare type.'

I asked him how old the tooth was.

'It was 10,300 years old – the oldest skeletal remains in the Americas from which they've successfully obtained DNA.

'But the remarkable thing is that this rare type of DNA – which is only found in 2 per cent of Native Americans overall – matched 20 per cent of the samples I had taken from Chumash Indians, here in California.

'And when we looked at a wider database of all Native Americans, we found that this type was also in southern Alaska, in north-west Mexico, in coastal Ecuador, in southern Chile, in southern Patagonia, and in prehistoric burials in Tierra del Fuego. All the way down the Pacific coast.'

'So do you think this is evidence for a coastal dispersal?' I asked.

'I think it is the best evidence to date for an ancient coastal migration: it suggests that this particular group gradually moved down the Pacific coast, taking advantage of coastal resources, and left behind descendants who still live in all of these areas.'

John and his colleagues had looked at samples from 584 Native Americans, from a geographical perspective. Haplogroup A was more common along the coast, but there was also a rare subgroup of haplogroup D that seemed to echo a migration into the Americas all the way down the Pacific coast of North and South America.[5, 6] This was the same rare type that had been found in DNA extracted from the ancient tooth discovered on Prince of Wales Island, in Alaska.[7] A recent study of Native American autosomal DNA markers – within chromosomes in the nucleus – also revealed a pattern, not only of decreasing diversity from north to south, indicating colonisation in that direction, but also suggesting that the coastal route was important.[8] Just possibly Arlington Woman may have been a descendant of the first ocean-going, beachcombing Americans.

Hunting American Megafauna: La Brea Tar Pits, Los Angeles

Having had my appetite whetted by the remains of those diminutive mammoths on Santa Rosa, I wanted to find out more about the large animals that roamed the Americas during the Ice Age.

Soon after I started working at Bristol University, I met colleagues in Earth Sciences who showed me some of the collections held in that department. There was one room in particular that fascinated me: there were samples of rocks from around the world, and, in a large glass cabinet, the impressive skeleton of a sabre-toothed cat. The bones were dark brown, almost as though they had been carved out of ebony. The stuff that had turned the skeleton that colour was tar. The cat was a composite skeleton, made from bones that had been dug up from the La Brea tar pits in Los Angeles. It had been brought to Bristol University by the explorer and famous palaeontologist Bob Savage, who was a professor in the Department of Earth Sciences during the fifties and sixties.

I was, therefore, very excited to be visiting the tar pits themselves. I had driven down the Californian coast from Santa Barbara to Los Angeles (stopping off to watch surfers at Malibu Beach on the way), and now I was heading into the heart of the city, to get a close look at the amazing palaeontological treasures that had emerged from the tar. I approached La Brea Museum around the edge of a lake, filled mostly with water, but with tarry edges, and huge methane bubbles rising to burst on the surface every few seconds. There I found myself face to face with life-size – full-size – mammoths. A great bull mammoth had become mired in the tar pit, and, on the bank, a cow and her calf were looking on helplessly. They were good models: it was almost as if I'd been transported back 20,000 years in time.

Inside the museum itself were reconstructed skeletons of a range of animals that had met sticky ends – in the most literal sense – in the tar pits: giant sloths, horses, camels, mammoths and mastodons, sabre-toothed cats, lions and dire wolves, an extinct species of wolf that used to live in both North and South America. In a glass-walled laboratory, palaeontologists worked away, cleaning and preparing the specimens that were still emerging from the tar pits. I went through a door that took me from the public side of the museum into laboratories and store rooms, where I met the curator, John Harris. I quickly discovered that John had also been at Bristol University, studying for his Ph.D., and he remembered Bob Savage's sabre-toothed cat from La Brea.

John showed me some of the museum stores: they comprised what seemed like endless corridors lined with shelves and trays containing thousands – millions – of bones from the tar pits. The bones were not fossilised in the true sense of the word: they had not turned to stone but were still bone because the tar had preserved them. Then John took me outside, to a pit where excavation was still going on. The pit was almost 10m deep and about 10m square. Down at the bottom of the pit, working away in the black, sticky sludge, were three young palaeontologists – Andrea Thomer, Michelle Tabencki and Ryan Long – all of whom were themselves well covered in tar. I sensibly donned a forensic-like white suit and started to descend the ladder into the pit, where I could safely stand on boards to one side.

'This is a strange job,' I offered, gingerly making my way down the ladder into the tarry trench.

'It's basically the strangest job in the world … digging up the bones of dead animals out of tar,' said Andrea.

'And it doesn't look like very nice stuff to excavate.'

'No, it's terribly sticky. Especially in the summertime. As it gets hotter it gets stickier.'

Around her Wellington-booted feet I could see bones sticking up out of the tar. The more I looked, the more I saw. It was like a mass grave of Ice Age beasts.

'We've had visiting palaeontologists come by who are just astounded to see this many fossils in one place. It's probably one of the richest fossil deposits in the world,' Andrea told me.

'And what type of species have you found?' I asked her.

'Well, the most common animal we find is the dire wolf. And the second most common is the sabre-cat. We have a very strange predator–prey ratio – we have way more predators than we do animals that they eat. We think it's because of the way that the animals became trapped in these tar pits: a large herbivore would walk in, get stuck, and then all of the carnivores would come and try to eat him, and then they would get stuck. About 70 per cent of the animals are predators rather than prey, and that's completely backwards.'

'And how old are these fossils that you're getting out of this pit?'

'Well, down at the bottom, they're 40,000 years old. And they're not actually *fossilised*. They're still bone – you get incredible detail.'

Michelle was trowelling the surface of the tar, removing a wet, oily layer so that excavation would be easier.

'It's called glopping,' she said. 'We have to do this every day before we really start work excavating bones from the pit. The tar is constantly rising up. If we didn't do this every single day, the pit would start filling up with liquid asphalt again. We're fighting nature, and it's a losing battle.'

Climbing back up out of the pit, I asked John about the scale of the operation.

'Since 1969 we've taken more than 70,000 bones out of this particular locality. All together we've got something like three and a half million specimens in our collections – representing more than 650 species of animals and plants.'

I asked him if they excavated all year round.

'No. We now only excavate during the summer. We do a ten-week excavation season and in just ten weeks we take out between one and two thousand bones.'

It was clearly a huge job and the tar pits seemed almost bottomless. And it wasn't just the planned excavations at La Brea itself that kept the palaeontologists busy. Construction work in downtown LA often turned up tarry surprises, and a recent sewer trench had unearthed tonnes of tar-soaked sediment – full of bones. I looked over to where John was pointing, just behind the tar pit where excavations were ongoing, and there were masses of industrial containers, full to the brim of sediment, waiting to be excavated. John was going to have to pull his workers out of the pit so that they could focus their attention on this unexpected haul of tarry bones.

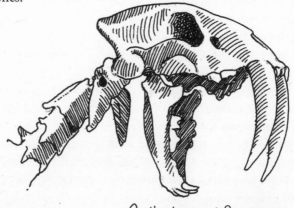

Smilodon californicus
at La Brea tar pits

Among all those Ice Age skeletons, only one human has ever been found.

'It's not that surprising,' said John. 'If you get stuck in a tar pit then you're likely to meet your end there – unless you have friends to pull you out.'

The tar pits gave a wonderful insight into the richness of animal life in ancient California. It was also obvious that the vast majority of these large animals were no longer with us. Just as in Eurasia and Australia, the arrival of humans in the Americas seemed to coincide with the demise of the megafauna.

'The peculiar thing,' John continued, 'is that about 13,000 years ago these large creatures disappeared – and it's one of the big mysteries of American palaeontology.

'Several people have come up with different ideas about how and why they disappeared. This was, after all, right at the end of the Ice Age; there would have been climatic change and environmental change. But that had happened at least ten times during the Pleistocene so that, in itself, climate change wouldn't have caused the extinction of the megafauna.

'Thirteen thousand years ago is about the time when humans arrived in North America, and some people think that it was humans hunting the megafauna, or introducing diseases against which the megafauna had no immunity, that was part of the cause.'

La Brea tar pits don't contain any evidence of human predation, but there are other sites in America that show, unequivocally, that humans were hunting mammoths. But, just as on other continents, whether or not humans hunted the mammoths and other megafauna to extinction, or whether changing climate and environment played a more significant role, it is difficult to know.

But John said that there was another, rather intriguing, theory about the disappearance of the American megafauna that had recently exploded on to the scene.

'The third suggestion that's come up recently is that, apparently, there was an explosion of an asteroid right over the Great Lakes region of North America at around 13,000 years ago …'

When I was in Arlington Springs and looking at the side of the gully (now eroded much further back than when Phil Orr discovered Arlington Woman), John Johnson had pointed out several black layers in the section. Most of them were probably due to forest fires that occasionally ripped through the island's vegetation. I had seen the results of recent forest fires in the hills above Santa Barbara: blackened trees and a thick layer of ash over the ground.

But it seems that one of these black layers represents something more than random and sporadic forest fires. Across North America, at more than fifty sites, a particular black layer has been found that dates to 12,900 years ago. The timing of this layer coincides with the onset of the Younger Dryas cool period, and with the extinction of the North American megafauna: mammoths, mastodons, ground sloths, horses and camels. Proponents of the asteroid hypothesis suggest that human overkill and climatic cooling are both inadequate explanations of megafaunal extinction. While mammoth and mastodon kill sites have been identified, there are none for the other thirty-three genera of megafauna that also disappeared. And previous cold spells similar in degree to the Younger Dryas had not resulted in mass extinctions.[1]

Geological analysis of the 12,900-year-old black layer from several sites revealed that it contained not only charcoal, carbon spherules and glass-like carbon indicative of intense forest fires, but also rather more strange components, like nanodiamonds and fullerenes containing extraterrestrial helium. These are very rare things indeed: they are found in meteorites and in the ground at extraterrestrial impact sites. So the suggestion is that the continent-wide wildfires were caused by a comet hit.[1]

No crater has yet been identified, but the researchers suggest that perhaps there is none to find: maybe the comet crashed into the Laurentide ice sheet, making very little mark on the earth beneath the ice – or perhaps it exploded in the air. In 1908, something from space – either a burnt-out comet or an asteroid, less than 150 metres in diameter, exploded in the atmosphere above Tunguska in Siberia. The resulting airburst set fire to 200km² of forest, and knocked down trees over a wider 2000km², while leaving no crater. Perhaps the widespread fires of 12,900 years ago were caused by multiple airbursts or impacts.

Maybe this proposed impact even set off the Younger Dryas, with clouds of soot and smoke from the forest fires blocking out sunlight. Perhaps the impact knocked off parts of the ice sheet, with rafting icebergs bringing down the surface temperature of the Arctic and North Atlantic oceans. It's an intriguing hypothesis, and the timing certainly works. An extraterrestrial impact around 12,900 years ago, sparking off continent-wide forest fires, environmental destruction, and subsequent cooling *could* have wiped out the American megafauna.

But other archaeologists argue that there is no need to look to the stars for an explanation of the disappearance of the large Pleistocene mammals of North America. For Gary Haynes of the University of Nevada the disappearance of thirty-three genera of large mammals at the same time as the appearance of distinctive Clovis spear points in the archaeological record is not a coincidence.[2] He argues that humans have been known to hunt herd animals to extinction without any 'help' from climate change or extraterrestrial impacts. Haynes also draws attention to the knock-on environmental consequences of losing large 'keystone' species, like mammoth and mastodon. These animals modify their environments, opening up areas of grazing for smaller herbivores. In Africa, large herbivores help to keep grassland savannah productive. Archaeologists have argued that the loss of these large grazers in Beringia may even have caused a shift from productive grassy steppe to moss-tundra. In other words, once humans had impacted particular species of megafauna, the loss of those animals would have further ecological consequences, spreading out to affect other species as well.[2]

Haynes argues that the climatic changes suggested by some to be instrumental in the demise of the megafauna did not happen at the right time, as the Younger Dryas cold spell succeeded rather than preceded some extinctions (although this is not necessarily at odds with the comet hypothesis, as the more immediate effects of the impact could have finished off the megafauna). And there are the archaeological kill sites that demonstrate that the Clovis hunters despatched mammoths. Haynes makes the case that mammoth and mastodon hunting would have been a sensible strategy for the Pleistocene hunters of North America.

The idea of Clovis people as 'big-game hunters', descendants of the Eurasian Upper Palaeolithic hunters of the steppe, has been around a long time. This image is very persuasive: it permeates our culture, fostered by artistic reconstructions and in film. The iconic image of plucky Ice Age hunters clad in furs, bravely bringing down a mammoth, is familiar to us all. But is it an accurate representation?

Clovis Culture: Gault, Texas

Driving east, I next made my way to Texas, in search of the mammoth hunters, to the site of Gault, near Austin. Leaving the freeway, I headed into the countryside and eventually turned off the road on to a track which led to a couple of sheds – these formed the headquarters of the Gault Archaeological Project. There I met Mike Collins, in blue jeans and Stetson the epitome of a Texan archaeologist. He was just putting the finishing touches to a wooden picnic table, which he loaded into his truck and then took me down to the site. It was in the valley below, in a field surrounded by woods.

Gault is one of the largest Clovis occupation sites, around 800m long and 200m wide. When I visited, the archaeologists had two trenches open and were painstakingly excavating and examining a small fraction of the site. The trenches were covered with white tents, to protect them – and the archaeologists at work inside them – from the elements. I followed Mike inside one tent where excavations were ongoing. It was blazing hot outside, but the tent was doing its job – inside it was shady and cool. Three archaeologists were busy in the bottom of a trench, digging down through dark black soil, and I could see stone tools sticking out of the walls of the trench.

'The site is still incredibly prolific,' said Mike. 'There have been people digging at this site for at least eighty years. We estimate that there are hundreds of thousands of artefacts out there, in the antiquities market and in private collections, that came from here.'

The site had a long history of being plundered. In fact, it was looters who initially brought the site to the attention of archaeologists, and the first excavations took place in the 1920s. Looting continued, though, and in the 1980s the then landowner was charging $25 a day for the chance to dig at Gault. Who knows how many precious artefacts were dug up, out of context, and flogged on the antiquities market? In 1991, one of the paid-up diggers dug a little deeper than most, and found two finely flaked, fluted Clovis spear points and a selection of engraved limestone pebbles. Incredibly responsibly, he reported these finds to Mike Collins, who then started scientific excavations at the site again. Even after all that plundering, Gault still held plenty of archaeological interest: the upper layers, containing 'Archaic' archaeology, had been heavily disturbed, but Mike was relieved to find that the deeper layers containing palaeoindian artefacts were left almost intact.

'Well, pretty much the entire prehistory of Texas is represented here,' Mike told me. 'It's amazing. From the artefacts which we've found and the dating we've been able to get, we can say that people have been here from five hundred years ago, back to at least 13,500 years ago.'

'Have you found any of those beautiful Clovis spear points here?'

'Yes – but Clovis points are actually in a very small minority. In our Clovis levels, we've got a million and a half artefacts, and just forty of them are projectile points. Every one of those is either broken, or resharpened down to the point that it was discarded. They're retooling here. They're making their spear points here, so what we find are the old worn-out ones that they're chucking away.'

So the Gault palaeoindian layers included Clovis artefacts, although most of them were waste flakes from making tools, like the ones I'd seen scattered on the ground. And there were animal bones, too. But there was no indication from these that the Clovis people at Gault were mammoth or big-game hunters. They seemed to be much more generalised, hunting and eating a great range of animals.[1]

And, in fact, this seems to be the picture that is now emerging as more Clovis sites are found or re-examined. The Clovis culture spread right across northern and central America, from coast to coast, and from southern Canada to Costa Rica. Recent radiocarbon dates have tightened its duration to a few hundred years, between 13,200 to 12,800 years ago.[2] Certainly, many of these sites have produced Clovis points in association with mammoth skeletons (for example, at Dent in Colorado and Naco in Arizona), which have led to the suggestion that the Clovis people were specialised big-game hunters – who may even have brought about the extinction of the mammoths with their deadly effective weaponry. Although some archaeologists have suggested that the sites with mammoth remains and Clovis tools may represent scavenging of carcasses, there are several sites where Clovis points have been found in among the bones of mammoth skeletons, and these seem to prove beyond doubt that hunting of these great beasts did occur. And why not? The existence of massive animals that offered a huge return in terms of energy (imagine how many people you could feed with a mammoth), as well as the convenience of long-term storage during the Ice Age.[3]

But maybe these mammoth kill sites have produced a skewed view of Clovis people as big-game hunters. Mike Collins argued that Gault is one of a growing number of sites that indicate a more generalised approach to hunting. He also believed that, on a wider scale, the spread of Clovis sites across America spoke of a generalised approach to subsistence – one that was flexible enough to facilitate survival across a large range of environments – rather than a highly specialised form of hunting. Perhaps there are a lot of Clovis sites with mammoth remains because the large bones are exactly what caught the eye of the archaeologists in the first place.[4]

The less 'spectacular' sites paint a different picture, and Mike argues that these show the hunters in their true colours, as generalists, going after mammals ranging in size from raccoons and badgers to bison and mammoths. The diet at Gault also included small mammals, frogs and birds. And it's not just the animal bones that made Mike doubt the myth of the mammoth hunter. The density of the occupation site at Gault suggests that these people were much less mobile than would be expected if they were (in the main) big-game hunters, tied in to following herd animals around a landscape. In contrast, the later Folsom culture of North America relates to highly mobile and highly specialised bison hunting.[1] The stone tools themselves also imply a less mobile lifestyle: unlike some of the Clovis hunting camps and kill sites, where stone tools have often been transported hundreds of miles from the source of the stone, Gault is sited close to a convenient source of chert. More than 99 per cent of stone tools and waste flakes are made from that local stone.[2,4]

More generally speaking, 'foraging theory', looking at the range and numbers of animals available at the end of the Ice Age, suggests that the early palaeoindians would have been much better off hunting whatever was available, rather than specialising in the large animals.[5]

But Mike doesn't claim that Clovis hunters never went after big game. Animals such as mammoth, horse and bison formed part of their diet, but plant materials and smaller animals were generally more important.[1] At the time of writing, there were at least twelve known Clovis mammoth or mastodon kill sites – compared with only six across the whole of Upper Palaeolithic Europe. Some archaeologists still argue that these proboscideans were important prey animals to Clovis people – albeit within a wider subsistence strategy – and that humans could have been instrumental in their extinction.[2,6] So, at the end of the day, it seems that labelling the Clovis people as mammoth hunters is not categorically *wrong*, but it's a narrow stereotype, like saying the French exist exclusively on frogs' legs (whereas, of course, they eat a much wider diet, including pâté de foie gras and snails as well …).

I felt I was getting closer to the truth about these early American hunters. At lunchtime we all sat in the shade of a tree, at the newly constructed picnic tables, to eat our sandwiches. After lunch, I went for a walk in the shady woods, beside a stream. The ground was littered with leaves and stone flakes. Next to the stream, lazing in the dappled sunlight, I saw a snake, a water moccasin, and walked very cautiously past it. Its eyes were open, but it didn't move as I passed by. It would probably have constituted a tasty snack for those Clovis people.

That afternoon, Mike wanted to show me another trench from previous years' excavations. Over the other side of the stream, just in the woods, was a deep pit.

'What we're really excited about, and the reason we're digging right here, is that previous tests in this part of the site tell us we also have a layer below Clovis. Older than Clovis …'

Mike pointed out layers near the bottom of the pit.

'Right there is where we find Clovis artefacts, but we continue to get artefacts down 25 to 30cm below that.'

He took a few bags of artefacts out of his pocket to show me.

'These,' he said, carefully passing the small stone flakes to me, 'are from that deep layer. They are pre-Clovis.'

'But for years and years archaeologists have been saying that Clovis is the oldest culture in the Americas,' I said.

'That's poppycock,' he replied. 'That's a paradigm that we spent seventy years honing and most of it is wrong. The paradigm is shifting. What happens when you break down a paradigm, a long-held theory, is that intellectual chaos follows because you don't know what the replacement theory is going to be.

'We have a luminescence date on this layer of 14,400 years ago. That's a thousand years before Clovis. It means we've got to rethink our ideas of the peopling of the Americas.'

For a long time, Clovis appeared to be the earliest archaeology in the Americas, so it seemed reasonable to assume that its makers were the first Americans. It also fitted rather beautifully with the timing of the melting back of the Laurentide and Cordilleran ice sheets, to form an ice-free corridor. This parting of the ice is thought to have happened around 14,000 to 13,500 years ago – and Clovis starts to appear around 13,200 years ago.[2] But a selection of sites across North and South America are now challenging the 'Clovis first' model. Gault was one of them. There were people in that valley long before the makers of the fine stone points arrived.

Meadowcroft Rockshelter in Pennsylvania has been proposed as another pre-Clovis site, where there is a suggestion of people having been present as early as 22,000 years ago, but the dates are controversial. It's possible that natural coal in the area may have contaminated the samples from the site, meaning that radiocarbon dating over-estimates the date.[7]

However, there are a growing number of sites with more reliable, pre-Clovis dates. At Schaefer and Hebior in Wisconsin, which would have been close to the southern edge of the Laurentide ice sheet, there's evidence of hunting or scavenging of mammoths long before the distinctive Clovis culture emerged, between 14,200 and14,800 years ago. There are also stone tools and the remains of extinct animals at Page-Ladson in Florida, dating to around 14,400 years ago, and at the Paisley Caves in Oregon, three human coprolites were found, dating to 14,100 years ago.[2] It's a patchy archaeological record, though. It would be nice if the archaeologists could find a well-dated occupation site, and maybe even some actual human remains from between 14,000 and 15,000 years ago. At the moment, there are just a few tantalising clues that humans were in North America that early: a handful of stone tools and flakes with animal bones, and some fossilised faeces.

Now, there are some sites that have been claimed to show a human presence earlier than 15,000 years ago, but the traces are very shadowy and the dates controversial. Four sites have recently come to the fore in the search for the earliest Americans: Cactus Hill in Virginia, La Sena in Nebraska, Lovewell in Kansas, and Topper in South Carolina. Topper, like Gault, has Clovis stone artefacts and older pre-Clovis stone tools in underlying sediments, dating to around 15,000 years ago. But in 2004, tools were found in even deeper layers at Topper, and the archaeologists have claimed that they may be in excess of – wait for it – *50,000* years old. But before we get too excited, these tools are not beautifully worked points or even things that are immediately recognisable as having been worked by human hand. They might just be natural. And with the archaeological and genetic data from Asia suggesting that the founding population there was not even in place before 40,000 years ago, this means we should view such a date for the Americas with profound scepticism.[2]

But it still seems likely that humans got into the Americas a long time before Clovis, maybe 15,000 years ago, perhaps even a bit earlier. In addition to the emerging archaeological clues, there's the genetic suggestion of an earlier colonisation along that coastal route into the Americas, and the circumstantial evidence relating to the environment along that north-west coast.

For Mike Collins, Clovis getting knocked off its pedestal as the first culture in the Americas didn't reduce its interest. It's still fascinating as a phenomenon that stretched across a whole continent, and there are still unanswered questions. Where did it come from? How did it develop? There is no clearly antecedent culture in north-east Asia or Beringia, although there are some common components (such as those ivory foreshafts).

In fact, the closest stone toolkit to Clovis comes from a *long* way away, in western Europe. The similarity between Solutrean and Clovis points and blades is quite remarkable. Some archaeologists have even suggested that this indicates a completely different origin and route for the peopling of the Americas: from France and Spain, spreading around the North Atlantic.[8] It's an interesting idea, but there is really no other evidence to back this up, and a lot of problems with the suggested route: the Solutrean ended some 5000 years before Clovis started, there's no evidence of any seafaring tradition in western Europe and the way would have been blocked by ice sheets.[9, 10] And then, of course, there are the genes of the Native Americans, which seem firmly to indicate an Asian origin of these people. Having said that, some have argued that the X haplogroup found among Native Americans indicates a European connection – as X is also found in Europe. Actually, this is probably right – but the connection is a very ancient one. The X lineages found in Europe and America split some 30,000 years ago, when the ancestors of those particular groups of Native Americans and Europeans were living on the Siberian steppe, before they went their separate ways to the west and east.[11] And recent genetic studies have picked up the X haplogroup among Mongolians, in the Altai region.[10] So it seems that the similarity between the Solutrean and Clovis industries may just be coincidental: an example of cultural convergence in these two widely separated (in space and time) groups of Ice Age hunters.

(Months later, in a field in Exeter, back home in England, I was to meet Professor Bruce Bradley and his masters student, Metin Eren. Bruce was quite clear that his hypothesis about the Solutrean-Clovis link and a North Atlantic route into the Americas had not yet been refuted. Indeed, although the best-known Clovis sites were in the western US, he pointed out that the oldest and densest scatter of sites was in the east. As he put it: 'It's thicker than the fleas on a dog in south-eastern US.' He was also much more convinced by the cultural connection between western Europe and North America, whereas, to him, the toolkit at Yana bore similarities to the Gravettian of Europe, but had no real links with American toolkits. This conversation made me realise that the first route into the Americas was still up for grabs – and Bruce and his colleagues were searching for evidence in the east. He was sure the Solutreans could have made their way,

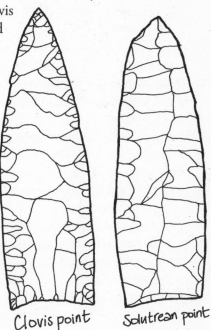

Clovis point   Solutrean point

by sea, around the North Atlantic. After all, those people had bows and arrows, and spear-throwers. 'The idea that they didn't have boats – give me a break!' exclaimed Bruce.)

The development and dispersal of Clovis *is* quite remarkable – in just a few hundred years it had spread through America. But if this wasn't the culture carried by the first colonisers, how was it disseminated? Mike's idea was that Clovis was a 'technocomplex' – a tool-making trend – taken up and shared by a wide range of ethnically diverse people whose ancestors had *already* dispersed across North America.[4] Clovis spread rapidly during the last centuries of the warmish Allerod interstadial, and the overlap between dates from sites widely spaced across North America makes it impossible to know where exactly it started and in which direction it advanced. It disappears from the archaeological record almost as swiftly as it appeared, at the onset of the Younger Dryas stadial, being replaced with another toolkit, named after Folsom, where it was first found.[2, 12] The date of the disappearance of Clovis takes us back to that comet impact idea. The black layer and the beginning of the Younger Dryas seem to mark the end of the megafauna in the fossil record of North America, but also the end of Clovis in the archaeological record. So did that comet strike indeed finish off the mammoths and mastodons and sloths and create such environmental destruction that human survivors were forced to reinvent their subsistence and devise new toolkits?

Later that afternoon I met archaeologist Andy Hemmings for a practical demonstration of the deadly efficiency of Clovis hunting technology. He showed me how the fluted base of a Clovis point enabled it to be hafted into a notch on a shaft. He also showed me bone points, another characteristic feature of Clovis toolkits. Three bone and ivory hooks have been found, of Clovis age, that suggest these hunters used spear-throwers – atlatls – to launch darts.[1] We had a go at throwing darts (both bone-tipped and metal-tipped – Andy wasn't keen on breaking the precious stone tips), using an atlatl, at our target: a car door. It took me a while to get the hang of aiming the contraption, but, when I did, I managed to make a decent hole in the door. I was suitably impressed. The atlatl allowed the dart to be launched with such force that it penetrated sheet steel. And, in Andy's capable hands, it was even more of a precise and vicious weapon. So the Clovis people certainly had the technology to kill massive animals like bison and mammoths – just not all the time.

It was now time for me to travel south and investigate the story in South America. There are no Clovis sites in Central or South America, although several of a similar age have yielded a basic flake technology which also includes some 'fishtail' points, some of which have fluted bases like Clovis points. But they are not Clovis: these sites represent different cultures, and adaptations to a very different environment, in South America – contemporary with Clovis sites in the north. One such site is Pedra Pintada, in the Brazilian Amazon, where excavations at the painted cave in the rainforest have yielded detailed information about the lifestyles of the South American palaeoindians. Sites like this constitute a significant challenge to any hangers-on of the 'Clovis first' theory, as it is hard to imagine how colonisers moving down through the ice corridor could have reached South America that quickly.

Over the decades, several sites in South America with apparently pre-Clovis dates have risen to fame, and then been knocked down, as the bastions of 'Clovis first' have picked holes in their archaeological procedures or dating methods. But there is one site in particular that has stood up to scientific scrutiny: Monte Verde in Chile.

On my way to the Amazon, and Pedra Pintada, I stopped off in Rio, where one of the oldest skulls in the Americas is kept in a box in the museum.

Meeting Luzia: Rio, Brazil

Upstairs, Rio Museum was like a palace that someone has forgotten to look after. The high ceilings were falling down, exposing beams and lath, and the paint was peeling on the stuccoed pilasters. Most of the huge rooms were divided into offices and corridors by the careful arrangement of filing cabinets and cupboards to form party walls.

In a large, dusty room somewhere in one corner of the museum, empty but for a huge wooden table, Walter Neves set down a metal case. He sprung the locks and carefully opened the lid. Inside was an almost complete skull. He lifted it out and set it down on a ring-doughnut-shaped piece of foam on the table. I was looking at the face – albeit the skeletal face – of an ancient American.

This was just one of the skulls found in the cave of Sumidouro (meaning swallet or sinkhole) in Lagoa Santa, in Brazil, in the nineteenth century. The Danish naturalist Peter Lund had discovered the human remains in association with the fossilised bones of megafauna, and had

therefore proposed that people had been
in the Americas at the same time that
the giant beasts were around. This
was long before any Clovis sites were
discovered in North America, and
Lund's contemporaries found it
impossible to accept his assertions
about the antiquity of those bones.
But archaeologists returning to
the caves in the twentieth century
found more human skeletons, in
the same layers as megafauna. In the
1970s, a skeleton was excavated from
sediments that contained charcoal.

Radiocarbon dating of this charcoal showed the skeleton to date back
to the end of the Pleistocene, around 13,000 years ago. Lund was at last
vindicated.[1] The skeleton became known as 'Luzia' – and it was Luzia's
skull that Walter Neves had placed on the table before me.

'I am proud to introduce you to Luzia,' said Walter.

'She is probably the oldest human skeleton ever found in the
Americas. And what I realised back in 1989 was that the morphology of
these skulls of the first Americans was very, very different from the skulls
of nowadays Native Americans.'

It *was* an odd looking skull. It didn't really look Asian. Everything
I'd found out so far about the origin of the first Americans pointed to
a homeland in East Asia – but this skull didn't appear to have much in
common with modern East Asians. I said so.

'No, and it's not just this skull,' said Walter. 'We have recently
published a paper with eighty skulls from Lagoa Santa and also like
seventy skulls from Colombia and they all show the same trends.'

I knew the paper. It was the biggest sample of early American
skulls that had ever been studied. After taking a suite of measurements
from the Lagoa Santa skulls, Walter had then compared this data with
Howell's database of measurements from more than 2500 skulls from
around the world. He had used powerful multivariate statistics – that
allow many different measurements to be compared all at the same time
– to look at how close the early Brazilian skulls were to crania from other
populations. And the results showed that the Lagoa Santa skulls were
closer to Australians, Pacific Islanders and even Africans than they were
to north-east Asians or modern Native Americans.[2]

Present-day Native Americans have a similar skull shape to modern north-eastern Asians: the braincase is short and wide, the face is broad and wide, with very little protrusion of the jaw, the nasal opening is high and narrow and the eye sockets are roundish. In contrast, Luzia's skull was long and narrow, with a quite pronounced, projecting jaw, and broad, rectangular eye sockets and a wide nasal opening – more like Australians, Melanesians and sub-Saharan Africans.

But how does this fit with the idea of early Americans coming from north-east Asia, through Beringia, into the Americas? Could it even be taken to suggest a trans-Pacific crossing from Australo-Melanesia to South America? Walter was quick to dismiss this idea.

'We never suggested that. We never thought that,' he said firmly. 'If you go and see the few human skeletons of the final Pleistocene in Asia they also look like this – not like today's East Asians. So these populations showing Australo-Melanesian, or even an African morphology, they were in East Asia by the final Pleistocene. So we have no doubt that Luzia's people also came from Asia using the Bering Strait as the entry to the Americas.'

He thought his results were entirely compatible with a colonisation of the Americas from the north, but that the early Americans were much more morphologically diverse than Native Americans are today, and included a lot of people – like Luzia – who didn't look at all East Asian.

Walter's argument was that these Brazilian skulls suggested that a model of just one population expansion into the Americas was too simple. He advocated that some people had come into the Americas before what we know as typical East Asian features had developed, or at least while there was a more diverse range of morphologies in East Asia: with more people who still looked like the original beachcombers. I was reminded of the Upper Cave skulls from Zhoukoudian: they also lacked classic 'East Asian' features.[3] Then, according to Walter, there was a later arrival of people with faces more like modern East Asians and Native Americans. Other physical anthropologists looking at skull shape have also found evidence for a two-wave colonisation of the Americas. Marta Lahr, of Cambridge University, found that skull shapes of modern Native Americans were similar to those of East Asians, but that skulls from archaeological populations from Tierra del Fuego and Patagonia showed a more 'generalised morphology'.[4] In 1996, a skeleton was discovered in Kennewick, Washington State. The skull was dated to about 9300 years old, and also appeared to be closer in shape to the Ainu of Japan and to Pacific Ocean islanders than to Native Americans.

Walter was aware that this idea of two migrations into the Americas didn't fit with the single migration revealed by genetics,[5] but he suggested that this discrepancy could be explained if some of the original genetic lineages have been lost, just as people who looked like Luzia had disappeared over time.[1, 2]

Combining the fossil and genetic data, Stephen Oppenheimer has suggested that there might have been *three* genetically, morphologically and culturally distinct populations that moved from Beringia into the Americas. This harks back to that idea of Beringia as a staging post, with various, different-looking populations originating in various parts of Asia moving in, and then expanding into the Americas. Oppenheimer's three populations include a group descended from the original beachcombers of Asia (these would be robust people like Luzia); a group connected with later 'East Asian'-looking people; and another group from the Russian Altai (refugees moving east from Siberia during the peak of the Ice Age, and sharing the mitochondrial X lineage with northern Europeans, descendants of westward-moving refugees).[6]

But whether or not these two (or three) morphologically, and perhaps culturally, distinct populations came down into the Americas at the same time, or in different waves, is currently very difficult to pin down. If these populations all flowed into America at the same time, this does not seem to explain why none of those early Americans (or, at least, the ones discovered to date, including Luzia and Kennewick Man) look 'East Asian'.

But recently a large study of more than five hundred skulls from late Ice Age America all the way to the present, including skulls from Lagoa Santa, suggested a synthesis between the genetic single-wave model and Walter's 'Two-Components' model. The researchers proposed an initial wave of colonisation, by a variable population from Siberia and Beringia, where East Asian and robust, palaeoamerican types like Luzia formed opposite ends of a continuous spectrum of variation. After colonisation, the northern, circumarctic groups then maintained continuous contact, allowing diffusion of that extreme north-east Asian morphology with facial flatness and projection of the cheekbones to America. This could explain the similarity between Siberian and Aleut-Eskimos.[7]

Some anthropologists urge caution over reconstructing past migrations based on craniofacial shape: European Upper Palaeolithic skulls appear closer to non-European skulls in Howell's database than to modern European skulls, despite the genetic indications that most modern Europeans are descended from Upper Palaeolithic populations.

It seems that, all over the world, skull and face shapes have changed a lot since the first colonisers arrived in the various continents.[8]

What is clear, though, is that there is no longer anyone looking like Luzia in the Americas today. Stephen Oppenheimer has suggested that the Olmec statues have this 'African' look, indicating that there may still have been people of this type living 3000 years ago. Walter Neves suspected that the ultimate disappearance of this robust type may have been fairly recent – perhaps even since European contact with the New World. He showed me a reconstruction of Luzia that had been made by forensic artist Richard Neaves. As Luzia had a projecting jaw, he had given her broad lips. In the flesh she certainly looked much more African than East Asian.

It is clear that there are quite a few different opinions and different explanations for the variation seen in the early palaeoindians, compared with Native Americans today. I imagine the picture will get clearer and some kind of consensus will emerge as more fossils and artefacts are found, and more genes analysed. The complete story of the first Americans has yet to emerge.

Ancient Hunter-Gatherers in the Amazon Forest: Pedra Pintada, Brazil

From Rio I flew north-east – to the Amazon. I was bound for an archaeological site that held information about how the early Brazilians had lived.

We flew over the lower reaches of that largest of rivers, which broke into skeins of water that meandered and reunited, and over vast stretches of forest, to land at a small airstrip at Santarém. I alighted from the plane into a warm breeze. Clouds of lemon-yellow butterflies were streaming across the runway, flying with the wind.

From Santarém I caught a ferry to Monte Alegre (pronounced by the locals as 'monchalegray'). The trip took around five hours, and, like all the other passengers, I bought myself a hammock to hang up on the double-decked ferry. By the time we were ready to leave the quayside in Santarém, the decks were festooned with a bright rainbow of hammocks: stripy, checked, fringed, and embroidered. By mid-afternoon just about everyone was tucked up in their hammocks, gently swaying with the motion of the boat, in the shade, out of the dazzling glare of the overhead sun.

Arriving at Monte Alegre close to sunset, I was met by Nelsi Sadeck as I got off the ferry, still swaying slightly. The next morning, we set off from the small town on a road which quickly turned into a dusty, rutted track. We were heading for Pedra Pintada, the 'Painted Rock'. Nelsi was a local guide, and had dug at Pedra Pintada for archaeologist Anna Roosevelt, of the University of Illinois, who had run excavations at the site during the early nineties.

Throughout South America, stone tools, including triangular points, have been found which are quite distinct from the North American complexes. Most of them are undated, so it's hard to know how they fit into the story of the colonisation of the Americas. Some such points had been discovered in the lower Amazon region: very different from fluted Clovis points, these were long and thin, with downturned 'wings' and a stem for attachment to a shaft. Other archaeologists had assumed them to be Holocene, made some time in the last 10,000 years. But Anna Roosevelt wanted to pin down their age. In order to do this, she needed to find a site where the archaeological layers were well preserved: it was no good just looking at surface finds. Cave sites seemed ideal – there was a good chance that early sediments would have been preserved intact – and Anna knew that there were lots of caves and rockshelters in the sandstone outcrops around Monte Alegre. Roosevelt's team began their investigations by surveying these caves, mapping them and testing the deposits inside using an auger to remove cores of sediment for inspection.[1]

I was expecting to be venturing deep into the rainforest, but in fact the landscape around Monte Alegre was dominated by low-level woodland and shrubby pastureland, where skinny white cows and horses grazed. Nelsi and I rattled along in the Toyota Land Cruiser, stopping to machete back fallen trees on the track, and eventually coming to a halt at the base of a large outcrop. Nelsi led me up the slope to see a painted rockshelter, called 'El Painel'. There were strange, abstract shapes, some looking like animal forms, one like a woman giving birth. There were also geometric patterns, and handprints, all in red and yellow ochre. Up on the rocky outcrop, looking down on the Amazon, I could still see those yellow butterflies on their journey, with the wind. And, high above us, black vultures circled.

Then we went on to the Caverna da Pedra Pintada itself. We walked through a narrow cleft, which was open at the top, with tree roots hanging down like lianas inside the cave. There were bats roosting high above us, on an overhang. I pointed them out.

El Painel

'Yes … vampire bats,' said Nelsi.

There were also peculiar, long-bodied wasps flying in and out of small nests, suspended by papery stalks from the walls of the cave. I proceeded with caution. We scrambled down into the main cave, which opened into a wide mouth. We were looking down on to the Amazon flood plain. On the walls were more paintings in red ochre.

The augured samples from Pedra Pintada suggested that the archaeology in the cave was undisturbed, and, in the early 1990s, the archaeologists had started to excavate.

'This is where we dug in '91, '92 and '93,' explained Nelsi, gesturing to the floor of the cave entrance. 'We excavated down to two metres.'

There were plenty of remains in the more recent, shallower layers, and then the archaeologists found a 'sterile' layer of sediment

– containing no archaeological finds. But below that there was a deep layer full of animal bones, shells, burnt plant remains, stone tools – and pieces of ochre: traces of the earliest inhabitants of the cave. Among the thousands of stone flakes, there were twenty-four finished stone tools, including stemmed, triangular points made from quartz-like chalcedony, that may have been used as spear or harpoon points. There were bones from amphibians, tortoises and turtles, snakes and mammals – but most of the bones were from freshwater fish. The archaeologists found further evidence of the dietary breadth of the ancient Amazonians from plant remains: there were fragments of burnt wood but also fruits and seeds from forest trees like jutai, achua, brazil nut and various palms, that still grow in the Amazon rainforest today. Radiocarbon dating of the carbonised plant, as well as luminescence dating of burnt stone tools and sediment, placed the early occupation of the cave at around 13,000 years ago.[1,2]

Intriguingly, the chunks of ochre found in the archaeological layers were similar in colour and chemical composition to the pigment in the paintings on the wall of the cave. It's impossible to know for sure, but those paintings *could* have been made by the palaeoindian cave-dwellers 13,000 years ago.

Some scholars have argued that the rainforests of Central and South America would have constituted an ecological barrier to colonisation by palaeoindians (a strange contention really, given that we know people were living in the rainforest at Niah at least 40,000 years ago, and probably much earlier). And Pedra Pintada definitively demonstrates that this is a misconception: palaeoindians were happily living in the late Pleistocene rainforest.[3]

The further importance of Pedra Pintada to theories of the colonisation of the Americas is that it demonstrates that there were people living in the Amazon Basin at the same time as (or possibly earlier than) Clovis people in the plains of North America, but more than 5000 miles to the south.[1] As these two cultures were contemporary, this is clearly a problem for the 'Clovis first' theory. There *must* have been people in the Americas before Clovis. Long before Clovis, in fact, as my last destination would prove.

I caught the boat back to Santarém, and made my way to the airport, where clouds of lemon-yellow butterflies were still streaming across the runway. Then I left the tropical warmth of Brazil behind me as I headed south, to Chile in mid-winter.

Black Soil and Revelations: Monte Verde, Chile

It was quite a shock stepping off the plane in Chile, into a grey, chilly winter's day. I was a long way from the vibrant colours and warmth of the tropics, and I was back on the Pacific coast.

It was so cloudy when I arrived in the small city of Puerto Montt that I couldn't see the amazing setting I was in: a landscape of lakes and volcanoes. Later, when the clouds lifted, I started to make out the snowy slopes of Osorno and Calbuco in the distance.

My impression of Chile over the days I was there was of almost unremitting dampness. It poured with rain most of the time. Lichen and moss grew on the wooden boards and shingles of houses in the villages I drove through, and the bare trees were swathed in moss. It was apt; the archaeological site I was investigating was special because of its very wetness.

My first visit was not to the site of Monte Verde itself, but to Valdivia University, up the coast, where some of the marvellous artefacts from the excavations were stored. A long, low wooden building there was a repository for a huge range of objects from ancient archaeological specimens to historical artefacts. There were shelves of beautiful painted pots, clay animals, teapots, and a Madonna whose detached hands lay on the pedestal at her feet. The curator cleared a space on a table, and brought out Tupperware boxes full of Monte Verde treasures. As the archaeological site was waterlogged (a damp peat bog), all sorts of materials have been preserved that would normally rot away quickly in the ground. Instead of 'postholes', where darkened soil shows archaeologists that wooden stakes were once planted in the ground, the stakes themselves were preserved. And there were wooden spikes that looked a lot like tent pegs. I examined a block of wood with a cut-down hole that had been drilled down – perhaps by twisting a stick to make fire. There was also *string* – unmistakable, twined-together fibres. And there was a piece of thick, darkened animal hide, perhaps from mastodon or mammoth. All these organic finds were discovered alongside more standard archaeological fare such as stone tools and animal bones. As I was inspecting these precious objects, Mario Piño arrived. Mario was a geologist who had worked on the excavation at Monte Verde, and he was to be my guide to the site itself.

I would have been lost without Mario to show me where the archaeology had been uncovered. Driving to Monte Verde itself, we pulled up beside a wooden gate which led into an insignificant-looking field. The ground was boggy and mossy, sloping down to a fast-flowing stream, with sheep grazing on its banks. There was no sign of any archaeology at this place – possibly the most important archaeological site in the Americas.

Mario and I made our way down to the stream. It curved around in a wide meander: there had been quite a significant change in its course since the last excavations in the 1980s. Mario pointed out the original bank edge, which was much straighter. The creek had moved about 20m, from the north to the south side of the archaeological site, within thirty years. We stepped down and walked into the bend of the stream, Chinchihuapi Creek, where a large log had recently been washed up on the low, sandy bank.

'The site, like so many others, was discovered by chance,' said Mario. 'Local villagers were widening the creek. When they were removing the sediment, cutting the curves, they found these huge bones, and kept them. Two university students who were travelling round the place took the bones to Valdivia.'

The bones were from Pleistocene animals, and archaeologists at Valdivia University decided to investigate further, joined by Tom Dillehay of the University of Kentucky. At first, they thought this might just be a fossil site, but when patches of charcoal – evidence of old hearths –

Chinchihuapi Creek

battered cobbles, cores and flakes started to appear, it was clear that it was archaeological. They had found the remains of an ancient camp.[1]

'That is about where the large hut was,' said Mario, indicating the log.

The 'hut' was a structure some 20m long. Digging through the sandy terraces of the creek, the archaeologists found a collection of wooden posts that had collapsed down. Some of them seemed to form divisions inside the hut: perhaps they were different living spaces. Microscopic pieces of hide were recovered from the sediment among the wooden stakes: it looked as if the hut had been covered in animal skins.

'The hut was heated with braziers,' explained Mario. 'We found the remains of small holes in the sand, lined with clay, full of charcoal.' Outside the hut, they found larger hearths that had probably been used for cooking.

As archaeological excavation proceeded, more and more remains came to light: there were digging sticks, plant remains, animal bones and skin, and even a child's footprint in the clay next to a hearth.[2]

Some 30m away from the main hut there were the remains of another structure: a strange, 'wishbone'-shaped outline that Dillehay had believed to be all that was left of another, small hut. It seemed to me to have been, at just a few feet across, a bit too small for a hut. Mario described the finds around and within the wishbone shape: mastodon bones, with cut marks from butchering, and, preserved in the wet ground, numerous plant remains. These included nine varieties of seaweed. Seaweed is an excellent source of iodine and other minerals, but Dillehay thought that some of these species may also have been used as medicines, as they are today, by the local Mapuche Indians. Rather bizarrely, some of these seaweeds appeared, combined with other potentially medicinal plants, in the form of chewed-up and spat-out cuds, in the 'wishbone hut'.[2]

The plant remains from the site included nuts and berries, and showed that Monte Verde had been occupied all year round. 'We found food from all four seasons,' said Mario. So it seemed that this place was more permanent than just a seasonal camp.

There were also the earliest remains of potatoes ever discovered – the limp but still recognisable skins of wild potatoes (*Solanum maglia*) – implying that humans had developed a taste for the humble spud at least 14,000 years ago.[3]

The archaeological remains from Monte Verde show that the people living there were exploiting resources from a wide range of habitats: from inland forests, freshwater marshes – and, as the seaweed, salt and bitumen found at the site show – the coast. During the Pleistocene, the coast would have been further away – about 90km to the west, compared with 25km today – so Monte Verde was near the coast but not *on* it. The presence of seaweed at the site suggests that the people living there either made visits to the seashore or were in contact with palaeoindians living on the coast itself.[2] There is certainly evidence of people living along the coast of South America, eating seabirds, anchovies and molluscs, from Quebrada Tacahuay and Quebrada Jaguay in Peru.[4, 5] Those sites date to between 11,000 and 13,000 years ago, but radiocarbon dating of plant remains and charcoal from the hearths at Monte Verde place the occupation of the site even earlier: some time between 14,000 and 14,600 years ago.[2] Such dates would be the final nails in the coffin of 'Clovis first'.

'Monte Verde means we have to rethink the moment when people came to America,' said Mario. 'From the findings in Monte Verde, and other investigations, it's proposed that the human entered America between 16,000 and 20,000 years ago. Any relation with the ice corridor does not make sense. The migration probably happened earlier, and along the Pacific coast.'

But Monte Verde has been a controversial site since its first discovery. By 1997, Tom Dillehay was so fed up with its detractors that he invited a crew of eminent palaeoindian archaeologists to visit the site, see the artefacts and make up their minds about it. They all agreed that the site was indeed archaeological, and that there was no reason to doubt the (pre-Clovis) radiocarbon dates.[6] It means that Monte Verde is the oldest – generally accepted – site in the whole of the Americas.

So it was agreed that Monte Verde was earlier than Clovis, and indeed earlier even than the opening of the ice-free corridor. People may have moved down into North America by that route, but they can't have been the first wave of colonisers. It seems much more likely, looking at the archaeological, geological and genetic evidence,[7] that the ancestors of the people who ended up at Monte Verde had entered the Americas, perhaps along the newly ice-free coastal route, some time around 15,000 years ago, and then spread down the Pacific coast into South America. (But at this point in time we can't ignore the possibility that the first Americans might – just might – have come across the North Atlantic. It will be interesting to 'watch this space' and see what evidence emerges from both the east and west coasts of Canada and the US in the future.) Once in North America, some colonisers may have spread inland, along the southern margin of the ice sheets, perhaps following dwindling herds of mammoth and mastodon to Wisconsin. The Clovis culture could have been developed by descendants of these people, or alternatively carried down later from Beringia (by people with the same genetic lineages as the first colonisers) through the ice-free corridor.[8]

People are still eating seaweed in Chile. I went for a rainy walk along the coast near Monte Verde, and met a man, kitted out in yellow waterproofs and sou'wester and carrying a sack: he was gathering stalks from the kelp that had washed up at the last high tide. When I took refuge from the driving rain in a small restaurant at the top of the cliffs, I sampled the local delicacy: seaweed empanadas.

Journey's End

At this point, my own journey of discovery was over. And, in a very small way, I was going to relive prehistory. Having been a nomad for half a year, I was going to settle down. And I was looking forward to going home.

I had ended my journey on the coast, as I had begun it. My travels had taken me all around the world, from our homeland in Africa all the way to the last continents to be populated: the Americas. I had endured extremes of temperature, from the icy north of Siberia to the searing heat of Australia. And, wherever I'd gone, I'd met people like me. Very often, we couldn't communicate directly in spoken language, but smiles and gestures were universal. And however different we all looked on the surface, those differences were superficial.

The 'science' of palaeoanthropology has been so misused in the past, to justify or emphasise differences between ethnic groups, to 'rank' people by head shape and size, skin colour and culture, but looking at the evidence objectively reveals a quite different truth, and carries a very positive message.

We're all members of a young species, going back less than 200,000 years. When we trace our ancestry back, we find that we're all related, on a great family tree of humanity. You can't rank *people* any more than you could rank twigs on a tree. We all have the same, very great-grandmother in African Eve. So wherever we've ended up, all over the world, we are all Africans under the skin.

I had also seen how our ancestors had spread over the world and survived, while huge fluxes in climate and environment transformed the face of the earth. Climate change is a feature of our world, although the length of our lifespan in comparison to the grand scale of geological time gives us a false impression of stability. But having said that, the last 11,000 years have been relatively stable, allowing us to settle down, start farming and achieve a huge population size.[1] There is no doubt that the world will alter around generations of our offspring, and indeed, anthropogenic climate change may produce much more dramatic fluctuations than any experienced by recent generations of our ancestors.

It seems that we may have been storing up trouble for ourselves since the time we invented agriculture and populations really took off. We have cleared vast swathes of woodland for agriculture, flooded huge areas to grow rice, and released ever-increasing quantities of carbon dioxide into the atmosphere. If it weren't for the man-made emission of these greenhouse gases, and the removal of carbon sinks, we could reasonably expect the world to start cooling down into a major glaciation within the next 50,000 years. But the scientific evidence for anthropogenic global warming is now irrefutable, and we just don't know what effect this disruption, or, as Chris Stringer put it, this 'tinkering with the Earth's climate machine', will have on the natural climate cycles in the longer term.[1]

Some of the changes we have to face may be catastrophic; some may even threaten to wipe us out in certain places, as they have done in the past. Some assessments of what will happen to the world and our place in it in the future are incredibly gloomy, but, taking the long view and looking at how early humans managed to survive and colonise the globe, it appears that we are a flexible and adaptable species. Chris Stringer points to the example of the Gravettians, who prevailed despite the deteriorating climate in Europe at the LGM, through new technology and extensive social networks.[1]

But in some ways we're different from the Gravettians. Very few hunter-gatherers remain in the world today. In the developed and the developing world, most people are settled, and there are billions of us on the planet. There's not much room for people displaced by rising sea levels or failing crops or lack of water,[2] but while our settled existence may make us less flexible, we surely have the capacity to come up with global solutions to the challenges ahead.

For instance, we can each, individually, aim for more 'low-tech', less energy-hungry lifestyles, but we need a worldwide, cooperative effort to tackle the problems of climate change. And any such plan needs to make economic sense. We could end up spending a vast amount trying to shave a fraction off the global temperature increase, whereas that money could be better spent now in developing countries, which are also likely to be hit hardest by the effects of climate change. Political scientist and 'Skeptical Environmentalist' Bjørn Lomborg points out that every person in the developing world could be guaranteed access to clean water and education for half the predicted cost of implementing the Kyoto Protocol. Cutting back hard on CO_2 emissions may not be the most beneficial approach, for us and future generations; we may be better off investing in research and development into renewable forms of energy, and in supporting developing countries.[3]

I think it would be fascinating, but probably quite scary, to come back in 200,000 years' time and see how our descendants are doing. I do hope we manage not to wipe ourselves out, and I'd like to think that we'll find a way to mitigate the damage caused by climate change and develop new technologies that mean we're not still pumping out such vast amounts of CO_2. It will require far-sighted and magnanimous politicians to achieve this. I hope that we'll learn to look after our environment better, and our own Palaeolithic bodies. And, of course, it would be lovely to think that all our achievements in literature, music, art and science will be passed on and built upon by future generations. I think the lessons of the past give us grounds for optimism. We are, after all, survivors. But perhaps the near future will be less rosy, and our civilisations will crumble. Our descendants might eventually be forced to go back to the ways of the ancients, to become hunter-gatherers once again.

Who knows? Stephen Jay Gould said, 'Life is a copiously branching bush continually pruned by the grim reaper of extinction.'[4] But I don't think the human lineage is about to get pruned just yet.

I have a vision of the Songlines stretching across the continents and ages; that wherever men have trodden they have left a trail of song (of which we may, now and then, catch an echo); and that these trails must reach back, in time and space, to an isolated pocket in the African savannah, where the First Man opening his mouth in defiance of the terrors that surrounded him, shouted the opening stanza of the World Song, 'I AM!'

Bruce Chatwin, *The Songlines*

References

Prologue

1. Cohen, D. J. New perspectives on the transition to agriculture in China. In: Yasuda, Y. (ed.), *The Origins of Pottery and Agriculture*, Roli Books, New Delhi, pp. 217–27 (2002).

Introduction

1. Foley, R. Adaptive radiations and dispersals in hominin evolutionary ecology. *Evolutionary Anthropology* 11: 32–7 (2002).
2. Stringer, C. Modern human origins: progress and prospects. *Philosophical Transactions of the Royal Society of London* 357: 563–79 (2002).
3. Lahr, M. M. The Multiregional Model of modern human origins: a reassessment of its morphological basis. *Journal of Human Evolution* 26: 23–56 (1994).
4. Field, J. S., & Lahr, M. M. Assessment of the Southern Dispersal: GIS-based analyses of potential routes at Oxygen Isotopic Stage 4. *Journal of World Prehistory* 19: 1–45 (2006).
5. Mithen, S. *After the Ice. A Global Human History*, Harvard University Press, Cambridge, Massachusetts (2003).
6. Stringer, C. *Homo Britannicus. The Incredible Story of Human Life in Britain*, Penguin Books, London (2006).
7. Lambeck, K., Esat, T. M., & Potter, E-K. Links between climate and sea levels for the past three million years. *Nature* 419: 199–206 (2002).
8. Pope, K. O., & Terrell, J. E. Environmental setting of human migrations in the circum-Pacific region. *Journal of Biogeography* 35: 1–21 (2008).
9. McBrearty, S., & Brooks, A. S. The revolution that wasn't: a new interpretation of the origin of modern human behaviour. *Journal of Human Evolution* 39: 453–563 (2000).
10. Klein, R. G. Archaeology and the evolution of human behaviour. *Evolutionary Anthropology* 9: 17–36 (2000).
11. Shea, J. I. The origins of lithic projectile point technology: evidence from Africa, the Levant and Europe. *Journal of Archaeological Science* 33: 823–46 (2006).
12. Oppenheimer, S. *Out of Eden. The Peopling of the World*, Constable & Robinson, London (2003).
13. Bouzouggar, A., Barton, N., Vanhaeren, M., *et al.* 82,000-year-old shell beads from North Africa and implications for the origins of modern human behaviour.

Proceedings of the National Academy of Sciences of the United States of America 104: 9964–9 (2007).

14.   Mellars, P. A new radiocarbon revolution and the dispersal of modern humans in Eurasia. *Nature* 439: 931–5 (2006).

15.   Lian, O. B., & Roberts, R. G. Dating the quaternary: progress in luminescence dating of sediments. *Quaternary Science Reviews* 25: 2449–68 (2006).

16.   Schwarcz, H. P., & Grun, R. Electron spin resonance (ESR) dating of the origin of modern man. *Philosophical Transactions of the Royal Society of London* 337: 145–8 (1992).

17.   Cann, R. L., Stoneking, M., & Wilson, A. C. Mitochondrial DNA and human evolution. *Nature* 325: 31–6 (1987).

18.   Cavalli-Sforza, L. L. The Human Genome Diversity Project: past, present and future. *Nature Reviews: Genetics* 6: 333–40 (2005).

1. African Origins

Meeting Modern-Day Hunter-Gatherers: Nhoma, Namibia

1.   Knight, A., Underhill, P. A., Mortensen, H. M., *et al.* African Y chromosome and mtDNA divergence provides insight into the history of click languages. *Current Biology* 13: 464–73 (2003).

2.   Marshall, L. *The !Kung of Nyae Nyae*, Harvard University Press, Cambridge, Massachusetts (1976).

3.   Smith, A. B. Ethnohistory and archaeology of the Ju/'hoansi bushmen. *African Study Monographs*, supplement 26: 15–25 (2001).

4.   Marino, F. E., Lambert, M. I., & Noakes, T. D. Superior performance of African runners in warm humid, but not in cool environmental conditions. *Journal of Applied Physiology* 96: 124–30 (2003).

5.   Bramble, D. M., & Lieberman, D. E. Endurance running and the evolution of *Homo. Nature* 432: 345–52 (2004).

6.   Lieberman, D. E., Bramble, D. M., Raichlen, D. A., & Shea, J. J. The evolution of endurance running and the tyranny of ethnography: a reply to Pickering and Bunn (2007). *Journal of Human Evolution* 53: 434–7 (2007).

African Genes: Cape Town, South Africa

1.   Tishkoff, S. A., & Williams, S. M. Genetic analysis of African populations: human evolution and complex disease. *Nature Reviews: Genetics* 3: 611–21 (2002).

2.   Richards, M., Macaulay, V., Hickey, E., *et al.* Tracing European founder lineages in the Near Eastern mtDNA pool. *American Journal of Human Genetics* 67: 1251–76 (2000).

3.   Cann, R. L., Stoneking, M., & Wilson, A. C. Mitochondrial DNA and human evolution. *Nature* 325: 31–6 (1987).

4.   Jorde, L. B., Watkins, W. S., Bamshad, M. J., *et al.* The distribution of human genetic diversity: a comparison of mitochondrial, autosomal and Y-chromosome data. *American Journal of Human Genetics* 66: 979–88 (2000).

5.   Jakobsson, M., Scholz, S. W., Scheet, P., *et al.* Genotype, haplotype and copy-number variation in worldwide human populations. *Nature* 451: 998–1003 (2008).

The Earliest Remains of Our Species: Omo, Ethiopia

1. White, T. D., Asfaw, B., DeGusta, D., *et al.* Pleistocene *Homo sapiens* from Middle Awash, Ethiopia. *Nature* 423: 742–7 (2003).
2. McDougall, I., Brown, F. H., & Fleagle, J. G. Stratigraphic placement and age of modern humans from Kibish, Ethiopia. *Nature* 433: 733–6 (2005).
3. Leakey, R. E. F. Early *Homo sapiens* remains from the Omo River Region of South-West Ethiopia. Faunal remains from the Omo Valley. *Nature*: 222, 1132–3 (1969).
4. Day, M. H. Early *Homo sapiens* remains from the Omo River Region of South-West Ethiopia. Omo human skeletal remains. *Nature* 222: 1135–8 (1969).
5. Johanson D., & Edgar B. *From Lucy to Language*, Simon & Schuster, New York (1996).
6. Schwartz, J. H., & Tattersall, I. Craniodental Morphology of Genus Homo (Africa and Asia), *The Human Fossil Record*, vol. 2, Wiley Liss, New Jersey, pp 235–40 (2003).

Modern Human Behaviour: Pinnacle Point, South Africa

1. Henshilwood C., & Sealy, J. Bone artefacts from the Middle Stone Age at Blombos Cave, South Africa. *Current Anthropology* 38: 890–95 (1997).
2. Minichillo, T. Raw material use and behavioural modernity: Howiesons Poort lithic foraging strategies. *Journal of Human Evolution* 50: 359–64 (2006).
3. Mellars, P. Going east: new genetic and archaeological perspectives on the modern human colonization of Eurasia. *Science* 313: 796–800 (2006).
4. D'Errico, F., Henshilwood, C., Vanhaeren, M., *et al. Nassarius kraussianus* shell beads from Blombos Cave: Evidence for symbolic behaviour in the Middle Stone Age. *Journal of Human Evolution* 48: 3–24 (2005).
5. Henshilwood, C. S., d'Errico, F., Yates, R., *et al.* Emergence of modern human behaviour: Middle Stone Age engravings from South Africa. *Science* 295: 1278–80 (2002).
6. Mellars, P. Why did modern human populations disperse from Africa *ca.* 60,000 years ago? A new model. *Proceedings of the National Academy of Sciences* 25: 9381–6 (2006).
7. Marean, C. W., Bar-Matthews, M., Bernatchez, J., *et al.* Early human use of marine resources and pigment in South Africa during the Middle Pleistocene. *Nature* 449: 905–9 (2007).

The First Exodus: Skhul, Israel

1. Tishkoff, S. A., & Williams, S. M. Genetic analysis of African populations: human evolution and complex disease. *Nature Reviews: Genetics* 3: 611–21 (2002).
2. Oppenheimer, S. The Great Arc of dispersal of modern humans: Africa to Australia. *Quaternary International* doi:10.1016/j.quaint.2008.05.015 (2008).
3. Flemming, N. C., Bailey, G. N., Courtillot, V. *et al.* Coastal and marine palaeo-environments and human dispersal points across the Africa-Eurasia boundary. In: *The Maritime and Underwater Heritage*, Wessex Institute of Technology, Southampton, pp 61–74 (2003).
4. Smith, T. M., Tafforeau, P., Reid, D. J. *et al.* Earliest evidence of modern human life history in North African early *Homo sapiens*. *Proceedings of the National Academy of Sciences* 104: 6128–33 (2007).

5.   Stringer, C. B., & Barton, N. Putting North Africa on the map of modern human origins. *Evolutionary Anthropology* 17: 5–7 (2008).

6.   Bouzouggar, A., Barton, N., Vanhaeren, M., *et al.* 82,000-year-old shell beads from North Africa and implications for the origins of modern human behaviour. *Proceedings of the National Academy of Sciences* 104: 9964–9 (2007).

7.   Lahr, M. M., & Foley, R. Multiple dispersals and modern human origins. *Evolutionary Anthropology* 3: 48–60 (1994).

8.   Oppenheimer, S. *Out of Eden. The Peopling of the World*, Constable & Robinson, London (2003).

9.   Field, J. S., & Lahr, M. M. Assessment of the Southern Dispersal: GIS-based analyses of potential routes at Oxygen Isotopic Stage 4. *Journal of World Prehistory* 19: 1–45 (2006).

10. Pope, K. O., & Terrell, J. E. Environmental setting of human migrations in the circum-Pacific region. *Journal of Biogeography* 35: 1–21 (2008).

11. Richards, M., Bandelt, H-J., Kivisild, T., & Oppenheimer, S. A model for the dispersal of modern humans out of Africa. *Nucleic Acids and Molecular Biology* 18: 225–65 (2006).

12. Smith, P. J. Dorothy Garrod, first woman Professor at Cambridge. *Antiquity* 74: 131–6 (2000).

13. Stringer, C., Grun, R., Schwarcz, H. P., & Goldberg, P. ESR dates for the hominid burial site of Es Skhul in Israel. *Nature* 338: 756–8 (1989).

14. Grun, R., Stringer, C., McDermott, F., *et al.* U-series and ESR analyses of bones and teeth relating to the human burials from Skhul. *Journal of Human Evolution* 49: 316–34 (2005).

15. Vanhaeren, M., d'Errico, F., Stringer, C., *et al.* Middle Palaeolithic shell beads in Israel and Algeria. *Science* 312: 1785–8 (2006).

16. Johanson, D., & Edgar, B., *From Lucy to Language*, Simon & Schuster, New York, (1996).

17. Stringer, C. Modern human origins: progress and prospects. *Philosophical Transactions of the Royal Society of London* 357: 563–79 (2002).

18. Vermeersch, P. M., Paulissen, E., Stokes, S. *et al.* A Middle Palaeolithic burial of a modern human at Taramsa Hill, Egypt. *Antiquity* 72: 475–84 (1998).

19. Underhill, P. A., Passarino, G., Lin, A. A. *et al.* The phylogeography of Y chromosome binary haplotypes and the origins of modern human populations. *Annals of Human Genetics* 65: 43–62 (2001).

20. Forster, P. Ice Ages and the mitochondrial DNA chronology of human dispersals: a review. *Philosophical Transactions of the Royal Society of London B* 359: 255–64 (2004).

21. Kivisild, T. Complete mtDNA sequences – quest on 'Out-of-Africa' route completed? In: Mellars, P., Boyle, K., Bar-Yosef, O., & Stringer, C. (eds), *Rethinking the Human Revolution: New Behavioural and Biological Perspectives on the Origin and Dispersal of Modern Humans*, McDonald Institute for Archaeological Research, Cambridge, pp 33–42 (2007).

22. Macaulay, V., Hill, C., Achilli, A., *et al.* Single, rapid coastal settlement of Asia revealed by analysis of complete mitochondrial genomes. *Science* 308: 1034–6 (2005).

23. Mellars, P. Going east: new genetic and archaeological perspectives on the modern human colonization of Eurasia. *Science* 313: 796–800 (2006).

24. Walter, R. C., Buffler, R. T., Bruggemann, J. H., *et al.* Early human occupation of the Red Sea coast of Eritrea during the last interglacial. *Nature* 405: 65–9 (2000).

25. Rose, J. The Arabian Corridor migration model: archaeological evidence for hominin dispersals into Oman during the Middle and Upper Pleistocene. *Proceedings of the Seminar for Arabian Studies* 37: 1–19 (2007).

An Arabian Mystery: Oman

1. Rose, J. The Arabian Corridor migration model: archaeological evidence for hominin dispersals into Oman during the Middle and Upper Pleistocene. *Proceedings of the Seminar for Arabian Studies* 37: 1–19 (2007).

2. Field, J. S., & Lahr, M. M. Assessment of the Southern Dispersal: GIS-based analyses of potential routes at Oxygen Isotopic Stage 4. *Journal of World Prehistory* 19: 1–45 (2006).

3. Petraglia, M. D., & Alsharekh, A. The Middle Palaeolithic of Arabia: implications for modern human origins, behaviour and dispersals. *Antiquity* 77: 671–84 (2003).

4. Parker, A. G., & Rose, J. I. Climate change and human origins in southern Arabia. *Proceedings of the Seminar for Arabian Studies* 38: 25–42 (2008).

5. Pope, K. O., & Terrell, J. E. Environmental setting of human migrations in the circum-Pacific region. *Journal of Biogeography* 35: 1–21 (2008).

6. Rose, J. The question of Upper Pleistocene connections between East Africa and South Arabia. *Current Anthropology* 45: 551–5 (2004).

7. Stringer, C. Coasting out of Africa. *Nature* 405: 24–7 (2000).

2. Footprints of the Ancestors: From India to Australia

Archaeology in the Ashes: Jwalapuram, India

1. Rampino, M. R., & Self, S. Volcanic winter and accelerated glaciation following the Toba super-eruption. *Nature* 359: 50–52 (1992).

2. Petraglia, M., Korisettar, R., Boivin N., *et al.* Middle Palaeolithic assemblages from the Indian subcontinent before and after the Toba super-eruption. *Science* 317: 114–16 (2007).

3. Oppenheimer, C. Limited global change due to the largest known Quaternary eruption, Toba E74 kyr BP? *Quaternary Science Reviews* 21: 1593–609 (2002).

4. James, H. V. A., & Petraglia, M. D. Modern human origins and the evolution of behaviour in the Later Pleistocene record of south Asia. *Current Anthropology* 46: S3–S16 (2005).

5. Gibbons, A. Pleistocene population explosions. *Science* 5130: 27–8 (1993).

6. Rampino, M. R., & Self S. Bottleneck in Human Evolution and the Toba eruption. *Science* 262: 1955 (1993).

7. Pope, K. O., & Terrell, J. E. Environmental setting of human migrations in the circum-Pacific region. *Journal of Biogeography* 35: 1–21 (2008).

8. Louys, J. Limited effect of the Quaternary's largest super-eruption (Toba) on land mammals from Southeast Asia. *Quaternary Science Reviews* 26: 3108–17 (2007).

9. Macaulay, V., Hill, C., Achilli, A., *et al.* Single, rapid coastal settlement of Asia revealed by analysis of complete mitochondrial genomes. *Science* 308: 1034–6 (2005).

10.   Mellars, P. Going east: new genetic and archaeological perspectives on the modern human colonization of Eurasia. *Science* 313: 796–800 (2006).

11.   Field, J. S., & Lahr, M. M. Assessment of the Southern Dispersal: GIS-based analyses of potential routes at Oxygen Isotopic Stage 4. *Journal of World Prehistory* 19: 1–45 (2006).

12.   Field, J. S., Petraglia, M. D., & Lahr, M. M. The southern dispersal hypothesis and the South Asian archaeological record: examination of dispersal routes through GIS analysis. *Journal of Anthropological Archaeology* 26: 88–108 (2007).

Hunter-Gatherers and Genes in the Rainforest: Lenggong, Perak, Malaysia

1.   Flint, J., Hill, A. V. S., Bowden, D. K., *et al*. High frequencies of α-thalassaemia are the result of natural selection by malaria *Nature* 321: 744–50 (1986).

2.   Oppenheimer, S. J., Higgs, D. R., Weatherall, D. J., *et al*. Alpha thalassaemia in Papua New Guinea. *Lancet* 25: 424–6 (1984).

3.   Oppenheimer, S. J., Hill, A. V. S., Gibson, F. D., *et al*. The interaction of alpha thalassaemia with malaria. *Transactions of the Royal Society of Tropical Medicine and Hygiene* 81: 322–6 (1987).

4.   Isa, H. M. Material culture transformation and its impact on cultural ecological change: the case study of the Lanoh in Upper Perak (2007).

5.   Carey, I., *Orang Asli. The Aboriginal Tribes of Peninsular Malaysia*. Oxford University Press, Oxford (1976).

6.   Rabett, R., & Barker, G. Through the looking glass: new evidence on the presence and behaviour of late Pleistocene humans at Niah Cave, Sarawak, Borneo. In: Mellars, P., Boyle, K., Bar-Yosef, O., & Stringer, C. (eds) *Rethinking the Human Revolution: New Behavioural and Biological Perspectives on the Origin and Dispersal of Modern Humans*, McDonald Institute for Archaeological Research, Cambridge, pp 411–24 (2007).

7.   Lahr, M. M. *The Evolution of Modern Human Diversity*, Cambridge University Press, Cambridge (1996).

8.   Oppenheimer, S. *Out of Eden. The Peopling of the World*. Constable & Robinson, London (2003).

9.   Hill, C., Soares, P., Mormina, M., *et al*. Phylogeography and Ethnogenesis of Aboriginal Southeast Asians. *Molecular Biology and Evolution* 23: 2480–91 (2006).

10.   Jablonski, N. G., & Chapman, G. The evolution of human skin coloration. *Journal of Human Evolution* 39: 57–106 (2000).

11.   Jablonski, N. G. The evolution of human skin and skin colour. *Annual Review of Anthropology* 33: 585–623 (2004).

12.   Norton, H. L., Kittles, R. A., Parra, E., *et al*. Genetic evidence for the convergent evolution of light skin in Europeans and East Asians. *Molecular Biology & Evolution* 24: 710–22 (2007).

13.   Thangaraj, K., Chaubey, G., Singh, V. K., *et al*. *In situ* origin of deep rooting lineages of mitochondrial Macrohaplogroup 'M' in India. *BMC Genomics* 7: 151 (2006).

14.   O'Connell, J. F., & Allen, J. Dating the colonization of Sahul (Pleistocene Australia-New Guinea): a review of recent research. *Journal of Archaeological Science* 31: 835–53 (2004).

Headhunting an Ancient Skull: Niah Cave, Borneo

1. *Tom Harrison, The Barefoot Anthropologist*, BBC Four (2006).
2. Barker, G., Barton, H., Bird, M., *et al.* The 'human revolution' in lowland tropical Southeast Asia: the antiquity and behaviour of anatomically modern humans at Niah Cave (Sarawak, Borneo). *Journal of Human Evolution* 52: 243–61 (2007).
3. Rabett, R. & Barker, G. Through the looking glass: new evidence on the presence and behaviour of late Pleistocene humans at Niah Cave, Sarawak, Borneo. In: Mellars, P., Boyle, K., Bar-Yosef, O., & Stringer, C. (eds), *Rethinking the Human Revolution: New Behavioural and Biological Perspectives on the Origin and Dispersal of Modern Humans*, McDonald Institute for Archaeological Research, Cambridge, pp. 411–24 (2007).
4. Detroit, F., Dizon, E., Falgueres, C., *et al.* Upper Pleistocene *Homo sapiens* from the Tabon Cave (Palawan, The Philippines): description and dating of new discoveries. *Comptes Rendus Palevol* 3: 705–12 (2004).
5. Storm, P. The evolution of humans in Australasia from an environmental perspective. *Palaeogeography, Palaeoclimatology, Palaeoecology* 171: 363–83 (2001).
6. Pope, K. O., & Terrell, J. E. Environmental setting of human migrations in the circum-Pacific region. *Journal of Biogeography* 35: 121 (2008).
7. Cattelain, P. Hunting during the Upper Palaeolithic: bow, spearthrower, or both? In: H. Knecht (ed.), *Projectile Technology*, Plenum Press, New York, pp. 213–40 (1997).
8. Hunt, C. O., Gilbertson, D. D., & Rushworth, G. Modern humans in Sarawak, Malaysian Borneo, during Oxygen Isotope Stage 3: palaeoenvironmental evidence from the Great Cave of Niah. *Journal of Archaeological Science* 34: 1953–69 (2007).

The Hobbit: Flores, Indonesia

1. Morwood, M., & Oosterzee, P. V. *The Discovery of the Hobbit. The Scientific Breakthrough that Changed the Face of Human History*, Random House Australia, Sydney (2007).
2. Brown, P., Sutikna, T., Morwood, M. J., *et al.* A new small-bodied hominin from the Late Pleistocene of Flores, Indonesia. *Nature* 431: 1055–61 (2004).
3. Jakob, T., Indriati, E., Soejono, R. P., *et al.* Pygmoid Australomelanesian *Homo sapiens* skeletal remains from Liang Bua, Flores: population affinities and pathological abnormalities. *Proceedings of the National Academy of Science* 103: 13421–6 (2006).
4. Falk, D., Hildebolt, C., Smith, K. *et al.* Brain shape in human microcephalics and *Homo floresiensis*. *Proceedings of the National Academy of Sciences* 104: 2513–18 (2007).
5. Obendorf, P. J., Oxnard, C. E., & Kefford, B. J. Are the small human-like fossils found on Flores human endemic cretins? *Proceedings of the Royal Society B* e-publication doi:10.1098/rspb.2007.1488 (2008).
6. Argue, D., Donlon, D., Groves, C., & Wright, R. *Homo floresiensis*: microcephalic, pymoid, *Australopithecus*, or *Homo*? *Journal of Human Evolution* 51: 360–74 (2006).
7. Larson, S. G., Jungers, W. L., Morwood, M. J., *et al. Homo floresiensis* and the evolution of the hominin shoulder. *Journal of Human Evolution* 6: 718–31 (2007).

8. Tocheri, M. W., Orr, C. M., Larson, S. G., *et al.* The primitive wrist of *Homo floresiensis* and its implications for hominin evolution. *Science* 317: 1743–5 (2007).

9. Moore, M. W., & Brumm, A. Stone artifacts and hominins in island Southeast Asia: new insights from Flores, eastern Indonesia. *Journal of Human Evolution* 52: 85–102 (2007).

10. O'Connor, S. New evidence from East Timor contributes to our understanding of earliest modern colonisation east of the Sunda Shelf. *Antiquity* 81: 523–35 (2007).

11. Morwood, M. J., Brown, P., Jatmiko, *et al.* Further evidence for small-bodied hominins from the Late Pleistocene of Flores, Indonesia. *Nature* 437: 1012–17 (2005).

A Stone Age Voyage: Lombok to Sumbawa, Indonesia

1. Macaulay, V., Hill, C., Achilli, A., *et al.* Single, rapid coastal settlement of Asia revealed by analysis of complete mitochondrial genomes. *Science* 308: 1034–6 (2005).

2. Oppenheimer, S. *Out of Eden. The Peopling of the World*, Constable & Robinson, London (2003).

3. Oppenheimer, S. The Great Arc of dispersal of modern humans: Africa to Australia. *Quaternary International* doi:10.1016/j.quaint.2008.05.015 (2008).

4. Ingman, M. & Gyllensten, U. Mitochondrial genome variation and evolutionary history of Australian and New Guinean aborigines. *Genome Research* 13: 1600–1606 (2003).

5. Van Holst Pellekan, S., Ingman, M., Roberts-Thomson, J., & Harding, R. M. Mitochondrial genomics identifies major haplogroups in aboriginal Australians. *American Journal of Physical Anthropology* 131: 282–94 (2006).

6. O'Connell, J. F., & Allen, J. Dating the colonization of Sahul (Pleistocene Australia-New Guinea): a review of recent research. *Journal of Archaeological Science* 31: 835–53 (2004).

7. O'Connor, S. New evidence from East Timor contributes to our understanding of earliest modern colonisation east of the Sunda Shelf. *Antiquity* 81: 523–35 (2007).

8. Bulbeck, D. Where river meets sea. A parsimonious model for *Homo sapiens* colonization of the Indian Ocean rim and Sahul. *Current Anthropology* 48: 315–21 (2007).

9. Bird, M. I., Taylor, D., & Hunt, C. Palaeoenvironments of insular Southeast Asia during the Last Glacial Period: a savanna corridor in Sundaland? *Quaternary Science Reviews* 24: 20–21 (2005).

10. Mulvaney, J., & Kamminga, J. *Prehistory of Australia*, Allen & Unwin Australia, Sydney (1999).

11. Bednarik, R. G. Maritime navigation in the Lower and Middle Palaeolithic. *Comptes Rendus de l'Académie des Sciences: Earth and Planetary Sciences* 328: 559–63 (1999).

12. Bednarik, R. G. Seafaring in the Pleistocene. *Cambridge Archaeological Journal* 13: 41–66 (2003).

13. Balter, M. In search of the world's most ancient mariners. *Science* 318: 388–9 (2007).

14. O'Connell, J. F., & Allen, J. Pre-LGM Sahul (Pleistocene Australia-New Guinea) and the archaeology of early modern humans. In *Rethinking the Human Revolution: New Behavioural and Biological Perspectives on the Origin and Dispersal of Modern Humans*, Mellars, P., Boyle, K., Bar-Yosef, O., & Stringer, C. (eds), McDonald Institute for Archaeological Research, Cambridge, pp. 395–410 (2007).

15. Pope, K. O., & Terrell, J. E. Environmental setting of human migrations in the circum-Pacific region. *Journal of Biogeography* 35: 121 (2008).

Footprints and Fossils: Willandra Lakes, Australia

1. Webb, S., Cupper, M. L., & Robins, R. Pleistocene human footprints from the Willandra Lakes, southern Australia. *Journal of Human Evolution* 50: 405–13 (2006).

2. Roberts, R. G., Flannery, T. F., Ayliffe, L. K., *et al.* New ages for the last Australian megafauna: continent-wide extinction about 46,000 years ago. *Science* 292: 1888–92 (2001).

3. Miller, G. H., Fogel, M. L., Magee, J. W., *et al.* Ecosystem collapse in Pleistocene Australia and a human role in megafaunal extinction. *Science* 309: 287–90 (2005).

4. Pope, K. O., & Terrell, J. E. Environmental setting of human migrations in the circum-Pacific region. *Journal of Biogeography* 35: 121 (2008).

5. Webb, S. Further research of the Willandra Lakes fossil footprint site, southeastern Australia. *Journal of Human Evolution* 52: 711–15 (2007).

6. Bowler, J. M., Jones, R., Allen, H., & Thorne, A. G. Pleistocene human remains from Australia: a living site and human cremation from Lake Mungo, Western New South Wales. *World Archaeology* 2: 39–60.

7. Thorne, A., Grun, R., Mortimer, G., *et al.* Australia's oldest human remains: age of the Lake Mungo 3 skeleton. *Journal of Human Evolution* 36: 591–612 (1999).

8. Bowler, J. M. & Magee, J. W. Redating Australia's oldest human remains: a sceptic's view. *Journal of Human Evolution* 38: 719–26 (2000).

9. Bowler, J. M., Johnston, H., Olley, J. M., *et al.* New ages for human occupation and climatic change at Lake Mungo, Australia. *Nature* 421: 837–40 (2003).

10. Mulvaney, J., & Kamminga, J. *Prehistory of Australia.* Allen & Unwin Australia, Sydney (1999).

11. Westaway, M. The Pleistocene human remains collection from the Willandra Lakes World Heritage Centre Area, Australia, and its role in understanding modern human origins. In: Tomida, Y. (ed.), *Proceedings of the 7th and 8th Symposia on Collection Building and Natural History Studies in Asia and the Pacific Rim, National Science Museum Monographs* 34: 127–38 (2006).

12. Brown, P. Australian Pleistocene variation and the sex of Lake Mungo 3. *Journal of Human Evolution* 38: 743–9 (2000).

13. Wolpoff, M. H., Hawks, J., Frayer, D. W., & Hunley, K. Modern human ancestry at the peripheries: a test of the replacement theory. *Science* 291: 293–7 (2001).

14. Schwartz, J. H., & Tattersall, I. *The Human Fossil Record*, vol. 2, *Craniodental Morphology of Genus Homo (Africa and Asia)*, Wiley Liss, New Jersey (2003).

15. Yokoyama, Y., Falgueres C., Semah F., *et al.* Gamma-ray spectromagnetic dating of late *Homo erectus* skulls from Ngandong and Sambungmacan, Central Java, Indonesia. *Journal of Human Evolution* 55: 274–7 (2008).

16. Stringer, C. B. A metrical study of the WLH-50 calvaria. *Journal of Human Evolution* 34: 327–32 (1998).

17. Webb, S. Cranial thickening in an Australian hominid as a possible palaeo-epidemiological indicator. *American Journal of Physical Anthropology* 82: 403–12 (1990).

18. Brown, P. Recent human evolution in East Asia and Australia. *Philosophical Transactions: Biological Sciences* 337: 235–42 (1992).

19. Thorne, A., & Curnoe, D. Sex and significance of Lake Mungo 3: reply to Brown 'Australian Pleistocene variation and the sex of Lake Mungo 3'. *Journal of Human Evolution* 39: 587–600 (2000).

20. Stone, T., & Cupper, M. L. Last Glacial Maximum ages for robust humans at Kow Swamp, southern Australia. *Journal of Human Evolution* 45: 99–111 (2003).

21. Hudjashov, G., Kivisild, T., Underhill, P. A., *et al.* Revealing the prehistoric settlement of Australia by Y chromosome and mtDNA analysis. *Proceedings of the National Academy of Sciences* 104: 8726–30 (2007).

22. Van Holst Pellekan, S., Ingman, M., Roberts-Thomson, J., & Harding, R. M. Mitochondrial genomics identifies major haplogroups in aboriginal Australians. *American Journal of Physical Anthropology* 131: 282–94 (2006).

23. Roberts, R. G., Jones, R., & Smith, M.A. Thermoluminescence dating of a 50,000-year-old human occupation site northern Australia. *Nature* 345: 153–6 (1990).

24. Roberts, R. G., Jones, R., Spooner, N. A., *et al.* The human colonisation of Australia: optical dates of 53,000 and 60,000 years bracket human arrival at Deaf Adder Gorge, Northern Territory. *Quaternary Geochronology (Quaternary Science Reviews)* 13: 575–83 (1994).

25. O'Connell, J. F., & Allen, F. J. When did humans first arrive in Greater Australia, and why is it important to know? *Evolutionary Anthropology* 6: 132–46 (1998).

26. O'Connell, J. F., & Allen, F. J. Dating the colonization of Sahul (Pleistocene Australia-New Guinea): a review of recent research. *Journal of Archaeological Science* 31: 835–53 (2004).

27. Bulbeck, D. Where river meets sea. A Parsimonious model for *Homo sapiens* colonization of the Indian Ocean rim and Sahul. *Current Anthropology* 48: 315–21 (2007).

28. Bird, M. I., Turney, C. S. M., Fifield, L. K., *et al.* Radiocarbon analysis of the early archaeological site of Nauwalabila I, Arnhem Land, Australia: implications for sample suitability and stratigraphic integrity. *Quaternary Science Reviews* 21: 1061–75 (2002).

29. Fullagar, R. L. K., Price, D. M., & Head, L. M. Early human occupation of northern Australia: archaeology and thermoluminescence dating of Jinmium rock-shelter, Northern Territory. *Antiquity* 70: 751–73 (1996).

30. Roberts, R., Bird, M., Olley, J., *et al.* Optical and radiocarbon dating at Jinmium rock shelter in northern Australia. *Nature* 393: 358–62 (1998).

Art in the Landscape: Gunbalanya (Oenpelli), Northern Territory, Australia

1. Morwood, M., & Oosterzee, P. V. *The Discovery of the Hobbit. The Scientific Breakthrough that Changed the Face of Human History*, Random House Australia, Sydney (2007).

2. Chatwin, B. *The Songlines*, Vintage, London (1987).

3. Hamby, L. *Twined Together: Kunmadj Njalehnjaleken* Injala, Arts and Crafts, Gunbalanya (2005).

3. Reindeer to Rice: The Peopling of North and East Asia

Trekking Inland: Routes into Central Asia

1. Derenko, M., Malyarchuk, B. A., Grzybowski, T., *et al.* Phylogeographic analysis of mitochondrial DNA in Northern Asian populations. *The American Journal of Human Genetics* 81: 1025–41 (2007).
2. Oppenheimer, S. *Out of Eden. The Peopling of the World*, Constable & Robinson, London (2003).
3. Derenko, M. V., Malyarchuk, B. A., Denisova, G. A., *et al.* Molecular genetic differentiation of the ethnic populations of south and east Siberia base on mitochondrial DNA polymorphism. *Russian Journal of Genetics* 38: 1196–1202 (2002).
4. Goebel, T. Pleistocene human colonization of Siberia and peopling of the Americas: an ecological approach. *Evolutionary Anthropology* 8: 208–27 (1999).
5. Goebel, T., Derevianko, A. P., & Petrin, V. T. Dating the Middle-to-Upper Paleolithic transition at Kara-Bom. *Current Anthropology* 34: 452–8 (1993).
6. Brantingham, P. J. The initial Upper Paleolithic in Northeast Asia. *Current Anthropology* 42: 735–46 (2001).
7. Krause, J., Orlando, L., Serre, D., *et al.* Neanderthals in central Asia and Siberia. 449: 902–4 (2007).
8. Vasil'ev, S. A. The Upper Palaeolithic of Northern Asia. *Current Anthropology* 34: 82–92 (1993).

On the Trail of Ice Age Siberians: St Petersburg, Russia

1. Pitulko, V. V., Nikolsky, P. A., Girya, E. Y., *et al.* The Yana RHS site: humans in the Arctic before the Last Glacial Maximum. *Science* 303: 52–6 (2004).
2. Vasil'ev, S. A., Sergey, A., Kuzmin, L. A., *et al.* Radiocarbon-based chronology of the Paleolithic in Siberia and its relevance to the peopling of the New World. *Radiocarbon* 44: 403–630 (2002).
3. Guthrie, R. D. Origin and causes of the mammoth steppe: a story of cloud cover, woolly mammoth tooth pits, buckles, and inside-out Beringia. *Quaternary Science Reviews* 20: 549–74 (2001).
4. Goebel, T. The 'microblade adaptation' and recolonization of Siberia during the Late Upper Pleistocene. In Elston, R. G., & Kuhn, S. L. (eds), *Thinking Small: Global Perspectives on Microlithization*, Archaeological Papers of the American Anthropological Association no. 12 (2002).
5. Goebel, T. Pleistocene human colonization of Siberia and peopling of the Americas: an ecological approach. *Evolutionary Anthropology* 8: 208–27 (1999).
6. Schlesier, K. H. More on the 'Venus' figurines. *Current Anthropology* 42: 410 (2001).
7. Soffer, O., Adovasio, J. M., & Hyland, D. C. More on the 'Venus' figurines: Reply. *Current Anthropology* 42: 410–12 (2001).
8. Hoffecker, J. F. Innovation and technological knowledge in the Upper Palaeolithic of Northern Eurasia. *Evolutionary Anthropology* 14: 186–8 (2005).
9. Vasil'ev, S. A. Man and mammoth in Pleistocene Siberia. *The World of Elephants. Proceedings of the 1st International Conference*, pp. 363–6, Rome (2001).
10. Ugan, A., & Byers, D. A global perspective on the spatiotemporal pattern of the

Late Pleistocene human and woolly mammoth radiocarbon record. *Quaternary International* doi: 10.1016/j.quaint.2007.09.035 (2008).

11. Lister, A. M., & Sher, A. V. Ice cores and mammoth extinction. *Nature* 378: 23–4 (1995).

12. Pushkina, D., & Raia, P. Human influence on distribution and extinctions of the late Pleistocene Eurasian megafauna. *Journal of Human Evolution* 54: 769–82 (2008).

13. Stuart, A. J. The extinction of the woolly mammoth (*Mammuthus primigenius*) and straight-tusked elephant (*Palaeoloxodon antiquus*) in Europe. *Quaternary International* 126–8: 171–7 (2005).

14. Stuart, A. J., Sulerzhitsky, L. D., Orlova, L. A., *et al.* The latest woolly mammoths (*Mammuthus primigenius* Blumenbach) in Europe and Asia: a review of the current evidence. *Quaternary Science Reviews* 21: 1559–69 (2002).

Meeting with the Reindeer Herders of the North: Olenek, Siberia

1. Vitebsky, P. *Reindeer People. Living with Animals and Spirits in Siberia*, HarperCollins, London (2005).

2. Ingold, T. On reindeer and men. *Man* 9: 523–38 (1974).

3. Pakendorf, B., Wiebe, V., Tarskaia, L. A., *et al.* Mitochondrial DNA evidence for admixed origins of central Siberian populations. *American Journal of Physical Anthropology* 120: 211–14 (2003).

4. Pakendorf, B., Novgorodov, I. N., Osakovskij, V. L., & Stoneking, M. Mating patterns amongst Siberian reindeer herders: inferences from mtDNA and Y-chromosomal analyses. *American Journal of Physical Anthropology* 133: 1013–27 (2007).

5. Uinuk-ool, T., Takezaki, N., Sukernik, R. I., *et al.* Origin and affinities of indigenous Siberian populations as revealed by HLA class II gene frequencies. *Human Genetics* 110: 209–26 (2002).

6. Burch, E. S. The caribou/wild reindeer as a human resource. *American Antiquity* 37: 339–68 (1972).

7. Galloway, V. A., Leonard, W. R., & Ivakine, E. Basal metabolic adaptation of the Evenki reindeer herders of Central Siberia. *American Journal of Human Biology* 12: 75–87 (2000).

8. Leonard, W. R., Galloway, V. A., Ivakine, E., *et al.* Nutrition, thyroid function and basal metabolism of the Evenki of central Siberia. *International Journal of Circumpolar Health* 58: 281–95 (1999).

9. Ebbesson, S. O. E., Schraer, C., Nobmann, E. D., & Ebbesson, L. O. E. Lipoprotein profiles in Alaskan Siberian Yupik Eskimos. *Artic Medical Research* 55: 165–73 (1996).

10. Steegman, A. T. Cold adaptation and the human face. *American Journal of Physical Anthropology* 32: 243–50 (1970).

11. Shea, B. T. Eskimo craniofacial morphology, cold stress and the maxillary sinus. *American Journal of Physical Anthropology* 47: 289–300 (1977).

12. Wallace, D. C. A mitochondrial paradigm of metabolic and degenerative diseases, aging, and cancer: a dawn for evolutionary medicine. *Annual Review of Genetics* 39: 359–407 (2005).

The Riddle of Peking Man: Beijing, China

1. Sautman, B. Peking Man and the politics of palaeoanthropological nationalism in China. *The Journal of Asian Studies* 60: 95–124 (2001).

2. Pope, G. G. Craniofacial evidence for the origin of modern humans in China. *Yearbook of Physical Anthropology* 35: 243–98 (1992).

3. Tattersall, I., & Sawyer, G. J. The skull of 'Sinanthropus' from Zhoukoudian, China: a new reconstruction. *Journal of Human Evolution* 31: 311–14 (1996).

4. Kamminga, J., personal correspondence.

5. Brown, P. Chinese Middle Pleistocene hominids and modern human origins in East Asia. In: *Human Roots. Africa and Asia in the Middle Pleistocene*, Barham, L., & Robson-Brown, K. (eds), Western Academic and Specialist Press, Bristol (2001).

6. Shen, G., Teh-Lung, K., Cheng., H., *et al.* High-precision U-series dating of Locality 1 at Zhoukoudian, China. *Journal of Human Evolution* 41: 679–88 (2001).

7. Lieberman, D. E. Testing hypotheses about recent human evolution from skulls: integrating morphology, function, development and phylogeny. *Current Anthropology* 36: 159–97 (1995).

8. Stringer, C. B. Reconstructing recent human evolution. *Philosophical Transactions: Biological Sciences* 337: 217–24 (1992).

9. Stringer, C. Modern human origins: progress and prospects. *Philosophical Transactions of the Royal Society of London* 357: 563–79 (2002).

10. Lieberman, D. E., Krovitz, G. E., Yates, F. W., *et al.* Effects of food processing on masticatory strain and craniofacial growth in a retrognathic face. *Journal of Human Evolution* 46: 655–77 (2004).

11. Macaulay, V., Hill, C., Achilli, A., *et al.* Single, rapid coastal settlement of Asia revealed by analysis of complete mitochondrial genomes. *Science* 308: 1034–6 (2005).

12. Shang, H., Tong, H., Zhang, S., *et al.* An early modern human from Tianyuan Cave, Zhoukoudian, China. *Proceedings of the National Academy of Sciences* 104: 6573–8 (2007).

An Archaeological Puzzle: Zhujiatun, China

1. Gao, X., & Norton, C. J. A critique of the Chinese 'Middle Palaeolithic'. *Antiquity* 76: 397–412 (2002).

2. Wu, X. On the origin of modern humans in China. *Quaternary International* 117: 131–40 (2004).

3. West, J. A., & Louys, J. Differentiating bamboo from stone tool cut marks in the zooarchaeological record, with a discussion on the use of bamboo knives. *Journal of Archaeological Science* 34: 512–18 (2007).

4. Shen, G., Wang, W., Cheng, H., & Edwards, R. L. Mass spectrometric U-series dating of Laibin hominid site in Guangxi, southern China. *Journal of Archaeological Science* 34: 2109–14 (2007).

5. Jian, L., & Shannon, C. L. Rethinking early Palaeolithic typologies in China and India. *Journal of East Asian Archaeology* 2: 9–35 (2000).

6. Shea, J. L. Lithic microwear analysis in archaeology. *Evolutionary Anthropology* 1: 143–50 (2005).

East Asian Genes to the Rescue: Shanghai, China

1. Ke, Y., Su, B., Song, X., *et al.* African origin of modern humans in East Asia: a tale of 12,000 chromosomes. *Science* 292: 1151–3 (2001).

2. Su, B., Xiao, J., Underhill, P., *et al.* Y-chromosome evidence for a northward migration of modern humans into Eastern Asia during the last Ice Age. *American Journal of Human Genetics* 65: 1718–24 (1999).

3. Li, H., Cai, X., Winograd-Cort, E. R., *et al.* Mitochondrial DNA diversity and population differentiation in southern East Asia. *American Journal of Physical Anthropology* 134: 481–8 (2007).

4. Kivisild, T., Tolk, H-V., Parik, J., *et al.* The emerging limbs and twigs of the East Asian mtDNA tree. *Molecular and Biological Evolution* 19: 1737–51 (2002).

5. Oppenheimer, S. *Out of Eden. The Peopling of the World*, Constable & Robinson, London (2003).

6. Yao, Y-G., Kong, Q-P., Bandelt, H-J., *et al.* Phylogeographic differentiation of mitochondrial DNA in Han Chinese. *American Journal of Human Genetics* 70: 635–51 (2002).

7. Pope, K. O., & Terrell, J. E. Environmental setting of human migrations in the circum-Pacific region. *Journal of Biogeography* 35: 1–21 (2008).

Pottery and Rice: Guilin and Long Ji, China

1. Diamond, J., & Bellwood, P. Farmers and their languages: the first expansions. *Science* 300: 597–603 (2003).

2. Matsumura, H., & Hudson, M. J. Dental perspectives on the population history of Southeast Asia. *American Journal of Physical Anthropology* 127: 182–209 (2005).

3. Oppenheimer, S. *Out of Eden. The Peopling of the World*, Constable & Robinson, London (2003).

4. Kuzmin, Y. V. Chronology of the earliest pottery in East Asia: progress and pitfalls. *Antiquity* 80: 362–71 (2006).

5. Pearson, R. The social context of early pottery in the Lingnan region of south China. *Antiquity* 79: 819–28 (2005).

6. Shelach, G. The earliest Neolithic cultures of Northeast China: recent discoveries and new perspectives on the beginning of agriculture. *Journal of World Prehistory* 14: 363–413 (2000).

7. Cohen, D. J. New perspectives on the transition to agriculture in China. In Yasuda, Y. (ed.), *The Origins of Pottery and Agriculture*, Roli Books, New Delhi, pp. 217–27 (2002).

8. Lu, T. & L-D. The occurrence of cereal cultivation in China. *Asian Perspectives* 45: 129–58 (2006).

9. Jiang, L., & Liu, L. New evidence for the origins of sedentism and rice domestication in the Lower Yangtzi River, China. *Antiquity* 80: 355–61 (2006).

10. Underhill, P. A., Passarino, G., Lin, A. A., *et al.* The phylogeography of Y chromosome binary haplotypes and the origins of modern human populations. *Annals of Human Genetics* 65: 43–62 (2001).

4. *The Wild West: The Colonisation of Europe*

On the Way to Europe: Modern Humans in the Levant and Turkey

1. Oppenheimer, S. *Out of Eden. The Peopling of the World*, Constable & Robinson, London (2003).

2.   Olszewski, D. I., & Dibble, H. L. The Zagros Aurignacian. *Current Anthropology* 35: 68–75 (1994).

3.   Kuhn, S. L., Stiner, M. C., Reese, D. S., & Gulec, E. Ornaments of the earliest Upper Palaeolithic: new insights from the Levant. *Proceedings of the National Academy of Sciences* 98: 7641–6 (2001).

4.   Mellars, P. Archaeology and the dispersal of modern humans in Europe: deconstructing the 'Aurignacian'. *Evolutionary Anthropology* 15: 167–82 (2006).

5.   Bar-Yosef, O., Arnold, M., Mercier, N., *et al.* The dating of the Upper Palaeolithic Layers in Kebara Cave, Mt Carmel. *Journal of Archaeological Science* 23: 297–306 (1996).

6.   Kuhn, S. L. Palaeolithic archaeology in Turkey. *Evolutionary Anthropology* 11: 198–210 (2002).

7.   Vanhaeren, M., d'Errico, F., Stringer, C., *et al.* Middle Palaeolithic shell beads in Israel and Algeria. *Science* 312: 1785–8 (2006).

8.   Mellars, P. Neanderthals and the modern human colonization of Europe. *Nature* 432: 461–5 (2004).

9.   Otte, M., & Derevianko, A. The Aurignacian in Altai. *Antiquity* 75: 44–8 (2001).

Crossing the Water into Europe: the Bosphorus, Turkey

1.   Kerey, I. E., Meric, E., Tunoglu, C., *et al.* Black Sea–Marmara Sea Quaternary connections: new data from the Bosphorus, Istanbul, Turkey. *Palaeogeography, palaeoclimatology, palaeoecology* 204: 277–95 (2004).

2.   Mellars, P. Neanderthals and the modern human colonization of Europe. *Nature* 432: 461–5 (2004).

3.   Mellars, P. A new radiocarbon revolution and the dispersal of modern humans in Eurasia. *Nature* 439: 931–5 (2006).

4.   Mellars, P. Archaeology and the dispersal of modern humans in Europe: deconstructing the 'Aurignacian'. *Evolutionary Anthropology* 15: 167–82 (2006).

5.   Underhill, P. A., Passarino, G., Lin, A. A., *et al.* The phylogeography of Y chromosome binary haplotypes and the origins of modern human populations. *Annals of Human Genetics* 65: 43–62 (2001).

Face to Face with the First Modern European: Oase Cave, Romania

1.   Zilhao, J. E., Trinkaus, E., Constantin, S., *et al.* The Peştera cu Oase people, Europe's earliest modern humans. In: Mellars, P., Stringer, C., Bar-Yosef, O., Boyle, K. (eds), *Rethinking the Human Revolution: New Behavioural and Biological Perspectives on the Origins and Dispersal of Modern Humans*, McDonald Institute of Archaeology Monographs, Cambridge (2007).

2.   Trinkaus, E., Moldovan, O., Milota, S., *et al.* An early modern human from the Peştera cu Oase, Romania. *Proceedings of the National Academy of Sciences* 100: 11231–6 (2003).

3.   Gibbons, A. A shrunken head for African *Homo erectus*. *Science* 300: 893 (2003).

4.   Carbonell, E., Bermudez de Castro, J. M., Pares, J. M., *et al.* The first hominin of Europe. *Nature* 452: 465–70 (2008).

5.   Brauer, G. The origin of modern anatomy: by speciation or intraspecific evolution. *Evolutionary Anthropology* 17: 22–37 (2008).

6.   Stringer, C. Modern human origins: progress and prospects. *Philosophical Transactions of the Royal Society of London* 357: 563–79 (2002).

7.   Campbell, B. The Centenary of Neanderthal Man: Part I. *Man* 56: 156–8 (1956).
8.   Schmitz, R. W., Serre, D., Bonani, G., *et al.* The Neanderthal type site revisited: interdisciplinary investigations of skeletal remains from the Neander valley, Germany. *Proceedings of the National Academy of Sciences* 99: 13342–7 (2002).
9.   Stringer, C. *Homo Britannicus. The Incredible Story of Human Life in Britain*, Penguin Books, London (2006).
10.   Bischoff, J. L., & Shamp, D. D. The Sima de los Huesos hominids date to beyond U/Th equilibrium (>350kyr) and perhaps to 400–500 kyr: new radiometric dates. *Journal of Archaeological Science* 30: 275–80 (2003).
11.   Klein, R. G. Whither the Neanderthals? *Science* 299: 1525–7 (2003).
12.   Krause, J., Orlando, L., Serre, D., *et al.* Neanderthals in central Asia and Siberia. 449: 902–4 (2007).
13.   Mellars, P. A new radiocarbon revolution and the dispersal of modern humans in Eurasia. *Nature* 439: 931–5 (2006).

Neanderthal Skulls and Genes: Leipzig, Germany
1.   Harvati, K., Gunz, P., & Grigorescu, D. Cioclovina (Romania): affinities of an early modern European. *Journal of Human Evolution* 53: 732–46 (2002).
2.   Stringer, C. Modern human origins: progress and prospects. *Philosophical Transactions of the Royal Society of London* 357: 563–79 (2002).
3.   Caramelli, D., Lalueza-Fox, C., Vernesi, C., *et al.* Evidence for a genetic discontinuity between Neanderthals and 24,000-year-old anatomically modern humans. *Proceedings of the National Academy of Sciences* 100: 6593–7 (2003).
4.   Currat, M., & Excoffier, L. Modern humans did not admix with Neanderthals during their range expansion into Europe. *PLoS Biology* 2: e2264–74 (2004).
5.   Kahn, P., & Gibbons, A. DNA from an extinct human. *Science* 277: 176–8 (1997).
6.   Green, R. E., Krause, J., Ptak, S. E., *et al.* Analysis of one million base pairs of Neanderthal DNA. *Nature* 444: 330–36 (2006).
7.   Noonan, J. P., Coop, G., Kudaravalli, S., *et al.* Sequencing and analysis of Neanderthal genomic DNA. *Science* 314: 1113–18 (2006).
8.   Wall, J. D., & Kim, S. K. Inconsistencies in Neanderthal genomic DNA sequences. *PLoS Genetics* 3: 1862–6 (2007).
9.   Dalton, R. DNA probe finds hints of human. *Nature* 449: 7 (2007).
10.   Krause, J., Orlando, L., Serre, D., *et al.* Neanderthals in central Asia and Siberia. 449: 902–4 (2007).
11.   Lalueza-Fox, C., Rompler, H., Caramelli, D., *et al.* A melanocortin 1 receptor allele suggests varying pigmentation among Neanderthals. *Science* 318: 1453–5 (2007).
12.   Trinkaus, E. Human evolution: Neanderthal gene speaks out. *Current Biology* 17: R917–19 (2007).
13.   Krause, J., Lalueza-Fox, C., Orlando, L., *et al.* The derived FOXP2 variant of modern humans was shared with Neanderthals. *Current Biology* 17: 1908–12 (2007).
14.   Morgan, J. Neanderthals 'distinct from us.' www.news.bbc.co.uk/1/hi/health/7886477.stm (12 February 2009).
15.   Bocquet-Appel, J-P., & Demars, P. Y. Neanderthal contraction and modern human colonization of Europe. *Antiquity* 74: 544–52 (2000).

Treasures of the Swabian Aurignacian: Vogelherd, Germany

1. Conard, N. J. Palaeolithic ivory sculptures from southwestern Germany and the origins of figurative art. *Nature* 426: 830–32 (2003).

2. Conard, N. J., Grootes, P. M., & Smith, F. H. Unexpectedly recent dates for human remains from Vogelherd. *Nature* 430: 198–201 (2004).

3. Conard, N. J., & Bolus, M. Radiocarbon dating the appearance of modern humans and timing of cultural innovations in Europe: new results and new challenges. *Journal of Human Evolution* 44: 331–71 (2003).

4. Kuhn, S. L. Palaeolithic archaeology in Turkey. *Evolutionary Anthropology* 11: 198–210 (2002).

5. Mellars, P. Archaeology and the dispersal of modern humans in Europe: deconstructing the 'Aurignacian'. *Evolutionary Anthropology* 15: 167–82 (2006).

6. Svoboda, J., van der Plicht, J., & Kuzulka, V. Upper Palaeolithic and Mesolithic human fossils from Moravia and Bohemia (Czech Republic): some new ^{14}C dates. *Antiquity* 76: 957–62 (2002).

7. Wild, E. M., Teschler-Nicola, M., Kutshera, W., *et al*. Direct dating of Early Upper Palaeolithic human remains from Mladec. *Nature* 435: 332–5 (2005).

8. Eren, M. I., Greenspan, A., & Sampson, C. G. Are Upper Paleolithic blade cores more productive than Middle Paleolithic discoidal cores? A replication experiment. *Journal of Human Evolution* 55: 952–61 (2008).

9. Bar-Yosef, O. The Upper Paleolithic revolution. *Annual Reviews in Anthropology* 31: 363–93 (2002).

10. Shea, J. I. The origins of lithic projectile point technology: evidence from Africa, the Levant and Europe. *Journal of Archaeological Science* 33: 823–46 (2006).

11. Oppenheimer, S. *Out of Eden. The Peopling of the World*, Constable & Robinson, London (2003).

Tracking Down the Last Neanderthals: Gibraltar

1. Barton, R. N. E., Currant, A. P., Fernandez-Jalvo, Y., *et al*. Gibraltar Neanderthals and results of recent excavations in Gorham's, Vanguard and Ibex Caves. *Antiquity* 73: 13–23 (1999).

2. Finlayson, C. *Neanderthals and Modern Humans. An Ecological and Evolutionary Perspective*. Cambridge University Press, Cambridge (2004).

3. Finlayson, C., Fa, D. A., Espejo, F. J., *et al*. Gorham's Cave, Gibraltar – the persistence of a Neanderthal population. *Quaternary International* 181: 64–71 (2008).

4. Finlayson, G., Finlayson, C., Pacheco, F. G., *et al*. Caves as archives of ecological and climatic changes in the Pleistocene – the case of Gorham's Cave, Gibraltar. *Quaternary International* 181: 55–63 (2008).

5. Stringer, C. Modern human origins: progress and prospects. *Philosophical Transactions of the Royal Society of London* 357: 563–79 (2002).

6. Hublin, J-J., Spoor, F., Braun, M., *et al*. A late Neanderthal associated with Upper Palaeolithic artefacts. *Nature* 381: 224–6 (1996).

7. Gilligan, I. Neanderthal extinction and modern human behaviour: the role of climate change and clothing. *World Archaeology* 39: 499–514 (2007).

A Cultural Revolution: Dolní Věstonice, Czech Republic

1.   Svoboda, J. A. The archaeological framework. In: Trinkaus, E., & Svoboda, J. (eds), *Early Modern Human Evolution in Central Europe*, Oxford University Press, Oxford, pp. 6–8 (2006).

2.   Straus, L. G. The Upper Palaeolithic of Europe: an overview. *Evolutionary Anthropology* 4: 4–16 (2005).

3.   Hoffecker, J. F. Innovation and technological knowledge in the Upper Palaeolithic of Northern Eurasia. *Evolutionary Anthropology* 14: 186–98 (2005).

4.   Vandiver, P. B., Soffer, O., Klima, B., & Svoboda, J. The origins of ceramic technology at Dolní Věstonice, Czechoslovakia. *Science* 246: 1002–8 (1989).

5.   Stevens, personal correspondence. When I discussed the Dolní Věstonice pottery fragments with my husband, field archaeologist Dave Stevens, he remarked that he had seen broken fragments of clay pipes that had been reused in kiln furniture, such as muffles, in post-medieval kilns, and that this was a well-known phenomenon.

6.   Formicola, V., Pontrandolfi, A., & Svoboda, J. The Upper Paleolithic triple burial of Dolní Věstonice: pathology and funerary behaviour. *American Journal of Physical Anthropology* 115: 372–9 (2001).

7.   Alt, K. W., Pichler, S., Vach, W., *et al.* Twenty-five-thousand-year-old triple burial from Dolní Věstonice: an Ice Age family? *American Journal of Physical Anthropology* 102: 123–31 (1997).

8.   Svoboda, J. A. The archaeological context of the human remains. In: Trinkaus, E., & Svoboda, J. (eds), *Early Modern Human Evolution in Central Europe*, Oxford University Press, Oxford, pp. 6–8 (2006).

9.   Pettitt, P. B., & Bader, N. O. Direct AMS radiocarbon dates for the Sunghir mid Upper Palaeolithic burials. *Antiquity* 74: 269–70 (2000).

10.   Vandiver, P. B., Soffer, O., Klima, B., & Svoboda, J. The origins of ceramic technology at Dolní Věstonice, Czechoslovakia. *Science* 246: 1002–8 (1989).

11.   Forster, P. Ice Ages and the mitochondrial DNA chronology of human dispersals: a review. *Philosophical Transactions of the Royal Society of London B* 359: 255–64 (2004).

12.   Metspalu, E., Kivisild, T., Kaldma, K., *et al.* The trans-Caucasus and the expansion of the Caucasoid-specific human mitochondrial DNA. In Papiha, S. S., *et al.* (eds), *Genomic Diversity. Applications in Human Population Genetics*, Kluwer Academic/ Plenum Publishers, New York, pp. 121–34 (1999).

13.   Oppenheimer, S. *Out of Eden. The Peopling of theWorld*, Constable & Robinson, London (2003).

Sheltering from the Cold: Abri Castanet, France

1.   Straus, L. G. The Upper Palaeolithic of Europe: an overview. *Evolutionary Anthropology* 4–16 (1995).

2.   Blades, B. Aurignacian settlement patterns in the Vézère Valley. *Current Anthropology* 40: 712–35 (1999).

3.   White, R. Systems of personal ornamentation in the Early Upper Palaeolithic: methodological challenges and new observations. In: Mellars, P., Boyle, K., Bar-Yosef, O., & Stringer C. (eds), *Rethinking the Human Revolution: New Behavioural and Biological Perspectives on the Origin and Dispersal of Modern Humans*, McDonald Institute for Archaeological Research, Cambridge, pp. 287–302 (2007).

4. White, R. Beyond art: toward an understanding of the origins of material representation in Europe. *Annual Reviews of Anthropology* 21: 537–64 (1992).

5. Mellars. Cognition and climate: why is Upper Palaeolithic cave art almost confined to the Franco-Cantabrian region? In Renfrew, C., & Morley, I. (eds), *Becoming Human. Innovation in Prehistoric Material and Spiritual Culture*, Cambridge University Press, Cambridge, chapter 14 (2009).

6. Straus, L. G. South-western Europe at the Last Glacial Maximum. *Current Anthropology* 32: 189–99 (1991).

7. Straus, L. G. The Upper Palaeolithic of Europe: an overview. *Evolutionary Anthropology* 4: 4–16 (2005).

Visiting the Painted Caves: Lascaux, Pech Merle and Cougnac, France

1. Pettitt, P., & Bahn, P. Current problems in dating Palaeolithic cave art: Candamo and Chauvet. *Antiquity* 77: 134–41 (2003).

2. Straus, L. G. The Upper Palaeolithic of Europe: an overview. *Evolutionary Anthropology* 4: 4–16 (1995).

3. Straus, L. G. South-western Europe at the Last Glacial Maximum. *Current Anthropology* 32: 189–99 (1991).

4. Barton, M., Clark, G. A., & Cohen, A. E. Art as information: Explaining Upper Palaeolithic art in Western Europe. *World Archaeology* 26: 185–207 (1994).

5. Lamason, R. L., Mohideen, M.-A. P. K., Mest, J. R., *et al.* SLC24A5, a putative cation exchanger, affects pigmentation in zebrafish and humans. *Science* 310: 1782–6 (2005).

6. Frost, P. European hair and eye color. A case of frequency-dependent sexual selection? *Evolution and Human Behaviour* 27: 85–103 (2006).

7. Gamble, C., Davies, W., Pettitt, P., & Richards, M. Climate change and evolving human diversity in Europe during the Last Glacial. *Philosophical Transactions: Biological Sciences* 359: 243–54 (2004).

8. Pereira, L., Richards, M., Goios, A., *et al.* High-resolution mtDNA evidence for the late-glacial resettlement of Europe from an Iberian refugium. *Genome Research* 15: 19–24 (2005).

9. Torroni, A., Bandelt, H.-J., Macaulay, V., *et al.* A signal, from human MtDNA, of postglacial recolonisation in Europe. *American Journal of Human Genetics* 69: 844–52 (2001).

10. Underhill, P. A., Passarino, G., Lin, A. A., *et al.* The phylogeography of Y chromosome binary haplotypes and the origins of modern human populations. *Annals of Human Genetics* 65: 43–62 (2001).

New Age Mesopotamia: Göbekli Tepe, Turkey

1. Mithen, S. *After the Ice. A Global Human History*, Harvard University Press, Cambridge, Massachusetts (2003).

2. Peters, J., & Schmidt, K. Animals in the symbolic world of pre-pottery Neolithic Göbekli Tepe, south-eastern Turkey: a preliminary assessment. *Anthropozoologica* 39: 179–218 (2004).

3. Byrd, B. F. Reassessing the emergence of village life in the Near East. *Journal of Archaeological Research* 13: 231–90 (2005).

4.   Bar-Yosef, O. The Upper Paleolithic revolution. *Annual Reviews in Anthropology* 31: 363–93 (2002).

5.   Lev-Yardun, S., Gopher, A., & Abbo, S. The cradle of agriculture. *Science* 288: 1602–3 (2000).

6.   Larsen, C. S. Biological changes in human populations with agriculture. *Annual Reviews in Anthropology* 24: 185–213 (1995).

7.   Papathanasiou, A. Health status of the Neolithic population of Alepotrypa Cave, Greece. *American Journal of Physical Anthropology* 126: 377–90 (2005).

8.   Armelagos, G. J., Goodman, A. H., & Jacobs, K. H. The origins of agriculture: population growth during a period of declining health. *Population and Environment* 13: 9–22 (1991).

9.   Pena-Chocarro, L., Zapata, L., Iriarte, M. J., *et al.* The oldest agriculture in northern Atlantic Spain: new evidence from El Miron Cave (Ramales de la Victoria, Cantabria). *Journal of Archaeological Science* 32: 579–87 (2005).

10. Balter, M. Ancient DNA yields clues to the puzzle of European origins. *Science* 310: 964–5 (2005).

11. Underhill, P. A., Passarino, G., Lin, A. A., *et al.* The phylogeography of Y chromosome binary haplotypes and the origins of modern human populations. *Annals of Human Genetics* 65: 43–62 (2001).

12. Richards, M., Macaulay, V., Hickey, E., *et al.* Tracing European founder lineages in the Near Eastern mtDNA pool. *American Journal of Human Genetics* 67: 1251–76 (2000).

13. Haak, W., Forster, P., Bramanti, B., *et al.* Ancient DNA from the first European farmers in 7500-year-old Neolithic sites. *Science* 310: 1016–18 (2005).

5. *The New World: Finding the First Americans*

Bridging the Continents: Beringia

1.   Taylor, R. E., Haynes, C. V. Jr & Stuiver, M. Clovis and Folsom age estimates: stratigraphic context and radiocarbon calibration. *Antiquity* 70: 515–25 (1996).

2.   Morlan, R. E. Current perspectives on the Pleistocene archaeology of eastern Beringia. *Quaternary Research* 60: 123–32 (2003).

3.   Zazula, G. D., Schweger, C. E., Beaudoin, A. B., & McCourt, G. H. Macrofossil and pollen evidence for full-glacial steppe within an ecological mosaic along the Bluefish River, eastern Beringia. *Quaternary International* 142–3: 2–19 (2006).

4.   Goebel, T. The Late Pleistocene dispersal of modern humans in the Americas. *Science* 319: 1497–502 (2008).

Mapping Native American Genes: Calgary, Canada

1.   Fiedel, S. J. Quacks in the Ice. Waterfowl, Paleoindians, and the discovery of America. In: Walker, R. B., & Driskell, B.N. (eds), *Foragers of the Terminal Pleistocene in North America*, University of Nebraska Press, Lincoln & London (2007).

2.   Vines, G. Genes in black and white. *New Scientist*, 8 July (1995).

3.   Starikovskaya, E. B., Sukernik, R. I., Derbeneva, O. A., *et al.* Mitochondrial DNA diversity in indigenous populations of the southern extent of Siberia, and the origins of Native American haplogroups. *Annals of Human Genetics* 69: 67–89 (2003).

4. Zegura, S. L., Karafet, T. M., Zhivotovsky, L. A., & Hammer, M. F. High-resolution SNPs and microsatellite haplotypes point to a single, recent entry of Native American Y chromosomes into the Americas. *Molecular Biology and Evolution* 21: 164–75 (2004).

5. Fagundes, N. J. R., Kanitz, R., Eckert, R., *et al.* Mitochondrial population genomics supports a single pre-Clovis origin with a coastal route for the peopling of the Americas. *American Journal of Human Genetics* 82: 583–92 (2008).

6. Wang, S., Lewis, C. M. Jr, Jakobsson, M., *et al.* Genetic variation and population structure in Native Americans. *PLoS Genetics* 3: 2049–67 (2007).

7. Oppenheimer, S. *Out of Eden. The Peopling of the World*, Constable & Robinson, London (2003).

8. Goebel, T. The Late Pleistocene dispersal of modern humans in the Americas. *Science* 319: 1497–502 (2008).

9. Shang, H., Tong, H., Zhang, S., *et al.* An early modern human from Tianyuan Cave, Zhoukoudian, China. *Proceedings of the National Academy of Sciences* 104: 6573–8 (2007).

10. Kitchen, A., Miyamoto, M. M., & Mulligan, C. J. A three-stage colonization model for the peopling of the Americas. *PLoS One* 3: e1596 (2008).

Exploring the Coastal Corridor: Vancouver, Canada

1. Hume, S. Tracking Simon Fraser's route. *Vancouver Sun* (2007).

2. Lacourse, T., Mathewes, R. W., & Fedje, D. W. Late-glacial vegetation dynamics of the Queen Charlotte Islands and adjacent continental shelf, British Columbia, Canada. *Palaeogeography, Palaeoclimatology, Palaeoecology* 226: 36–57 (2005).

3. Ramsey, C. L., Griffiths, P. A., Fedje, D. W., *et al.* Preliminary investigation of a late Wisconsinian fauna from K1 cave, Queen Charlotte Islands (Haida Gwaii), Canada. *Quaternary Research* 62: 105–9 (2004).

4. Leonard, J. A., Wayne, R. K., & Cooper, A. Population genetics of Ice Age brown bears. *Proceedings of the National Academy of Sciences* 97: 1651–4 (2000).

5. Erlandson, J. M., Graham, M. H., Bourque, B. J., *et al.* The Kelp Highway hypothesis: marine ecology, the coastal migration theory, and the peopling of the Americas. *The Journal of Island and Coastal Archaeology* 2: 161–74 (2007).

6. Fedje, D. W., & Josenhans, H. Drowned forests and archaeology on the continental shelf of British Columbia, Canada. *Geology* 28: 99–102 (2000).

7. Mandryk, C. A. S., Josenhans, H., Fedje, D. W., & Mathewes, R. W. Late Quaternary paleoenvironments of Northwestern North America: implications for inland versus coastal migration routes. *Quaternary Science Reviews* 20: 310–14 (2001).

Finding Arlington Woman: Santa Rosa Island, California

1. Johnson, J. R., Stafford, T. W., Ajie, H. O., & Morris, D. P. Arlington Springs revisited. In: *Proceedings of the Fifth California Islands Symposium*, pp. 541–5 (2000).

2. Waguespack, N. M. Why we're still arguing about the Pleistocene occupation of the Americas. *Evolutionary Anthropology* 16: 63–74 (2007).

3. Agenbroad, L. D., Johnson, J. R., Morris, D., *et al.* Mammoths and humans as late Pleistocene contemporaries on Santa Rosa Island. *Proceedings of the Sixth California Islands Symposium*, pp. 3–7 (2005).

4. Dixon, E. J. Human colonization of the Americas: timing, technology and process. *Quaternary Science Reviews* 20: 277–99 (2001).

5.   Eshleman, J. A., Malhi, R. S., Johnson, J. R., *et al.* Mitochondrial DNA and prehistoric settlements: native migrations on the western edge of North America. *Human Biology* 76: 55–75 (2004).

6.   Johnson, J. R., & Lorenz, J. G. Genetics, linguistics and prehistoric migrations: an analysis of California Indian mitochondrial DNA lineages. *Journal of California and Great Basin Anthropology* 26: 33–64 (2006).

7.   Kemp, B. M., Malhi, R. S., McDonough, J., *et al.* Genetic analysis of early Holocene skeletal remains from Alaska and its implications for the settlement of the Americas. *American Journal of Physical Anthropology* 132: 605–21 (2007).

8.   Wang, S., Lewis, C. M. Jr, Jakobsson, M., *et al.* Genetic variation and population structure in Native Americans. *PLoS Genetics* 3: 2049–67 (2007).

Hunting American Megafauna: La Brea Tar Pits, Los Angeles

1.   Firestone, R. B., West, A., Kennett, J. P., *et al.* Evidence for an extraterrestrial impact 12,900 years ago that contributed to the megafaunal extinctions and the Younger Dryas cooling. *Proceedings of the National Academy of Sciences* 104: 16016–21 (2007).

2.   Haynes, G. The catastrophic extinction of North American mammoths and mastodonts. *World Archaeology* 33: 391–416 (2002).

Clovis Culture: Gault, Texas

1.   Collins, M. B. Discerning Clovis subsistence from stone artifacts and site distributions on the southern plains periphery. In: Walker, R. B., & Driskell, B. N. (eds), *Foragers of the Terminal Pleistocene in North America*, University of Nebraska Press: Lincoln and London (2007).

2.   Goebel, T. The Late Pleistocene dispersal of modern humans in the Americas. *Science* 319: 1497–502 (2008).

3.   Haynes, G. The catastrophic extinction of North American Mammoths and Mastodonts. *World Archaeology* 33: 391–416 (2002).

4.   Collins, M. B. The Gault Site, Texas, and Clovis research. *Athena Review* 3: 31–41 (2002).

5.   Byers, D. A., & Ugan, A. Should we expect large game specialization in the late Pleistocene? An optimal foraging perspective on early Paleoindian prey choice, *Journal of Archaeological Science* 32: 1624–40 (2005).

6.   Koch, P. L., & Barnosky, A. D. Late quaternary extinctions: state of the debate. *Annual Review of Ecology, Evolution and Systematics* 37: 215–50 (2006).

7.   Mithen, S. *After the Ice. A Global Human History*, Harvard University Press, Cambridge, Massachusetts (2003).

8.   Bradley, B., & Stanford, D. The Solutrean-Clovis connection: reply to Straus, Meltzer and Goebel. 38: 704–14 (2006).

9.   Straus, L. G. Solutrean settlement of North America? A review of reality. *American Antiquity* 65: 219–26 (2006).

10.   Straus, L. G., Meltzer, D. J., & Goebel, T. Ice Age Atlantis? Exploring the Solutrean-Clovis 'connection'. *World Archaeology* 37: 507–32 (2006).

11.   Oppenheimer, S. *Out of Eden. The Peopling of the World*, Constable & Robinson, London (2003).

12.   Holliday, V. T. Folsom drought and episodic drying on the Southern High Plains from 10,900–10,200 C14 yr BP. *Quaternary Research* 53: 1–12 (2000).

Meeting Luzia: Rio, Brazil

1. Neves, W. A., Hubbe, M., & Pilo, L. B. Early Holocene human skeletal remains from Sumidouro Cave, Lagoa Santa, Brazil: History of discoveries, geological and chronological context, and comparative cranial morphology. *Journal of Human Evolution* 52: 16–30 (2007).

2. Neves, W. A., and Hubbe, M. Cranial morphology of early Americans from Lagoa Santa, Brazil: implications for the settlement of the New World. *Proceedings of the National Academy of Sciences* 102: 18309–14 (2005).

3. Neves, W. A., Prous, A., Gonzalez-Jose, R., *et al*. Early Holocene human skeletal remains from Santana do Riacho, Brazil: Implications for the settlement of the New World. *Journal of Human Evolution* 45: 19–42 (2003).

4. Lahr, M. M. Patterns of modern human diversification: implications for Amerindian origins. *Yearbook of Physical Anthropology* 38: 163–98 (1995).

5. Wang, S., Lewis, C. M. Jr, Jakobsson, M., *et al*. Genetic variation and population structure in Native Americans. *PLoS Genetics* 3: 2049–67 (2007).

6. Oppenheimer, S. *Out of Eden. The Peopling of the World*, Constable & Robinson, London (2003).

7. Gonzalez-Jose, R., Bortolini, M. C., Santos, F. R., *et al*. The peopling of the Americas: craniofacial shape variation on a continental scale and its interpretation from an interdisciplinary view. *American Journal of Physical Anthropology* 137: 175–87 (2008).

8. van Vark, G. N., Kuizenga, D., & Williams, F. L. Kennewick and Luzia: Lessons from the European Upper Palaeolithic. *American Journal of Physical Anthropology* 121: 181–4 (2003).

Ancient Hunter-Gatherers in the Amazon Forest: Pedra Pintada, Brazil

1. Roosevelt, A. C., Lima da Costa, M., Machado, C. L., *et al*. Paleoindian cave dwellers in the Amazon: the peopling of the Americas. *Science* 272: 373–84 (1996).

2. Roosevelt, A. C. Clovis in context: new light on the peopling of the Americas. *Human Evolution* 17: 95–112 (2002).

3. Roosevelt, A. C. Ancient and modern hunter-gatherers of lowland South America: an evolutionary problem. In Plew, M. (ed.), *Advances in Historical Ecology*, Columbia University Press, New York, pp. 165–92 (1998).

Black Soil and Revelations: Monte Verde, Chile

1. Dillehay, T. D., & Collins, M. B. Early cultural evidence from Monte Verde in Chile. *Nature* 332: 150–52 (1988).

2. Dillehay, T. D., Ramfrez, C., Piño, M., *et al*. Monte Verde: seaweed, food, medicine, and the peopling of South America. *Science* 320: 784–6 (2008).

3. Ugent, D., Dillehay, T., & Ramirez, C. Potato remains from a late Pleistocene settlement in southcentral Chile. *Economic Botany* 41: 17–27 (1987).

4. Keefer, D. K., deFrance, S. D., Moseley, M. E., *et al*. Early maritime economy and El Niño events at Quebrada Tacahuay, Peru. *Science* 281: 1833–5 (1998).

5. Sandweiss, D. H., McInnis, H., Burger, R. L., *et al*. Quebrada Jaguay: early South American maritime adaptations. *Science* 281: 1830–32 (1998).

6. Meltzer, D. J., Grayson, D. K., Ardila, G., *et al*. On the Pleistocene antiquity of Monte Verde, southern Chile. *American Antiquity* 62: 659–63 (1997).

7.   Dixon, E. J. Human colonization of the Americas: timing, technology and process. *Quaternary Science Reviews* 20: 277–99 (2001).

8.   Goebel, T. The Late Pleistocene dispersal of modern humans in the Americas. *Science* 319: 1497–502 (2008).

Journey's End

1.   Stringer, C. *Homo Britannicus. The Incredible Story of Human Life in Britain*, Penguin Books, London (2006).

2.   Mithen, S. *After the Ice. A Global Human History*, Harvard University Press, Cambridge, Massachusetts (2003).

3.   Lomborg, B. *The Skeptical Environmentalist. Measuring the Real State of the World*, Cambridge University Press, Cambridge (2001).

4.   Gould, S. J. *Wonderful Life.* Norton & Co., New York (1989).

References and Sources for Figures and Maps

p. 5   *The Human Family Tree: 'Splitters' Hominin Taxonomy.* Based on figure 1 in Wood, B., Lonergan, N. The hominin fossil record: taxa, grades and clades. *Journal of Anatomy* 212: 354–76 (2008).

p. 11   *Ages and Stages.* Adapted from figure, p. 300 in: Stringer, C., *Homo Britannicus. The Incredible Story of Human Life in Britain*, Penguin Books, London (2006).

p. 15   *Basic Guide to Toolkits.* Sourced mainly from Klein, R. G. Archaeology and the evolution of human behaviour. *Evolutionary Anthropology* 9: 17–36 (2000).

p. 53   *Bodo and Omo crania.* Based on photographs and descriptions in: Schwartz, J. H., and Tattersall, I., *The Human Fossil Record*, vol. 2, *Craniodental Morphology of Genus Homo (Africa and Asia)*, Wiley Liss, New Jersey, pp. 235–40 (2003).

p. 63   *Routes out of Africa.* Based partly on figure 1 in Bulbeck, D. Where river meets sea. A parsimonious model for *Homo sapiens* colonization of the Indian Ocean rim and Sahul. *Current Anthropology* 48: 315–21 (2007).

p. 101   *Mitochondrial DNA ivy branches.* Based on Stephen Oppenheimer's description and figure in box 2: Shriver, M. D., & Kittles, R. A. Genetic ancestry and the search for personalized genetic histories. *Nature Reviews Genetics* 5: 611–18 (2004).

p. 119   *Routes from Sunda to Sahul.* Based on figure 1 in Bulbeck, D. Where river meets sea. A Parsimonious model for *Homo sapiens* colonization of the Indian Ocean rim and Sahul. *Current Anthropology* 48: 315–21 (2007).

p. 152   *Routes into central and northern Asia.* Based on figures 5.5, 5.7 and 5.9 in Oppenheimer, S. *Out of Eden. The Peopling of the World*, Constable & Robinson, London (2003).

p. 207   *Routes into Europe.* Based on figure 1 in: Mellars, P. Neanderthals and the modern human colonization of Europe. *Nature* 432: 461–5 (2004); figure 1 in Bar-Yosef, O. The Upper Paleolithic revolution. *Annual Reviews in Anthropology* 31: 363–93 (2002); figure 3.4 in Oppenheimer, S. *Out of Eden. The Peopling of the World*, Constable & Robinson, London (2003).

p. 208   *Upper Palaeolithic Artefacts from Üçağizli Cave.* Redrawn from Kuhn, S. L., Stiner, M. C., Reese, D. S., & Gulec, E. Ornaments of the earliest Upper Palaeolithic: new insights from the Levant. *Proceedings of the National Academy of Sciences* 98: 7641–6 (2001).

p. 209   *Pierced Nassarius shells from Üçağizli Cave*. Redrawn from Kuhn, S. L., Stiner, M. C., Reese, D. S., & Gulec, E. Ornaments of the earliest Upper Palaeolithic: new insights from the Levant. *Proceedings of the National Academy of Sciences* 98: 7641–6 (2001).

p. 214   *Simplified plan of Peştera cu Oase*. Redrawn from Zilhao, J. E., Trinkaus, E., Constantin, S., *et al*. The Peştera cu Oase people, Europe's earliest modern humans. In: Mellars, P., Stringer, C., Bar-Yosef, O., Boyle, K. (eds), *Rethinking the Human Revolution: New Behavioural and Biological Perspectives on the Origins and Dispersal of Modern Humans*, McDonald Institute of Archaeology Monographs, Cambridge (2007).

p. 262   *Stages in the manufacture of Aurignacian beads*. Based on figures in White, R. Systems of personal ornamentation in the Early Upper Palaeolithic: methodological challenges and new observations. In: Mellars, P., Boyle, K., Bar-Yosef, O., & Stringer, C. (eds), *Rethinking the Human Revolution: New Behavioural and Biological Perspectives on the Origin and Dispersal of Modern Humans*, McDonald Institute for Archaeological Research, Cambridge (2007).

p. 300   *Routes into the Americas*. Based on figure 7.1 in Oppenheimer, S. *Out of Eden. The Peopling of the World*, Constable & Robinson, London (2003); figure 1 in Goebel, T. The Late Pleistocene dispersal of modern humans in the Americas. *Science* 319: 1497–502 (2008).

ACKNOWLEDGEMENTS

I want to express my thanks to the huge team of people who made this book and the BBC2 television series possible.

The producers, Michael Moseley, Kim Shillinglaw and Paul Bradshaw, made the journey happen and kept it on track!

I was away filming for around 26 weeks, through the spring and summer of 2008, and I couldn't have done it without the love and support of my husband, Dave Stevens, who stayed at home while I roamed the world. I'm extremely grateful to the originators of Skype and Facebook for helping me stay in contact with family and friends when I was far, far away.

I must also thank my Head of Department, Jeremy Henley, the Department of Anatomy and the University of Bristol, for granting me a leave of absence, allowing me to take the wonderful opportunity of making this journey.

Many friends and colleagues have kindly assisted me with this book. I am massively indebted to Stephen Oppenheimer, Colin Groves and Jo Kamminga, for their careful reading of early drafts and wise advice and guidance. Many of the producers, directors and researchers on the series, especially Kim Shillinglaw, Paul Bradshaw, Dave Stewart, Pete Oxley, Naomi Law, Mags Lightbody and Sam Cronin, also provided me with valuable feedback and fact checks. Many thanks are due to Martha Sullivan and Jodie Pashley for furnishing me with transcripts of interviews.

I am very grateful to Chris Stringer (Natural History Museum) and Peter Forster (Anglia Ruskin University) for their thoughtful advice. And to Paul Valdes and Joy Singarayer (Bristol Research Initiative for the Dynamic Global Environment – BRIDGE – University of Bristol), climate consultants for the series, for sharing the palaeoclimate maps with me. Enormous thanks to Dave Stevens who took my rough sketches and made them into beautiful maps and diagrams.

Thank you to everyone involved with filming the series. Each of the five programmes – corresponding with the chapters in this book – was filmed by a different team, with a cast of contributors, some of whom I have written about in the book, others whom I must remember here.

In Africa: thanks to Dave Stewart, producer/director extraordinaire (thank you for recommending *The Songlines*!); Mags Lightbody, who saved my laptop in Dubai, and carried Omo II 'home' in a shawl; Graham Smith ('the bear'), cameraman, often to be seen hanging out the side of a helicopter or small plane; Rob McGregor, camera assistant/ cameraman – thank you for yoga on the beach and for hauling me off the rocks in Israel; Andrew Yarme, soundman – thank you for driving to Cape Town and for the Sounds of Omo!

Thanks also to Arno and Estelle Oostuysen, for looking after us at Nhoma camp; all the Bushmen of Nhoma, and Theo for keeping guard on our night in the bush; to Raj Ramesar, for an introduction to Capetonian genetics, and all the study participants who shared their results with us; Kyle Brown, for the tour around Pinnacle Point; Jeff Rose, for introducing me to Omani archaeology; and Yoel Rak, for showing me Skhul Cave.

In India, Southeast Asia and Australia: thanks to Ed Bazalgette, producer/director (and outback-barbie guitarist); Naomi Law, researcher and macarena teacher, goddess of organisation and serenity (remember the night when the lights went out in Mungo?); Chris Titus King, cameraman with an excellent sense of the absurd; Freddie Claire, soundman, with a joke for every occasion (Poppadom preach and Indian otters); Alex Byng, camera assistant; Phil Dow, camera assistant in Oz; Toby Sinclair, fixer extraordinaire in India (thank you for the jasmine garlands!); and Alan d'Cruz, fixer in Malaysia.

Many thanks to Michael Petraglia and Ravi Korisettar, for giving up time to talk to me during their excavation at Jwalapuram; Bert Roberts, for fascinating insight into luminescence dating and the Hobbit controversy; Stephen Oppenheimer, for talking to me about genetics and phylogeography, and for reviewing the India-to-Australia chapter – huge thanks; Hamid Isa, for his knowledge of the Semang people; Ipoi Datan, for bringing the Niah skull back to its findspot; Tony Djubiantono, for letting me examine the bones from Flores; Robert Bednarik, for masterminding the construction of a Stone Age raft; Sally May, fixer and Australian rock art expert in Gunbalanya; Anthony Murphy and all the artists at the Injalak Arts & Crafts Centre; Michael Westaway, archaeologist at Mungo; Alan Thorne, for introducing me to Mungo Man; and Sheila van holst Pellekaan, for an insight into Australian genetics.

In Siberia and China: thanks to Fiona Cushley, assistant producer, for her Russian expertise and walking the wall with me; Tim Cragg, cameraman,

Adam Prescod, soundman and Jack Burton, second cameraman, for keeping going in the c-c-cold; and Qian Hong, fixer in China.

I am also grateful to Svetlana Demeshchenko, head curator at the Hermitage, for letting me see the beautiful artefacts from Mal'ta; Vladimir Pitulko, for talking to me about Yana – and for trying to get me there!; Piers Vitebsky and Anatoly Alekseyev, for introducing me to Arctic culture; Jo Kamminga, for teaching me to make a bamboo knife, and for reviewing the manuscript of this book – and for my copy of *Prehistory of Australia*; Xingzhi Wu, for introducing me to Zhoukoudian and Peking Man; Wei Jun, Wang Hao Tian, Liu Cheng Jie and Liu Cheng Yi; and Fu Xianguo, for an introduction to early Chinese pottery.

In Europe: thanks to Phil Smith, producer/director, for his excellent direction and wry sense of humour (though I'm sad that River Euphrates was not on the soundtrack!); Finola Lang, assistant producer, for her girly company; Jonathan Partridge, cameraman and gentleman; Simon Farmer, soundman, for his beautiful watercolour postcards; Adrian O'Toole, camera assistant, 'they'll like that back at broadcasting hice'; (and thank you to all of the above for my massive Romanian birthday cake).

Thank you to Michael Pitts and John Chambers, diver cameraman and diving buddy in Gibraltar; Nathalie Cabrier, formidable fixer in France, with her infallible sense of direction!; Klaus Schmidt, director of the Göbekli Tepe site; Silviu Constantin, Mihai Bacin, Virgil Dragusin and Alexandra Hillebrand, for taking me to Peştera cu Oase, and for the book of caves; Clive and Gerry Finlayson, and Darren Fa, for conversations about Neanderthals and the sea caves of Gibraltar; Nick Conard, for showing me the site and the beautiful artefacts from Vogelherd, and for giving me such an insight into the Swabian Aurignacian; Wulf Hein, for showing me how to use an atlatl, and sorry for losing your arrows in the long grass; Katerina Harvati, for her knowledge of heads and hybrids; Ed Green, for explaining the Neanderthal Genome Project to me; Jiri Svoboda, for taking me to the vineyards of Dolní Věstonice and showing me the wonderful ivory carvings in the museum; Martina Laznickova, for helping me to make a reconstruction of the Dolní Věstonice Venus; Randall White, for an introduction to the Aurignacian and Abri Castanet; Michel Lorblanchet, for his demonstration of Palaeolithic stencilling technique, and for being my guide in Cougnac Cave; and The Musee Duyputren, for letting me examine rachitic skeletons there. And many thanks to Bruce Bradley and Metin Eren for the lesson in Palaeolithic stone tool manufacture.

In the Americas: thanks to Pete Oxley, producer/director, for not losing me in a tar pool and down a crevasse!; Clare Duncan, assistant producer; Paul Jenkins, cameraman (we were good up that glacier, weren't we?); Simon Farmer, soundman and catalogue model; and David McDowall, assistant cameraman (What! No beer at a folk festival?!).

Also: Karina Rehavia, fixer in Brazil – thank you for looking after us and for rescuing my books! Thanks to Mike Collins and the Gault team; John Johnson on Santa Rosa; John Harris, for getting me into a sticky mess at La Brea tar pits; Rolf Mathewes and his pollen at Simon Fraser University; Rob Toohey, for looking after me in the water and up on the glacier; Jim Orava, for teaching me to ice-climb; Quentin Mackie, for introducing me to some Canadian bears; Tracey Pierre and the Tsuu T'ina First Nation in Canada; Walter Neves and the Brazilian National Museum, for introducing me to Luzia; and Mario Piño for talking me around the site at Monte Verde.

(And thank you to everyone else I have mentioned in the book.)

The views and opinions expressed in this book, where I am not reporting on specific items of published research, are my own – as are any mistakes.

Many thanks to my agents, Hilary Murray and Luigi Bonomi. Finally, I am hugely grateful to my editors, Richard Atkinson and Natalie Hunt, to my patient copyeditor, Richard Collins, and to the whole team at Bloomsbury.

INDEX